Tourism Resilience and Adaptation to Environmental Change

In recent years, resilience theory has come to occupy the core of our understanding and management of the adaptive capacity of people and places in complex social and environmental systems. Despite this, tourism scholars have been slow to adopt resilience concepts, at a time when the emergence of new frameworks and applications is pressing.

Drawing on original empirical and theoretical insights in resilience thinking, this book explores how tourism communities and economies respond to environmental changes, both fast (natural hazard disasters) and slow (incremental shifts). It explores how tourism places adapt, change, and sometimes transform (or not) in relation to their environmental context, with an awareness of intersection with societal dynamics and links to political, economic and social drivers of change. Contributions draw on empirical research conducted in a range of international settings, including indigenous communities, to explore the complexity and gradations of environmental change encounters and resilience planning responses in a range of tourism contexts.

As the first book to specifically focus on environmental change from a resilience perspective, this timely and original work makes a critical contribution to tourism studies, tourism management and environmental geography, as well as environmental sciences and development studies.

Alan A. Lew is a professor in the Department of Geography, Planning, and Recreation at Northern Arizona University, USA where he teaches in geography, urban planning, and tourism. His research interests focus on tourism in the Asia-Pacific region, particularly in East and Southeast Asia. He is the founding editor-in-chief of the journal, *Tourism Geographies*, a Fellow of the International Academy for the Study of Tourism, and a member of the American Institute of Certified Planners.

Joseph M. Cheer is a lecturer at the National Centre for Australian Studies (NCAS), Monash University and directs the activities of the Australia and International Tourism Research Unit (AITRU). His research draws from transdisciplinary perspectives, especially human geography, cultural anthropology and political economy with a focus on the Asia-Pacific region. He is focused on research-to-practice with an emphasis on resilience building, sustainability and social justice.

Routledge Advances in Tourism
Edited by Stephen Page
School for Tourism, Bournemouth University

For a full list of titles in this series, please visit www.routledge.com/series/SE0258

Tourism Resilience and Adaptation to Environmental Change

Definitions and Frameworks

Edited by Alan A. Lew and
Joseph M. Cheer

Routledge
Taylor & Francis Group

LONDON AND NEW YORK

First published 2018
by Routledge
2 Park Square, Milton Park, Abingdon, Oxon OX14 4RN

and by Routledge
52 Vanderbilt Avenue, New York, NY 10017, USA

First issued in paperback 2020

Routledge is an imprint of the Taylor & Francis Group, an informa business

British Library Cataloguing-in-Publication Data
A catalogue record for this book is available from the British Library

Library of Congress Cataloging-in-Publication Data
A catalog record for this book has been requested

ISBN 13: 978-0-367-66766-5 (pbk)
ISBN 13: 978-1-138-20679-3 (hbk)

Typeset in Times New Roman
by Saxon Graphics Ltd, Derby

Contents

Figures

Tables

Contributors

L. Arifin Bakti is a graduate research assistant and PhD candidate in School of Forestry at Northern Arizona University. His research focus is on community resilience in small island tourism systems. He is also the Director of the Research Centre for Regional Planning, and a faculty member in the Department of Soil Science at the University of Mataram, Indonesia. He has ten years of experience as a senior adviser in urban planning for the United Nations Development Programme (UNDP) in Jakarta, Indonesia, and in early 2000 he was involved in the establishment of Gili EcoTrust on the island of Gili Trawangan.

Georgette Leah Burns is a Senior Lecturer at Griffith University in Queensland, where she is also a foundation member of the Environmental Futures Research Institute. Her research focuses on environmental ethics and the interactions between people and wildlife in nature-based tourism settings. In addition to numerous book chapters and journal articles on this topic, Leah is the author of *Dingoes, Penguins and People: Engaging Anthropology to Reconstruct the Management of Wildlife Tourism Interactions* (2010) and co-editor of *Engaging with Animals: Interpretations of a Shared Existence* (2014).

Joseph M. Cheer is lecturer at the National Centre for Australian Studies (NCAS), Monash University, and directs the activities of the Australia and International Tourism Research Unit (AITRU). His PhD is in Cultural Anthropology and his MA (Tourism and Development Geography) examines the intersection between aid and tourism. Joseph's research draws from transdisciplinary perspectives, especially human geography, cultural anthropology and political economy with a focus on the Asia–Pacific region. As a former practitioner in the international tourism industry, international development sector, and as a management consultant and business owner, he is focused on research-to-practice with an emphasis on resilience building, sustainability and social justice.

Fangfang Chen is a graduate student in Sun Yat-sen University in Guangzhou, China, with a special interest on resilience and small tourism entrepreneurs.

Jana Raadik Cottrell is faculty in the Honors Program at Colorado State University and tourism lecturer/researcher at Maritime Academy of Tallinn University of Technology, Estonia. She teaches courses in leisure, international travel and tourism development. She is a nature-based tourism specialist with expertise in island community development via responsible tourism. Her research focuses on sustainable tourism development on islands and sense of place. Present projects include second homeowners' studies in island communities and collaborative conservation efforts within protected areas in Europe.

Stuart Cottrell is an Associate Professor in human dimensions of natural resources at Colorado State University, teaching courses in ecotourism, sustainable tourism development, protected area management and research in human dimensions of natural resources and tourism. His research includes sustainable tourism, travel and tourism behaviour, visitor impact management and public perceptions of landscape disturbance issues. Present projects involve a National Science Foundation grant to examine natural resource management agency and public perceptions of the bark beetle as a land disturbance issue on natural resources, recreation and tourism.

Shanshan Dai is a Lecturer in Sun Yat-sen University in Guangzhou, China. She works in sustainable tourism and tourism resilience, tourism impact, tourism economics and tourism destination management.

Esther A. Duke is Director of Special Projects and Programs for the Human Dimensions of Natural Resources (HDNR) Department, Colorado State University (CSU). Esther has directed seven conference/training programs focused on environmental conservation and nature-based tourism (150–600 attendees). She teaches courses in tourism and human dimensions. She has published two edited books. Esther has experience working cross-culturally as director of conferences in Kenya and Italy and managing research networks in China, Central America and Europe. Esther received her MS from the HDNR at CSU in 2010. Her research focuses on ecosystem services. She received her BA from Illinois Wesleyan University.

Stephen Espiner's current research interests focus on the human dimensions of protected natural area management, with particular reference to nature-based recreation and tourism and associated conservation, community and visitor management issues. Key themes in his work include the use of natural resource settings for recreation and tourism, and implications for Protected Area management and local community development. Stephen also undertakes research on risk and resilience in nature-based tourism communities, and recreational value conflicts, social impacts, risk perception and communication.

Irina Herrschner holds an MA (Tourism) and has recently submitted her PhD at the University of Melbourne on German cultural diplomacy. She is a Teaching Associate in the Faculty of Arts at the University of Melbourne and in the School of Languages, Linguistics and Culture at Monash University. Her research and forthcoming publications focus on cultural diplomacy, cinematic diplomacy and other forms of global mobilities. Irina is the co-editor of the bookseries 'Global Germany in Transnational Dialogues'.

Kevin Hillmer-Pegram is Associate Professor of Sustainability Studies at Colorado Mountain College, in Glenwood Springs. Previously, Kevin taught Geography and completed an Interdisciplinary PhD at the University of Alaska Fairbanks, specializing in the resilience and adaptation of small arctic communities to climate change and economic globalization – including the expansion of arctic tourism. Through his ongoing teaching and research, Kevin strives to integrate critical social theory, environmental science, and activism to make the world a better place.

Tisha Holmes is an Assistant Professor in the Department of Urban and Regional Planning at Florida State University. Her research examines the social and institutional factors that influence resilience to environmental hazards in marginalized coastal communities. Her research also emphasizes active community participation in research, education and decision-making processes to address the present and potential impact of climate change risks. Tisha has a particular interest in exploring the intersections of climate change, adaptation planning, health equity and social welfare.

Phoebe Honey is a Marketing and Public Relations Executive with a Master's Degree in Tourism and six years' experience in destination marketing with Destination Gippsland, a Regional Tourism Board in Victoria, Australia. She was the recipient of the Victorian Tourism Industry Council 2014/2015 Local Government and Tourism Organisation Development Award, a scholarship which enabled her to conduct research into Tourism Crisis Management and Resilience in the Tourism Sector in Christchurch, New Zealand. She is also passionate about nature based and sustainable tourism.

Kumi Kato is a Professor at the Faculty of Tourism, and Deputy Director at the Center for Tourism Research at Wakayama University, Japan. Her interests include spiritual culture and its creative expressions applied to sustainable community development. Her current work includes community reconstruction in Fukushima, and slow tourism development in communities along the world heritage pilgrimage routes.

Dominic Lapointe is Assistant Professor in the Department of Urban and Tourism Studies at the University of Quebec in Montreal. He holds a doctoral degree in Regional Development. His research addresses development and environment issues with the use of critical theory, especially in the fields of tourism, conservation, environment and climate change.

Alan A. Lew is a Professor in the Department of Geography, Planning, and Recreation at Northern Arizona University where he teaches courses in geography, urban planning and tourism. His research interests focus on tourism in the Asia–Pacific region, particularly in east and south-east Asia. He has published over 25 books, including *Understanding and Managing Tourism Impacts: An Integrated Approach* (2009), *Tourism Geography* (2014), and *World Regional Geography* (2015). He is the founding editor-in-chief of the journal, *Tourism Geographies*, a Fellow of the International Academy for the Study of Tourism, and a member of the American Institute of Certified Planners.

Apisalome Movono is a PhD candidate in Tourism Management at the Department of International Business and Asian Studies, Griffith University, Brisbane, Australia. He has a background in Marine Affairs and Tourism Studies and gained his MA from the University of the South Pacific, where he also holds a position as an Assistant Lecturer. His passion for indigenous issues has influenced his work with indigenous communities in Fiji and the greater South Pacific. His previous publications and research interests examine issues related to tourism, climate change, resilience, vulnerability and more specifically on conceptualizing communities as complex adaptive social-ecological systems.

Maggie C. Miller completed her PhD in the Department of Recreation and Leisure studies at the University of Waterloo in April 2017. Her previous degrees include a MA in Tourism Policy and Planning from the University of Waterloo and a BSc in Tourism and Events Management from George Mason University. Much of her work focuses on socio-cultural dimensions of tourism and sustainable development, with a particular emphasis on engaging research to understand and enhance social justice and equity in international tourism contexts. In her dissertation research, Maggie employed visual methodologies to better understand Nepal's adventure tourism industry as perceived by Sherpa natives who inhabit and work within the Mt Everest region.

Pin T. Ng is a Professor of Economics at the W. A. Franke College of Business, Northern Arizona University. He received his Bachelor degree in Economics from the University of Minnesota at Duluth and PhD in Economics from the University of Illinois at Urbana-Champaign. He teaches mainly Business Statistics, Econometrics and Managerial Economics at the Northern Arizona University. His research interests span Econometrics, Computational Statistics, Urban and Rural Economics, International Finance, and Tourism and Leisure Study. The major theme of his research has been the computational aspects of quantile regression and applications of quantile regression technique to interdisciplinary research.

Chin-cheng (Nickel) Ni is a Professor in the Department of Environmental and Cultural Resources, National Tsing Hua University, Taiwan, where he teaches courses in tourism geography, island studies, and urban and rural analysis. His research interests focus on tourism and community development in marginal regions. The issues of sustainability, vulnerability and resilience at community level are core areas of his recent publications.

Leonardo Nogueira de Moraes holds a Bachelor of Tourism degree (University of São Paulo, 2003) a Specialization in Tourism and Hospitality Marketing Management (University of São Paulo, 2004) and a PhD in Architecture and Planning (University of Melbourne, 2015). Leonardo was a visiting researcher at Lund University Centre of Excellence for the Integration of Social and Natural Dimensions of Sustainability (LUCID, Sweden) in 2011 and is currently a Tenured Senior Lecturer at Instituto Federal de São Paulo (IFSP, Brazil), one of the founding members of the International Observatory of Public Policies for Sustainability (IOPPS) and a researcher on regional tourism resilience at the University of Melbourne.

Natalie Ooi is an Assistant Professor and Program Coordinator of the Ski Area Management Program within the Department of Human Dimensions of Natural Resources in the Warner College of Natural Resources at Colorado State University. She teaches within the Master of Tourism Management and Graduate Certificate in Ski Area Management programmes. Her areas of research include sustainable tourism, mountain resort development and impacts, social capital and backpacker tourism.

Caroline Orchiston is the Deputy Director of the Centre for Sustainability at the University of Otago, Dunedin, New Zealand. Caroline's research focuses on natural hazards (including earthquakes, tsunami and climate change) and community resilience, and how we can work with communities to build resilience before future hazard events. Her tourism research has focused on post-earthquake recovery and organizational resilience in Christchurch (New Zealand), as well as the issues faced by tourism-reliant peripheral communities in New Zealand and the Pacific Northwest of the United States.

Bruno Sarrasin is Associate Professor and chair of the Department of Urban and Tourism Studies at the University of Quebec in Montreal. He led the tourism and hospitality management programme and the French-speaking tourism research journal *Teoros*. Author of some 40 scientific and transfer publications, he has presented several lectures on international tourism issues. He is particularly interested in the socio-political analysis of nature tourism, especially in developing countries.

Tsung-Chiung (Emily) Wu, PhD, is a professor in the Department of Tourism, Recreation, and Leisure Studies at National Dong Hwa University in Taiwan. Her current research focuses on sustainable tourism development, community resilience, cultural tourism, rural tourism and tourism resource management.

Honggang Xu is in the School of Tourism Management, Sun Yat-sen University, Guangzhou, China, where she teaches Tourism System Analysis and Tourism Planning and Public Policy. She is an editor for the journal, *Tourism Geographies*, and has research experience in the areas of Tourism Geography, System Dynamics and Mobilities.

Part I
Introduction

1 Environmental change, resilience and tourism

Definitions and frameworks

Alan A. Lew and Joseph M. Cheer

Resilience and environmental change

The repercussions of global warming and changeable weather patterns such as El Niño on many of the world's coral reefs, as exemplified by Australia's Great Barrier Reef, demonstrate significant local-level effects of environmental change that are potentially severe for tourism communities (Norström et al., 2016). Indeed, natural hazard disasters are often 'glocalized', rather than limited to only one particular locality, as demonstrated in incidences of tsunamis, volcanic eruptions and hurricanes (Uitto, 2016). This renders local adaptive responses inescapably interdependent within global action contexts, with the success or failure of either linked to some degree.

That said, while environmental crises can be catastrophic in their outcomes, they can also present opportunities for change and innovation where feedback loops signal directional potentials, possibly resulting in more effective cooperation and better outcomes, and mitigating the 'tragedy of the commons' (Lindahl, Crépin & Schill, 2016). In a sense, this is illustrative of the aims of resilience where the new state condition can at times lead to far more resilient conditions. One of the benefits of a crisis, so long as the change does not lead to permanent destruction, is that tipping points may become obvious, enabling a clearer understanding of the causal mechanisms that generated the crisis situation (SRC, 2015). In identifying such tipping points, communities can 'self organise and respond' (Arctic Council, 2016), as well as spur innovation in a kind of creative–destruction manner (Schumpeter, 1943; Holling, 2001).

Adaptive responses that leverage knowledge, experience and innovation are underpinned by policy learning and paradigm changes (Lew, 2014). However, the development of optimal adaptive capacities rests on 'ensuring that a system is able to accept change and unpredictability, and is designed to be safe to fail, as well as being able to respond to the needs of the most vulnerable' (Hall, 2018, p.28).

Tourism and the natural environment

The experience of natural environments has probably always been a motivation for leisure, recreation, education and other forms of touristic travel (Meyer-Arendt,

2004; Williams & Lew, 2014; Hill, Curtain & Go, 2014). In response, tourism industries have become prominent in places that have the kinds of natural resources that people associate with their travel and tourism needs. Mountains, beaches, islands, and tropical climates are among the many natural features that tourists, and therefore tourism industries, are drawn to, especially when they are all found in the same place. Tourism economies in such places are largely dependent on their natural endowments and are vulnerable to any threats that might damage those resources.

Every tourism place, whether its attractions are nature-based or cultural, are defined by their location in terms that geographers refer to as *site* and *situation* (Lew, Hall & Timothy, 2015). *Site* characteristics are all the natural environmental features that are inherent and intrinsic to the place's location. This includes the topography of the land and water, as well as types of vegetation and soil, and climate and weather patterns. These offer opportunities for human settlement and development, which includes tourism development where such features meet the touristic interests of visitors. As these are developed, they become new topographic features added to the site characteristics of a place. (Thus, a topographic map not only shows ground elevation, but also vegetation, roads and built-up settlement areas.)

Human settlements tend to be created with an assumption that the site characteristics at the time of their development will remain largely unchanged into the future. In some instances, such as in floodplains, we know through experience that this is not true and we therefore use regulation and engineering to adapt land uses and to manage the potential changes that we can foresee. Unfortunately, it is not always possible to foresee all of the potential changes that nature can bring our way, and, even when potential nature-related hazards are known, there are often insufficient funds to address a community's vulnerabilities.

While some environmental change may be primarily associated with a site's characteristics (such as a landslide or the overuse of groundwater), most environmental changes are much broader in scope and origin. This points to the *situation* characteristics of a place, which is defined as the relationships that a place has with other places, as well as with its larger spatial context beyond its immediate site location. Climate change, for example, is driven more by regional and global processes than by local site changes. (Although there also are some local impacts from urban heat islands and air pollution caused by fossil fuel emissions and seasonal air temperature inversions.) Flooding, as mentioned above, is often caused by weather phenomena that are much broader than the individual site that experiences the flooding. However, how that site is naturally endowed and how humans have changed the site can determine whether the flood's impact reaches disaster proportions.

In their relationship to their environmental site and situation, tourism places are not separate from non-tourism places. They both share similar challenges (and opportunities) from changing environmental conditions. All places experience environmental changes that sometimes occur fast and unexpected, and at other times are slow and almost imperceptible (Lew, 2014). These changes have the potential to impact all residents and all economic sectors in a place. Some types of environmental changes, however, may impact tourism more than other economic sectors.

Where a tourism economy is built on environmental attractions, there are special concerns. These tourism places are primarily concerned about changes in their natural resource attractions. The biggest threat (or vulnerability) is the inability to bring tourists to the attractions. This can occur from either:

- the loss of access to a natural resource attraction, including the loss of the attraction itself; or
- a loss of access to tourist source areas, so that tourists cannot get to the tourism place.

From the supply side, tourism is an economic activity and both loss scenarios fundamentally impact the financial viability of the industry. Financial concerns serve as the bottom line for most tourism destination decision making because if tourism activities are not economically worthwhile, then it is likely that the tourism system will either completely collapse or transform into a lower-level form, both of which could be considered undesirable states.

Understanding how to maintain tourism activities, and a tourism community's overall quality of life, at a desirable level is generally what the study of tourism resilience is all about. Resilience is generally defined as how a system responds without succumbing to external drivers pressuring it to change (Folke 2016). Responses include resistance (sometimes referred to as 'resilience'), adaptation and transformation. Social-ecological resilience theory applies a systems approach understanding to how communities as integrated social and environment entities respond to change. Systems exist in a nested hierarchy of larger systems and smaller subsystems, known as a 'panarchy' (Gunderson & Holling, 2002; Allen et al., 2014). A tourism economy, for example, would be a subsystem of a larger local or regional economy. In reality, all systems are subsets of the global world panarchy system. Within this global panarchy, systems may be defined in different ways, depending on the focus of a research question. In addition to the overall tourism economy system of a destination (place or region), some of the key systems that tourism researchers tend to focus on include:

- the system of tourist attractions;
- the system of tourism infrastructure and workforce (which may be further subdivided into accommodations, food, transportation and travel services);
- the system of tourist markets (tourist origins and types of tourists);
- the system of drivers of change that impact the tourism economy (all the above); and
- a tourism community system (the larger community that a tourism economy is a part of).

Each of these defines a different system that is being impacted by different external pressures and is responding in a different way. They are all, however, legitimate topics within the study of tourism resilience.

Human- and nature-based change

The focus of this book is on how tourism places respond, adapt, change and sometimes transform (or not) in relation to changes in their environmental context. Sometimes these changes are primarily nature-driven, with tourism places being forced to respond to them. At other times the changes are mostly human-driven under social policies that modify natural environments to better exploit their resource potential. These two scenarios, however, are extremes on a continuum of human–environment interactions that is far less bifurcated than it may sometimes appear. What seems to start out as a human-driven or nature-driven event or process will quickly evolve into a dialectical discourse as mostly human systems respond to natural processes, and, in turn, mostly natural systems respond to human actions.

The world is a social-ecological system in which separations of nature- and human-driven changes are ontologically difficult to make (Wight, 2005). Humans, animals, plants, soils, water bodies and land masses are all open subsystems that interact and influence each other within an all-encompassing global system. Humans influence climate systems, but so does plant and ocean activity. Conversely, climates (as distinct from weather events) influence human settlement by defining the soils and types of organism and animals that are best adapted to an environment. It is important to recognize that these system relationships are deep and complex, even when, from a human perspective, it is more convenient to generalize the sources of change as being human-driven or nature-driven in their character. Disease epidemics, whether impacting humans or animals (e.g. the H5N1 avian bird flu), are an example of a crisis event that is difficult to categorize into simple human or environmental processes.

All three of the scenarios described here (human-driven, nature-driven, and nature–human discourse) are considered in the chapters of this book, through studies of community and tourism resilience responses to shifting environmental contexts. Nature-driven change is primarily seen in chapters related to climate change (which is a slow change process: Chapters 8, 9 and 16) and those discussing preparation and responses to natural hazard disaster events (mostly earthquakes, a very fast change process: Chapters 11, 12, 13 and 14; although Chapters 3 and 6 also cover hazardous weather events). A few chapters address both fast and slow environmental drivers (Chapters 3, 10 and 15).

Discounting the fact that climate change is, by most accounts, largely human-driven in its disaster proportions (Pachauri & Meyer, 2014), other human-caused changes also place significant pressures on natural ecosystems. These are discussed to varying degrees in all the chapters with respect to how humans respond to environmental changes through engineering and social modifications.

However, some chapters have a specific focus on, for example, government and other policies that impact specific natural sites (Chapters 3, 4, 5 and 7) or the impacts of private tourism development and tourists themselves on special environments and communities (Chapters 10 and 16).

Although not explicitly stated by all the authors, all of the chapters in this volume recognize an implicit interrelationship between human and natural realms. Traditional indigenous culture groups, however, have historically demonstrated a closer innate awareness of their integration with their natural environments than is often the case in modern, industrial societies (Lew & Kennedy, 2002). A section of the book, therefore, specifically covers potential resilience responses and lessons from indigenous populations to their changing social-ecological conditions (Chapters 16 and 17).

Fast and slow change

In addition to the issue of human-driven and nature-driven change, a second major variable among the chapters of this book is the intensity and speed of change (Table 1.1). Environmental change can happen very quickly and with great intensity, which is often considered a natural disaster, especially when human interests are impacted. Earthquakes, hurricanes and typhoons, landslides and sinkholes, and forest and grassland fires, are all examples of fast and intense environmental changes (see the natural hazard disaster chapters cited above). On the other hand, natural ecosystems are also continually undergoing slow shifts and changes, in almost imperceptible increments. Climate change has historically been a very slow process, although it appears to have sped up in recent years, at least from a human perspective, due to increasingly visible weather event impacts on human settlements and investments (Folke, 2016). Other slow drivers of change that impact natural ecosystems and human relations to those ecosystems include government policies and programmes (Chapters 3, 4, 5, and 7), economic globalization (Chapters 3, 7 and 9), and broader shifts in the social structure of places (Chapter 3, 9, 16 and 17).

Ultimately, however, what is fast, slow or intermediate in speed is dependent on the time scale against which it is measured (Lew, 2014). While most people will recognize sudden natural hazard disaster events as a fast change phenomena, there may be considerable disagreement on whether climate change and globalization related events are slow, intermediate or fast in nature. For example, the economic transformation of some communities following China's 2008 Wenchuan Earthquake has itself been a fast event, by many standards (Chapter 11). On the other hand, a fast change earthquake has many other impacts that are only recognized and addressed over a relatively long period. These may include psychological impacts (Chapter 13) and issues related to a community's spiritual resilience (Chapter 12).

Table 1.1 Major resilience themes of chapters in this book

Chapter	Author(s)	Nature Driven[1]	Human Driven[1]	Fast Change	Slow Change	Engineering Resilience[2]	General Topics
1	Lew & Cheer	N	H	F	S	E	Concepts
2	Duke, Cottrell & Cottrell	N	H	F	S		Concepts
3	Lew, Ni, Wu & Ng	N	H	F	S	E	Community Development
4	Burns		H		S		Environmental Sustainability
5	Nogueira de Moraes	N	H		S		Community Development
6	Holmes	N		F			Weather Hazards – Social Adaptations
7	Lew & Wu		H		S	E	Wetland Conservation
8	Ooi	N			S	E	Climate Change – Engineering Adaptations
9	Lapoint & Sarrasin	N			S		Climate Change – Social Adaptations
10	Bakti & Lew	N	H	F	S	E	Coral Reef Conservation
11	Xu, Chen & Dai	N		F			Earthquake Recovery – Economic
12	Miller	N		F			Earthquake Recovery
13	Herrschner & Honey	N		F			Earthquake Recovery – Psychological
14	Kato	N		F			Earthquake Recovery – Spiritual
15	Orchiston & Espiner	N		F	S	E	Environmental Change – Social Adaptation
16	Hilmer-Pegram	N	H		S		Climate Change – Traditional Systems
17	Movono	N	H		S		Social Change – Traditional Systems
18	Cheer & Lew	N	H	F	S	E	Concepts

Notes:
1 Human driven change and Nature driven change describe the initial locus of the driver of change and not the larger pre-event context or subsequent responses of mostly human and natural systems.
2 All chapters address Social-Ecological Resilience to a significant degree, but only a few address Engineering Resilience approaches.

Source: Authors

Engineering and social-ecological resilience

A third issue that is especially relevant to environmental change events and processes is the difference between engineering resilience responses and social-ecological resilience responses (Table 1.1). Engineering resilience is the use of engineering techniques and structures to harden and strengthen physical infrastructure and buildings against potential nature-based disaster events and other potential crisis events, with the goal of recovering pre-existing service levels as quickly as possible after such an event (Woods, 2006). Examples of engineering approaches include earthquake-resistant buildings, seawalls, redundant power grids and onsite electrical power generation facilities. Disaster preparedness, including public awareness campaigns, are also part of an engineering approach to building community resilience against these types of environmental change event. Dams are a good example of engineering resilience in many parts of the world where they have primarily been effective in reducing undesirable flood events, while also providing more consistent and dependable water and energy supplies for agriculture and human settlements.

Engineering resilience is the most common approach in preparing for, and responding to, natural hazard disasters, and, for many people, the concept of resilience is equated to engineering resilience (Hall, 2017). In tourism studies, crisis management planning by businesses or a destination area's tourism industry overall has mostly been undertaken from an engineering approach (Ritchie, 2004). The focus is on identifying potential threats, reducing system vulnerabilities to those threats, maintaining system viability through a crisis or disaster event, and returning to pre-event levels as quickly as possible.

A few chapters in this book specifically focus on engineering resilience responses to environmental change (Chapters 3, 7, 8, 10 and 15). These include snow-making technologies to address warming temperatures in ski areas (Chapter 8), technologies to speed the growth (and recovery) of coral reef organisms (Chapter 10), and the strengthening of transportation and other infrastructure in quake-prone regions (Chapter 15).

An engineering approach is also effective as a first response to a major natural hazard disaster event, such as a large earthquake. Governments may need to respond in an authoritarian manner, suspending the normal legal and market system protocols by banning access to some areas, forcing evacuations and providing emergency medical, housing and food supplies through the use of military personnel who are also charged with repairing key access and utility infrastructure lines. These issues, which fall more under the topic of natural hazards disaster management than community resilience, have not been directly addressed by the chapters in this book.

Slow, non-disaster environmental change processes are very different and often do not lend themselves to engineering resilience approaches (Folke, 2016). In many instances, research scientists or other experts are the first to notice slow environmental changes, and they may find it difficult to convince a skeptical public of their concerns. In other cases, environmental changes may occur in direct response to purposeful human actions, either as a policy initiative by a government, or as a group or people act in accordance with what they consider to be appropriate social behaviour. Much of resilience theory comprises an effort to model these kinds of slow change using an interrelated open systems approach (Brand & Jax, 2007), with the overall context being that of a coupled social-ecological system. As such, social-ecological resilience has come to be an encompassing concept that includes a deep understanding of how human and environmental systems move through cycles of organization, growth and collapse (the adaptive cycle), and impact one another through space and time (panarchy). It is generally considered quite separate from the engineering resilience approach.

All the chapters in this volume adopt a social-ecological resilience approach to a significant degree, even those that also discuss engineering resilience. This is because of the wide recognition of the complexity of our contemporary world and the challenges that we face. A complex adaptive systems approach lends itself to modelling and understanding those challenges, and social-ecological resilience theory has become the most widely adopted theory in this context (Folke, 2016).

Conclusion

As noted above, tourism resilience is about understanding how to maintain tourism activities and a tourism community's overall quality of life at desirable levels. Based on this definition, enhancing tourism resilience requires knowledge of and answers to these questions:

1   What is a 'desirable level' of tourism activities and of a tourism community's quality of life?
2   What are the social and ecological variables that need to be monitored and managed to achieve or maintain that desirable level?

The answer to the first of these two questions is addressed primarily through the political process, as residents and interest groups express their desires in multiple ways, including direct action, to those in leadership positions. Political issues (including political economy and political ecology) are touched upon to some degree in many of the chapters in this book, although often in a secondary or peripheral manner. (For a greater focus on social change issues and tourism, see Cheer & Lew, 2018.) This is because resilience theory and resilience thinking, with their origins in environmental and systems sciences (Strunz, 2012), tend to focus more on applied theories (question 2) rather than normative theories (question 1). Resilience theory seeks to build descriptive frameworks and models to explain how places respond to changing conditions over time. The goal is to better prepare communities (systems) to adapt to their continually changing contexts.

Based in resilience thinking and resilience theory (the former being a more open concept, the latter more specific), this volume mostly focuses on practical and applied methodologies of resilience. Politics are not completely ignored (see especially Chapters 9, 10 and 16), but are often only one component in a multifaceted discussion of resilience responses to changing environments. With that consideration in mind, this volume provides a broad range of insight into how resilience relates to tourism within the context of changing environmental conditions. The articles are applied, theoretical and at times critical in nature. Readers will benefit from the considerable depth of experience and knowledge presented, with a more coherent (if still complex) understanding of tourism resilience in the end.

References

Allen, C.R., Angeler, D.G., Garmestani, A.S., Gunderson, L.H. & Holling, C.S. (2014). Panarchy: Theory and Application. Nebraska Cooperative Fish and Wildlife Research Unit, Staff Publications. Paper 127. Retrieved from http://digitalcommons.unl.edu/ncfwrustaff/127

Arctic Council (2016). *Arctic Resilience Report, 2016.* (M. Carson & G. Peterson, eds.) Stockholm: Stockholm Environment Institute and Stockholm Resilience Centre.

Brand, F.S. & Jax, K. (2007). Focusing the meaning(s) of resilience: resilience as a descriptive concept and a boundary object. *Ecology and Society*, 12(1), 23. Retrieved from http://www.ecologyandsociety.org/vol12/iss1/art23.

Cheer, J. & Lew, A.A., eds. (2018). *Tourism, Resilience and Sustainability: Adapting to Social, Political and Economic Change.* London: Routledge.

Folke, C. (2016). Resilience. In *Oxford Research Encyclopedia of Environmental Science* (pp. 1–68). New York: Oxford University Press. DOI: 10. 1093/acrefore/9780199389414.013.8.

Gunderson, L. & Holling, C.S. (2002). *Panarchy: Understanding Transformations in Human and Natural Systems.* Washington, DC: Island Press.

Hall, C.M. (2018). Resilience in tourism: development, theory, and application. In J. Cheer and A.A. Lew, eds., *Tourism, Resilience and Sustainability: Adapting to Social, Political and Economic Change* (pp. 18–33). London: Routledge.

Hill, J., Curtin, S. & Go, G. (2014). Understanding tourist encounters with nature: a thematic framework. *Tourism Geographies*, 16(1), 68–87. DOI: 10.1080/14616688.2013.851265.

Holling, C.S. (2001). Understanding the complexity of economic, ecological, and social systems. *Ecosystems* 4: 390–405.

Lew, A.A. (2014). Scale, change and resilience in community tourism planning. *Tourism Geographies,* 16(1), 14–22. DOI:10.1080/14616688.2013.864325

Lew, A.A. & Kennedy, C.L. (2002). Tourism and culture clash in American Indian country. In S. Krakover and Y. Gradus, eds., *Tourism in Frontier Areas* (pp. 259–83). Lexington, Kentucky: Lexington Books.

Lew, A.A., Hall, C.M. & Timothy, D. (2015). *World Regional Geography: Tourism Destinations, Human Mobilities, Sustainable Environments*, 2nd edn. Des Moines, Iowa: Kendall-Hunt.

Lindahl, T., Crépin, A.-S., & Schill, C. (2016). "Potential disasters can turn the tragedy into success". Environmental and Resource Economics, 65(3), 657–76.

Meyer-Arendt, K. (2004). Tourism and the natural environment. In A.A. Lew, C.M. Hall and A.M. Williams, eds. *A Companion to Tourism*. London: Blackwell. (Blackwell's Companion to Geography series).

Norström, A.V., Nyström, M., Jouffray, J.B., Folke, C., Graham, N.A., Moberg, F., … & Williams, G. J. (2016). Guiding coral reef futures in the Anthropocene. *Frontiers in Ecology and the Environment*, 14(9), 490–98.

Pachauri, R.K. & Meyer, L.A., eds. (2014). *Climate Change 2014: Synthesis Report.* (Contribution of Working Groups I, II and III to the Fifth Assessment Report of the Intergovernmental Panel on Climate Change.) Geneva: IPCC.

Ritchie, B.W. (2004). Chaos, crises and disasters: a strategic approach to crisis management in the tourism industry. *Tourism Management*, 25, 669–83.

Schumpeter, J., 1943. *Capitalism, Socialism and Democracy.* New York: Harper and Row.

SRC (2015). *Applying Resilience Thinking: Seven Principles for Building Resilience in Social-Ecological Systems.* Stockholm: Stockholm Resilience Centre.

Strunz, S. (2012). Is conceptual vagueness an asset? Arguments from philosophy of science applied to the concept of resilience. *Ecological Economics,* 76, 112–18.

Uitto, J.I. (2016). Environmental hazards, climate change and disaster risk reduction: JK Mitchell's relevance to the global sustainable development agenda. *Journal of Extreme Events*, 1671006.

Wight, I. (2005). Placemaking as applied integral ecology: evolving an ecologically wise planning ethic. *World Futures*, 61: 127–37. DOI: 10.1080/02604020590902407

Williams, S. & Lew, A.A. (2014). *Tourism Geography: Critical understandings of place, space and experience, 3rd edition.* Oxford: Routledge.

Woods, D.D. (2006) Essential characteristics of resilience. In E. Hollnagel, D.D. Woods, N. Leveson, eds., *Resilience Engineering: Concepts and Precepts* (pp. 21–34). Aldershot: Ashgate.

2 Applying the adaptive capacity cycle to tourism development

An exploration of social-ecological resilience

Esther A. Duke, Stuart Cottrell and Jana Raadik Cottrell

Introduction

Tourism is a multifaceted adaptive system with non-linear dynamics, which can cause unpredictable complex and changing outcomes (Cochrane, 2010; Malanson, 1999; Miller & Twinning-Ward, 2005). Holling's (1986, 1987) adaptive cycle demonstrates how complex systems of ongoing transition work and this can be useful for understanding the tourism development process and planning for sustainability. The adaptive cycle shows how "sudden surprises," such as those impacting tourism destinations, may affect resilience and/or vulnerability (Liburd, 2010).

Linear disciplinary approaches to understanding tourism are limited and an emerging more holistic social-ecological systems (SES) view of tourism provides new avenues for exploring tourism:

> [T]he dependency of tourism on natural resources, its interlinked elements of economics, politics, psychology, anthropology and ecology, its cross-cultural, cross-sectorial and multi-scalar characteristics and its international linkages, mean that tourism systems constitute excellent examples of complex SES [social-ecological systems].
>
> (Cochrane, 2010, p. 173)

This new ecological view of tourism (Lóránt, 2011) allows for more informed adaptive management of tourism destinations, which can better build resilience. Every tourism destination is comprised of a unique network of socio-cultural features, natural features, and both natural and built infrastructure:

> Evidence makes it increasingly clear that a tourism system is an ecosystem, like an urban ecosystem or agro-ecosystem, in which tourism is merged with life support systems and related social systems which are likely to extend well beyond the recognized destination.
>
> (Farrell & Twining-Ward, 2005, p. 115).

To understand the tourism ecology of a destination, one must understand both the human and natural ecology of that place and how these two ecologies interact at different scales (Potts & Harrill, 1998, 2002; Tyler & Dangerfield, 1999).

Through an exploration of recent developments in conceptualizing "sustainability" inspired by ecosystem ecology (Folke, Carpenter, Elmqvist, Gunderson, Holling & Walker, 2002; Holling, Gunderson & Peterson, 2002; Holling, 1996) and ecosystem services (MA, 2005), we can begin to reveal the ecology of tourism: "the relationship of tourists, communities, managers, developers, and policymakers to each other, and especially to their environment…" (Farrell & Runyan, 1991, p. 27). Beginning in the 1990s, there were efforts to apply insights from ecological research to tourism studies (e.g., Tyler & Dangerfield, 1999; Farrell & Runyan, 1991). By exploring how contemporary social and ecological scientists think about SES systems, we can better understand the context and components of a tourism system. It is not necessary to develop an in-depth understanding of SES systems; introductory-level knowledge can greatly expand one's understanding of tourism systems. The key is to recognize the components and the relationships between them. In Chapter 17 of the *Millennium Ecosystem Assessment*, de Groot and Ramakrishnan (2005) attribute 30 per cent of global travel and tourism revenue related to cultural tourism and ecotourism. Tourism is a valuable industry that depends on intact biodiversity and ecosystem services, including cultural ecosystem services. As Farrell and Twining-Ward (2004, 2005) emphasize, tourism researchers (and one might argue tourism professionals, as well) are often well versed in the economic, business and sometimes social-cultural components, but know correspondingly little about the natural-resources or ecosystem-service components of tourism systems.

Elsewhere in this book, other authors delve into more human ecology when overviewing the economic, socio-cultural and political/institutional dimensions and drivers of change in tourism development (e.g., conservation initiatives, cultural preservation). Understanding and managing the relationship between ecosystem services, social systems and tourism systems allows one to potentially manage for resilience (Cochrane, 2010; Liburd, 2010). With human induced global change, comes greater uncertainty about the timing and spread of shocks and disturbances (ecological, health or economic crises) (Lew, 2013). This is a major concern for leaders in the tourism industry where success depends upon the flow of people and money on national and international scales.

In this chapter, we introduce the concept of an SES systems perspective to provide a framework for understanding the co-evolution of human and natural systems. We then explore the concepts of resilience and adaptive capacity as they apply to tourism SES systems and examine the utility of the adaptive cycle approach for understanding and managing tourism systems.

Resilience and SES

Resilience, initially couched in an ecological context (Holling, 1973; Gunderson, 2000), is the capacity for an ecological, social and linked social-ecological system

(SES) to absorb perturbation while maintaining fundamentally similar structure and function to its pre-disturbed condition (Engle, 2011; Folke, 2006). More specifically, resilience has been characterized as:

> (1) the amount of change the system can undergo and still retain the same controls on function and structure; (2) the degree to which the system is capable of self-organization; and (3) the ability to build and increase the capacity for learning and adaptation.
>
> (Benson & Craig, 2014, p. 779).

Resilience-thinking acknowledges disequilibrium and nonlinear change in social-ecological systems and can be a useful concept for tourism management and planning (Seidl, 2014; Stockholm Resilience Centre, 2014). In ecology, resilience is often defined as the capacity for a system to absorb a perturbation while maintaining fundamentally similar structure and function to its pre-disturbed condition (Gunderson, 2000). In a societal context, this definition is adapted to encompass the capability of a coupled social-ecological system to recuperate the environmental, economic and aesthetic properties that sustain a system prior to some disturbance (e.g., hurricane, fire, war, economic upheaval) (Folke, 2006). In recent years, resilience theories have been applied to tourism with great success (Lew, 2013; Lew, Ng, Ni & Wu, 2016; Liburd, 2010). However, much about how social and ecological factors and processes work in isolation and in combination to promote or erode resilience in social-ecological systems is currently unknown, as the degree to which management policies and market forces can help to mitigate undesirable social and ecological outcomes in tourism requires new research (Cochrane, 2010; Lew, 2013; Liburd, 2010). Meanwhile, the concept of resilience in SES is not well established and needs further clarification for it to enhance tourism development (Brand & Jax, 2007; Cochrane, 2010; Liburd, 2010).

Holling's adaptive cycle model (see Figure 2.1) was developed by and named after biologist and systems theorist C.S. Holling. Beginning with his seminal paper in 1973, and throughout his career, Holling's research furthered the scope, application and understanding of what is often labeled resilience-thinking. His work applied systems theory in ecology through simulation modeling and policy analysis. His most influential work focused on adaptive management, the adaptive cycle, resilience and panarchy. His work has revolutionized ecology and has also been applied to social systems with great success (Gunderson & Holling, 2002; Holling, 1973, 1986, 1996).

Adaptive capacity and SES

Working on ecosystem modeling, Holling (1973, 1986, 1996) argues that ecosystems and social systems do not have a single state. These systems do not follow a linear path along a single equilibrium, but actually move across multiple points within and even sometimes across a basin of attraction. There are multiple stable points which systems can move between in response to external (e.g., hurricane, flood, acts of war or terrorism) or internal (e.g., change in leadership) forces. This has important implications for tourism systems and our ability to actively manage a system for sustainability (Cochrane, 2010). For example, this model illustrates how it is possible to push a system beyond its basin of attraction into a new basin with a new set of equilibrium points. This ability to escape a basin of attraction is referred to in the resilience literature as a transformation or a measure of adaptive capacity (Walker, Holling, Carpenter & Kinzig, 2004).

Thirty years after Holling's seminal paper, Walker, Holling and colleagues continue to clarify the increasingly applied (and often misused) concept of resilience. In Walker, Holling et al. (2004) they couch the application of resilience within an SES framework and emphasize the importance of three attributes:

- resilience;
- adaptability; and
- transformability.

They explore a rangeland case study in Zimbabwe where an SES rangeland defined originally by the amount of grass, shrubs and cattle transforms to a new stability landscape through environmental degradation coupled with the introduction of new ways for earning a living, such as ecotourism, based on wildlife and rivers:

> At times societies or groups may find themselves trapped in an undesirable basin that is becoming so wide, and so deep, that movement to a new basin or sufficient reconfiguration of the existing basin becomes extremely difficult. At some point, it may prove necessary to configure an entirely new stability landscape—one defined by new state variables, or the old state variables supplemented by new ones.
>
> (Walker, Holling et al., 2004, p. 5)

Similar transformations have taken place elsewhere in Africa, including Kenya and Namibia. In these countries diverse players, including NGOs with an interest in using tourism as a development tool, and for-profit tourism operators, have come together to address impacts on wildlife stemming from the shift away from large swaths of communally managed lands and towards privatization and fenced parcels of land. These partners have launched an ecotourism model known as wildlife conservancies to mitigate the growing threat of habitat fragmentation (Jandreau & Berkes, 2016). Such capacity to create a new stable landscape is referred to as transformability: the capacity to start anew from and evolve new livelihoods when existing ecological, economic or social structures change dramatically with new emergent variables introduced. Such vital changes cascade through and possibly transform the entire panarchy with its multitude of constituent adaptive cycles. Many examples of SESs exist that are becoming locked in and unable to transform until it is too late (mass tourism destinations such as Cancun, floodplains and flood control; forest fire suppression at ever larger scales). In this context, how can individual tourism businesses and tourism destinations work together to develop transformability and avoid such lock-ins? These lock-in examples exemplify the importance of adaptive capacity.

Adaptive capacity and tourism

Adaptive capacity is the ability of a social-ecological system to adapt to disturbances and maintain ecological and/or social resilience (Clarvis & Engle, 2015). The United Nations Intergovernmental Panel on Climate Change (IPCC) defines adaptive capacity as "the ability to respond successfully to climate variability and change" (Adger et al., 2007, p. 727; Dilling et al., 2015). Within social systems, adaptive capacity presents opportunities for adaptation through flexibility and learning in the context of ecological disturbances and uncertainty (Armitage, 2005; Folke, Colding & Berkes, 2003; Smit & Wandel 2006). Adaptive capacity is a function of the levels of information exchange, resources and learning opportunities related to the attributes of the system, also known as knowledge transfer (Smit & Wandel, 2006). Thus, adaptive capacity is an integral component of an SES system and an important consideration for tourism development (Potts & Harrill, 1998, 2002).

A resilient SES is not necessarily always desirable. Resilience is a concept of how to cope and respond to a disturbance event that cannot be reversed or mitigated. Resilience involves coping without fundamentally changing structure and function (Walker, Carpenter et al., 2002; Walker, Holling et al., 2004). For instance, after Hurricane Katrina, the homeless population in New Orleans demonstrated resilience in that they quickly returned to their original state of being without a permanent shelter. In this case, resilience describes the lock-in of an unwanted social condition. The term resilience is often misused when what is really meant is adaptive capacity or transformability—the ability to adapt or transform to a new stable state. In such cases, what we seek is an improved social condition, or bounce-forward, and not a bounce-back to the previous stable state (Cochrane, 2010). There are some examples of this in the tourism industry, as well, including: coral reef tourism in Australia (Biggs, 2011); gorilla tourism in Uganda (Lepp, 2002, 2008a, 2008b); and tourism recovery in Thailand and Sri Lanka (Calgaro & Cochrane, 2009).

Both increasing adaptive capacity and resilience in tourism can have positive effects since "resilience thinking acknowledges disequilibrium and nonlinear change in SESs" (Benson & Craig, 2014, p. 779) and resilience, or the "the amount of change the system can undergo and still retain the same controls on function and structure" (ibid.) is desirable in places where tourism is well balanced. Tourism is an ongoing process gaining momentum globally as new markets (e.g., China) open to further development. Resilience thinking takes this into account because of the acknowledgment of disequilibrium and the inevitability of change and the necessity of adaptation. For instance, resilience allows for change and improvement so that local communities can see benefits. From a tourism illustrative perspective, tours of the disaster zone in New Orleans have brought work and opportunities for some people taking advantage of the aftermath as a form of adaptation to the changed environmental conditions. Very soon after the hurricane, disaster tourism became a new phenomenon in New Orleans with guided boat and bus tours to neighborhoods that were severely damaged (Gotham, 2007). Some tours even include a glimpse into the social state of the homeless and seeing how the poor live in that aftermath of the hurricane.

Adaptive capacity and resilience in a tourism system

Assuming the existence of multiple stable states, resilience relates to the likelihood of a system reorganizing or flipping from one state to another as the result of a disturbance or surprise event. In tourism systems, local to global economic linkages and flows of people and resources can result in the rapid and unexpected spread of impacts from shocks or disturbances. Major ecological (e.g., Deepwater Horizon oil spill, Hurricane Katrina), health (e.g., the SARS outbreak), economic (e.g., global and regional financial crises), political (e.g., the 9/11 terrorist attacks, the Gulf War) and other shocks and disturbances (Lew, 2013) impact the social-ecological system of tourism (Cochrane, 2010) and, depending on the resiliency of the system in a given place, can cause a system to reorganize or to flip to a different state. Holling's (1987) adaptive cycle model provides insight into how such events are viewed in resilience-thinking (Figure 2.1).

Holling's (1987) adaptive cycle was largely extended from work done by Gunderson and Holling (2002). Originally conceived as a way of thinking about ecosystems and the dynamic processes that occur within a system, the adaptive cycle model illustrates the resilience concept as a cycle or loop featuring four phases:

- growth/exploitation (r phase);
- conservation (K phase);
- collapse/release (Ω or omega phase); and
- reorganization (α or alpha phase).

The cycle is depicted as a figure eight with returns to an exploitation phase to begin again. This cycle offers a more complete (systems view) of what Butler's (1980) Tourism Area Life Cycle (TALC) model began to explore (Cochrane,

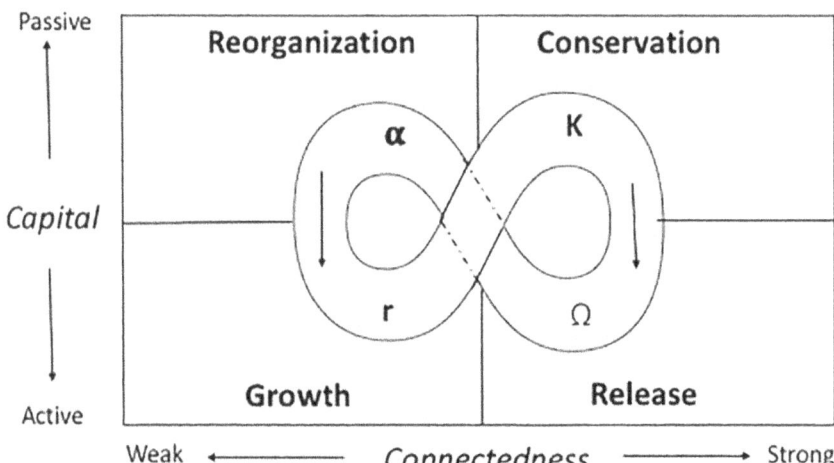

Figure 2.1 Adaptive cycle model
Source: Adapted from Holling, 1987

2010; Liburd, 2010; Farrell & Twining-Ward, 2004). Butler proposed three progressive phases of exploitation, development and conservation, leading up to one of three states:

- standstill;
- decline; and
- rejuvenation.

Meanwhile, Holling's adaptive cycle suggests that a destination would encounter decline, release and reorganization. Butler's cycle is driven by tourist numbers and does not clearly allow for reorganization after decline. Resilience might be beneficial in the introductory and growth stages of a tourism enterprise (in Butler's model), but, in the mature stage of business, sustainable tourism development including social, economic, environmental and institutional imperatives (Cottrell, Vaske & Roemer, 2013) would still be relevant as the growth rate will typically stabilize over time, thus allowing goals and benchmarks to be put in place and maintained.

The adaptive cycle can be split into two primary transitions. The r phase (growth) to K phase (conservation) represents the slow formation of structure, stability, networks and the accumulation of resources. The omega (release) and alpha phases (reorganization) refer to the rapid breakdown of the system and the manner in which it reassembles itself (Walker & Salt, 2006).

The phases of the adaptive cycle can further explain ideas that the TALC began to explore (Cochrane, 2010; Petrosillo et al. 2006; Farrell & Twining-Ward, 2005). Butler's model was based on the lifecycle of a product where products experience a period of growth followed by stability and then, if product improvements are not made, an eventual decline in appeal and sales. Butler's TALC has three fixed growth stages. The exploration stage is similar to Holling's reorganization stage; the involvement and development stages are similar to the growth stage while the consolidation stage mirrors the conservation stage (when the system becomes somewhat rigid). The three undetermined future trajectories of the TALC include stagnation, decline or rejuvenation, which correlate with the release stage of the adaptive cycle. Holling's cycle better recognizes that when faced with a release event a tourism system can reorganize or flip into another stability domain. Holling's adaptive cycle allows understanding of what could happen after the release phase, related to the decline phase in the TALC, and it helps to illustrate the lifecycle of a tourist resort in terms of connectedness and resilience from a systems approach versus the mere number of visitors (Petrosillo et al., 2006). Furthermore, while Butler's TALC implies the possibility of reorganization after the stagnation phase through rejuvenation, Hollings adaptive cycle better illustrates the possibility through the nested hierarchy of adaptive cycles resulting in a multitude of possible outcomes (panarchy).

Within a given system, multiple cycles, or figure eight loops, can exist as a nested hierarchy of adaptive cycles that are associated with subsystem elements. This is referred to as *panarchy* and adds additional layers of complexity, but also provides for a more accurate representation of system responses (Gunderson &

Holling, 2002). Cycles move at various rates across time and space. As a result, faster and smaller cycles can reorder themselves within a more stable system without destroying the greater entity. The Panarchy Cycle is self-similar. It repeats itself on many scales. The higher the spatial scale the greater the impact and the slower the change that normally takes place. For example, deep-seated cultural values usually change slowly across a larger social system and, as a result, they can moderate a more radical reorganization that might otherwise occur in a subsystem collapse. On the other hand, a sudden collapse in a subsystem can trigger a steamrolling process that leads to the collapse of the larger system, if its resilience is weak (Gunderson & Holling, 2002).

Can the adaptive cycle apply to tourism? We think it can serve as a valuable metaphor (Pendall, Foster & Cowell, 2010). A tourist destination begins in the alpha stage and "evolves continuously" in the process of ordering (van der Duim, Ren & Jóhannesson, 2012) while moving through the growth stage, where it takes shape and forms relationships in a way best suited to the point where it becomes a relatively stable environment, until a point (due to external or internal forces) when it will decline and reorganize either as a rejuvenated tourism destination following the current modes of ordering or perhaps transformed by new modes of ordering (van der Duim et al., 2012). Tourism developers and managers heavily influence the organization stage when developing infrastructure, engaging (or not engaging) local populations in decision making, marketing attractions and managing for certain types of experiences. This will affect how various actors in the network accumulate resources and position themselves in relation to each other. The case of gorilla tourism in the Bwindi Impenetrable Forest, Uganda, illustrates the various phases of the adaptive capacity cycle.

An example of adaptive capacity

The case of gorilla-viewing tourism in Bwindi Impenetrable Forest (Lepp, 2002; UNEP, 2014) is used to illustrate application of the adaptive capacity cycle and to explore the extent to which Holling's integrative theories of change may have practical utility. Bwindi Impenetrable National Park is located in southwestern Uganda in East Africa. It is most notable for the approximately 350 Bwindi gorillas, half the world's population of the critically endangered mountain gorillas. In the early to mid-1990s, the park became a popular tourist destination with gorilla-tracking being the main attraction. The park became a major tourist destination following the collapse of gorilla tourism in Rwanda due to civil war, and the absence of law and order in the Democratic Republic of Congo (UNEP, 2014). Uganda became the only safe country in which gorillas could be viewed in the wild. When applying the adaptive cycle to this case study, the late 1980s and early 1990s would be the rapid growth phase (r), ripe with opportunity for innovation as new opportunities and resource become available to be exploited (also referred to as the exploitation phase). Interconnections and regulation are relatively weak during this phase. A growth phase may or may not resemble a previous growth phase (r is the maximum rate of growth in growth models). In the case of gorilla-tracking in the

Bwindi Impenetrable Forest, this tourism was new, thus the cycle was different from previous cycles. Between 1991 and 1997 the park earned approximately US$1 million per year, visitor numbers were tightly controlled and a permit was required for entry (UNEP, 2014). This constituted a conservation phase (K) in the cycle, during which energy is stored and the system becomes more rigid (K is the parameter for carrying capacity or maximum population size in growth models). It is a phase of accumulation, and often monopolization. Connections and regulations increase in number and become stronger, such as the tightly controlled permitting process for seeing gorillas. Specialists predominate because they are more conservative and efficient. Resources are conserved and tend to be bound up in the system's structure. Growth rates slow as a loss of flexibility limits the system's ability to respond to changing external (e.g., market) and internal (e.g., demographic) conditions. Thus, as it becomes increasingly dependent on existing structures and processes, it becomes more vulnerable to disturbances.

In March 1999, a major shock to the system (or "disaster" event) took place with approximately 150 former Rwandan Interahamwe guerrilla fighters infiltrating Bwindi from the Democratic Republic of the Congo and kidnapping 14 foreign tourists and their Ugandan guide. The guerillas ultimately released 6 tourists and murdered the rest with machetes and clubs. The Ugandan guide was doused with gasoline and set on fire and the park warden was killed. The attack was reportedly intended to "destabilize Uganda" and frighten away tourists from the park, thereby depriving the Ugandan government of this important and growing stream of revenue (BBC, 1999). This would constitute a release phase (Ω) event, in which external geopolitical pressures destroy the system's network, connections and regulatory processes. It is a chaotic phase where uncertainty rules and the destruction forced the system to respond with creative reorganization, or otherwise collapse and disappear.

After the attack, the Bwindi Impenetrable National Park closed for four months. When it reopened, visitor numbers were low, many people canceled their trips and the popularity of the gorilla tours plunged, resulting in a huge loss of revenue for both local residents and the government. The Ugandan government invested significant funds to renovate the park, yet still faced the major challenge of reassuring foreign tourists that it was safe. It took several years, but investments in infrastructure, increased security measures, marketing efforts and improved revenue sharing with local Batwa people (many of whom were forced to relocate when the park was created) eventually paid off. Following the attack, park visitation dropped from 3,347 in 1998 to 2,111 in 1999. It took three years to regain pre-crisis visitation with 4,048 visitors recorded in 2002. The years immediately following the disaster exemplify the reorganization phase (α), during which a process of restoring the old system or creating an entirely new system emerges. The small event can influence and shape the larger event, as novelty, innovation and experimentation become possible. The released capital (energy) can regroup around new opportunities. This phase may or may not end with a new identity (in a new basin of attraction), but, even with the same identity, it will likely be different in some essential ways.

With greater geopolitical stability in the region, visitation reached 12,100 in 2009 helping to make tourism one of the leading foreign exchange earners for Uganda. A number of new campsites, hotels and lodges were built to cater for the growth in tourist numbers. The new infrastructure and revenue-sharing system increased community support for tourism and led to the development of improved facilities. Bwindi today provides a better quality tourist experience than before the 1999 incident. The years following the initial recovery illustrate a prolonged growth phase (r), with potential for a new adaptive cycle. Although this case is primarily a description of the phases of the adaptive cycle applied in retrospect, implications for management and application of the model follow.

Resilience and the adaptive cycle

Resilience and adaptive capacity is highest when the ecosystem begins to reorganize itself (α) because the structure is something of a blank slate and there is maximum opportunity for experimentation (Bennett, Dearden, Murray & Kadfak, 2014; Engle, 2011). The cost of failure is low since the system has not yet begun to accumulate and store resources. These conditions foster innovation and the most adaptable or well-suited actors and relationships will emerge. As a structure solidifies and the cycle moves into the growth stage, resilience remains high, but starts to decrease as patterns become locked-in. The potential for change and adaptation decreases. By the time the system reaches the conservation stage, resilience is at its lowest point. The system has reached its lock-in point and cannot adapt to change without triggering a release of energy and resources. When the system is unable to adequately respond to an external shock, it will collapse, at which point resilience is lowest. Assuming the system does not succumb to the shock, reorganization will begin a new cycle of growth in resilience, when and where an informed tourism system should play a facilitating role (Cochrane, 2010; Potts & Harrill, 2002).

Both tourists and tourism managers could identify the contributions and capabilities they bring to these adaptive processes. As one small example, at least from the hazards literature, social capital is often the most powerful force of resilience-building (Potts & Harrill, 2002). All else being equal, communities or neighborhoods with stronger and more numerous connections to persons and organizations both inside and beyond their group or community demonstrate greater resilience (Simonsen et al., 2014). Tourism operators and managers can greatly affect the way their companies organize and network with each other. Some challenging questions are:

- What types of connections are desired?
- Do we want everyone to be marginally connected to everyone else, or would it be better if a company developed fewer, but deeper connections?
- How does supply chain management or the way partnerships are structured shape these network patterns?

If the notion of panarchy is to be accepted, each tourism destination is nested within a hierarchy of systems, and these systems, along with the memories each actor brings with them of the old system, will all affect the potential to reorganize (Davoudi, 2012). The adaptive cycle model further illustrates the concept of resilience while helping to understand tourism in a specific place and time (e.g., Bwindi Impenetrable National Park).

Some of the challenges with the previous discussion is the uncertainty about knowing where a destination is located vis-à-vis Butler's TALC and adaptive capacity cycles. While it may be obvious when a major disaster takes place, even in a disaster scenario, other phases are more complicated and can only truly be known in retrospect (such as with the Bwindi National Park case study). Further, non-disaster disturbances and persistent pressures for change are difficult to identify, let alone address. Within a panarchy, some systems may fail, while others may remain robust. The following attempts to address these issues.

Implementing resilience-thinking in tourism systems

As Cochrane (2010, p. 182) suggests, "understanding how the resilience cycle works is interesting, but analysis needs to be taken forward to policy and intervention to be of practical use." She offers a diagram of the "Sphere of Tourism Resilience" in an attempt to apply complexity theory and resilience to tourism in practice and suggests that it might be used to identify "the factors of tourism system resilience and thus allows a focus on supporting these elements in order to work towards the elusive goals of sustainability" (ibid.). In addition, several existing planning frameworks provide principles to apply resilience-thinking to tourism development based on principles and process. A couple of notable examples are the Stockholm Resilience Centre's seven principles for building resilience in social-ecological systems (Simonsen et al., 2014) and the six principles included in the little-known Travel Ecology Framework (Potts & Harrill, 1998, 2002).

In 2014 the Stockholm Resilience Centre produced a booklet titled *Applying Resilience Thinking: Seven Principles for Building Resilience in Social-Ecological Systems*. The seven principles are:

- maintain diversity and redundancy;
- ʻmanage connectivity;
- manage slow variables and feedbacks;
- foster complex adaptive systems thinking;
- encourage learning;
- broaden participation; and
- promote polycentric governance systems.

Each principle is presented with an example of its application, which provides lessons for tourism development.

Cities are one type of tourism destination. In their 2013 article titled "Towards sustainable cities: Extending resilience with insights from vulnerability and transition theory," Seeliger and Turok posit that:

Cities at all stages of development need to provide jobs, food and services for their people. There is no formula that can unilaterally be applied in all urban environments to achieve this. The complex interaction of social, economic and ecological cycles within cities makes it impossible to predict outcomes. Resilience theory, with its engineering, multi-equilibria and social-ecological approaches, provides some of the foundations for understanding the full range of the complex social and ecological interactions that underpin sustainable cities. It is proposed that these insights could be extended by a sharper focus on the social and technological innovation that has traditionally been the emphasis of vulnerability and transition theories respectively.

(Seeliger & Turok, 2013, p. 5)

Like sustainable cities, all sustainable tourism destinations must be able to avoid hazardous conditions, where possible, or respond positively where risks are unavoidable and change inevitable. In some circumstances they need, above all, to withstand disasters and bounce back from adverse stress (engineering resilience) (Davoudi, 2012). Retaining essential visitor services is important to avoid systemic crises and breakdown. In other situations, it is vital that tourism destinations adapt to a new environment by adjusting certain structures and systems to match the altered conditions (multi-equilibria resilience). They may be experiencing stagnation or they may have reached a critical threshold, beyond which incremental change is insufficient and a major transformation is necessary to regain a steady state. In a third, highly dynamic situation, a process of persistent, ongoing change occurs, requiring tourism destinations to continually rearrange their structures and reorganize their institutions (social-ecological resilience). It is important to avoid becoming locked into a single development path based on outdated arrangements and inappropriate patterns of resource use.

Another approach to tourism planning that alludes to resilience-thinking is *Travel Ecology* introduced by Potts and Harrill in 1998. The relatively unknown and ideological Travel Ecology framework is a tourism planning approach uniting the themes of social development and ecological sustainability encouraging increased adaptability from within the community (Potts & Harrill, 1998; 2002). It is argued that sustainable tourism has failed thus far to account for the social and political inequalities inherent within the concept of sustainable development (Cochrane, 2010; Liburd, 2010), especially in the case of community involvement. The Travel Ecology framework poses six broad principles to consider for community sustainable tourism development (Potts & Harrill, 1998), including discovery, mutuality, locality, historicity, potentiality and enhancement. Largely adopted by authors based on their work in post-Soviet countries dealing with tourism development after rapid societal changes, travel ecology principles can be viewed as useful guidance to understand the potential for resilience and adaptive capacity within communities:

- *Discovery*—a process of community exploration, involving participation and self-awareness, and an inventory of resources (social, environment and economic). The knowing of what makes a place unique—its sense of place. The discovery principle would encourage learning and participation, thus better knowledge of resources for development and change.

- *Mutuality*—creating social capital from shared values, ideas and concerns through local engagement, respect and consensus over sustainable development, including relationship-building and capacity-building. Mutuality correlates with principles of broadening participation and promoting polycentric governance systems listed by the Stockholm Resilience Centre
- *Locality*—thinking and acting locally while being asked to think globally. All aspects of the community contribute, including shared and backyard knowledge. Community is constructed on social networks and ecological grounding, with awareness of mundane environments. Although locality might at first hand emphasize stability, it also refers to opportunities for adaptability based on local social-ecological systems.
- *Historicity*—historical and cultural knowledge as a source of self-esteem and sense of identity. Historicity as a principle couples well with the first three principles in the Stockholm Resilience Centre list. Maintaining diversity, slow variables and feedback is enabled through enhanced connectivity to cultural values and knowledge as potential for positive change and adaptability.
- *Potentiality*—reaching for optimum human achievement through integration of tourism, community and ecology as mechanisms to achieve quality of life, encouraging notions of growth and maturation, along with integration.
- *Enhancement*—going beyond conservation and preservation to enhance communities for the future through program planning and management, and implemented through political processes grounded in true democracy and dealing with resource dependency and the allocation of scarce resources through participatory deliberation. Potentiality and enhancement principles related to the Stockholm Resilience Centre list as fostering complex adaptive systems thinking keeping in mind the capacity of available ecological resources.

These principles are posed for developing tourism ideologically from a local community perspective with the potential to enhance resilience in tourism development locally. It is important to build sustainable communities resilient enough to survive in a highly competitive international environment (Ritchie & Crouch, 2003). The travel ecology approach, while stemming from aspects of political ecology, unites the themes of social development and ecological sustainability in the context of community-oriented tourism development. As Bookchin (1989, p. 33) states, "the basic problems which pit society against nature emerge from within social development itself-not between society and nature." This approach emphasizes the importance of building an adaptive social network and institutional resilience locally to address complex issues faced by societal and environmental changes (adaptive capacity and resilience in an SES).

Conclusion

The goal of this chapter is to apply resilience thinking to tourism through use of the adaptive cycle model and existing planning frameworks. With the existence of multiple stable states, resilience relates to the likelihood of a system reorganizing, or flipping, from one state to another, as the result of a major disturbance or surprise event. Holling's adaptive cycle model provided in Figure 2.1, may be useful to facilitate a dialog on the aspects of tourism development in phases of rapid growth with increasing efficiency and rigidity, versus periods of slow change exposing areas of vulnerability, versus phases of complete disorder and disintegration, versus rebuilding stages of innovation where innovation begins to emerge in pulses. The adaptive capacity cycle can be applied to tourism with success if considered in the early stages of a destination area planning process via a set of principles (e.g., Travel Ecology, principles for resilience in SES). In tourism systems, local to global economic linkages and flows of people and resources can result in the rapid and unexpected spread of impacts from shocks or disturbances. Major ecological, health, economic, political and other shocks and disturbances impact the SES of tourism and, depending on the resiliency of the system in a given place, can cause a system to reorganize, or to flip, to a different state. This chapter shows how Holling's adaptive cycle model can illustrate this resilience through a cyclical process with four phases—growth, conservation, release and reorganization—in a tourism context. Yet, there is much work to be done in the application of resilience thinking and its application to the slow change challenges of tourism development versus disaster incidents or as a retrospective academic exercise.

References

Adger, W.N., Agrawala, S., Mirza, M.M.Q., Conde, C., O'Brien, K., Pulhin, J., Pulwarty, R., Smit, B. & Takahashi, K. (2007). Assessment of adaptation practices, options, constraints and capacity. In: Parry, M.L., Canziani, O.F., Palutikof, J.P., van der Linden, P.J. & Hanson, C.E. (eds) *Climate Change 2007: Impacts, Adaptation and Vulnerability* (pp. 717–43). Cambridge: Cambridge University Press.

Armitage, D. (2005). Adaptive capacity and community-based natural resource management. *Environmental Management*, 35, 703–15.

BBC (1999). Uganda tourists "butchered." In BBC News World, March 3. Retrieved April 30, 2016, from http://news.bbc.co.uk/2/hi/africa/289196.stm

Bennett, N.J., Dearden, P., Murray, G. & Kadfak, A. (2014). The capacity to adapt? Communities in a changing climate, environment, and economy on the northern Andaman coast of Thailand. *Ecology and Society*, 19(2): 5.

Benson, M.H. & Craig, R.K. (2014). The end of sustainability. *Society & Natural Resources*, 27(7), 777–82.

Biggs, D. (2011). Understanding resilience in a vulnerable industry: the case of reef tourism in Australia. *Ecology & Society*, 16(1). Retrieved from www.ecologyandsociety.org/vol16/iss1/art30/

Bookchin, M. (1989). *Remaking Society*. New York: Black Rose Books.

Brand, F.S. & Jax, K. (2007). Focusing the meaning(s) of resilience: resilience as a descriptive concept and a boundary object. *Ecology and Society*, 12(1), 23.

Butler, R.W., (1980). The concept of a tourist area cycle of evolution: implications for management of resources. *Canadian Geography*, 24(1), 5–12.

Calgaro, E. & Cochrane, J. (2009). *Comparative Destination Vulnerability Assessment for Thailand and Sri Lanka*. Stockholm: Stockholm Environment Institute.

Clarvis, M.H. and Engle, N.L. (2015) Adaptive capacity of water governance arrangements: a comparative study of barriers and opportunities in Swiss and US states. *Regional Environmental Change*, 15, 517–27.

Cochrane, J. (2010). The sphere of tourism resilience. *Tourism Recreation Research*, 35(2), 173–85.

Cottrell, S.P., Vaske, J. & Roemer, J. (2013). Resident satisfaction with sustainable tourism: the case of Frankenwald Nature Park, Germany, *Tourism Management Perspectives*, 8, 42–48.

Davoudi, S. (2012). Resilience: a bridging concept of a dead end? *Planning Theory and Practice* 13(2), 299–333, http://dx.doi.org/10.1080/14649357.2012.677124

Dilling, L., Lackstrom, K., Haywood, B., Dow, K., Lemos, M.C., Berggren, J. & Kalafatis, S. (2015). What stakeholder needs tell us about enabling adaptive capacity: the intersection of context and information provision across regions in the United States. *Weather Climate Society*, 7, 5–17.

van der Duim, R., Ren, C. & Jóhannesson, G. (2012). Actor-Network Theory and Tourism: Ordering, Materiality and Multiplicity. London: Routledge.

Engle, N.L. (2011). Adaptive capacity and its assessment. *Global Environmental Change*, 21(2), 647–56.

Farrell, B.H. & Runyan, D. (1991). Ecology and tourism. *Annals of Tourism Research*, 8, 26–40.

Farrell, B H. & Twining-Ward, L. (2004). Reconceptualizing tourism. *Annals of Tourism Research*, 31(2), 274–95.

Farrell, B.H. & Twining-Ward, L. (2005). Seven steps towards sustainability: tourism in the context of new knowledge. *Journal of Sustainable Tourism*, 13(2), 109–22.

Folke, C. (2006). Resilience: the emergence of a perspective for social-ecological systems analyses. *Global Environmental Change*, 16, 253–67.

Folke, C., Carpenter, S., Elmqvist, T., Gunderson, L., Holling, C.S. & Walker, B. (2002). Resilience and sustainable development: building adaptive capacity in a world of transformations. *AMBIO: A journal of the human environment*, 31(5), 437–40.

Folke, C.J., Colding, J. & Berkes, F. (2003). Synthesis: building resilience and adaptive capacity in socio-ecological systems. In: Berkes F, Folke C, Colding, J (eds) *Navigating social-ecological systems: Building resilience for complexity and change*, Cambridge, MA: Cambridge University Press, pp 352–87.

Gotham, K.F. (2007). (Re) Branding the Big Easy tourism rebuilding in post-Katrina New Orleans. *Urban Affairs Review*, 42(6), 823–50.

de Groot, R. & Ramakrishnan, P.S. (2005). Cultural and amenity services. In: Hasan, R., Scholes, R. & Ash, N. (eds), Millennium Ecosystem Assessment. Washington, DC: Island Press, pp. 455–76.

Gunderson, L.H. (2000). Resilience in theory and practice. *Annual Review of Ecology Systematics*, 31, 425–39.

Gunderson, L.H. & Holling, C.S. (eds.) (2002). *Panarchy: Understanding Transformations in Human and Natural Systems*. Washington, DC: Island Press.

Holling, C.S. (1973). Resilience and stability of ecological systems. *Annual Review of Ecology and Systematics*, 1–23.

Holling, C.S. (1986). The resilience of terrestrial ecosystems: local surprise and global change, in W.C. Clark and R.E. Munn (eds.), *Sustainable Development of the Biosphere* (pp. 292–317). Cambridge: Cambridge University Press.

Holling, C.S. (1987). Simplifying the complex: the paradigms of ecological function and structure. *European Journal of Operational Research*, 30(2), 139–46.

Holling, C.S., (1996). Engineering resilience versus ecological resilience. In: Schulze, P.C. (Ed.), *Engineering within Ecological Constraints* (pp. 31–44). Washington, DC: National Academy Press.

Holling, C.S., Gunderson, L.H. & Peterson, G.D. (2002). Surprises and sustainability cycles of renewal in the Everglades in *Panarchy: Understanding Transformations in Human and Natural Systems*, L.H. Gunderson and C.S. Holling, eds., Washington, DC: Island Press, pp. 63–102.

Jandreau, C. & Berkes, F. (2016). Continuity and change within the social-ecological and political landscape of the Maasai Mara, Kenya. *Pastoralism*, 6(1), 1–15.

Lepp, A. (2002). Uganda's Bwindi Impenetrable national park: meeting the challenges of conservation and community development through sustainable tourism. In Harris, R., Griffin, T. & Williams, P. (eds.) *Sustainable Tourism: A Global Perspective* (pp. 210–20). New York: Elsevier.

Lepp, A. (2008a). Attitudes towards initial tourism development in a community with no prior tourism experience: the case of Bigodi, Uganda. *Journal of Sustainable Tourism*, 16(1), 5–22.

Lepp, A. (2008b). Tourism and dependency: an analysis of Bigodi Village, Uganda. *Tourism Management*, 29, 1206–14.

Lew, A.A. (2013). Scale, change and resilience in community tourism planning, *Tourism Geographies*, 16(1), 14–22.

Lew, A.A., Ng, P.T., Ni, C-C. & Wu, T-C. (2016). Community sustainability and resilience: similarities, differences and indicators. *Tourism Geographies*, 18(1):18–27.

Liburd, J.J. (2010). Introduction to sustainable tourism development. In J.J. Liburd & D. Edwards, *Understanding the Sustainable Development of Tourism* (pp. 1–18). Oxford: Goodfellow Publishers.

Lóránt, D. (2011). Tourism ecology: towards the responsible, sustainable tourism future, *Worldwide Hospitality and Tourism Themes*, 3(3), 210–16.

Malanson, J.P. (1999). Considering complexity, *Annals of the Association of American Geographers*, 89, 746–53.

MEA (2005). *Ecosystems and Human Well-being*. Vol. 5. Millennium Ecosystem Assessment. Washington, DC: Island Press.

Miller, G. & Twinning-Ward, L. (2005). *Monitoring for a Sustainable Tourism Transition: The Challenge of Developing and Using Indicators*, Wallingford: CABI Publishing.

Pendall, R., Foster, K.A. & Cowell, M. (2010). Resilience and regions: building understanding of the metaphor. *Cambridge Journal of Regions, Economy and Society*, 3(1), 71–84.

Petrosillo, I., Zurlini, G., Grato, E. & Zaccarelli, N. (2006). Indicating fragility of socio-ecological tourism-based systems. *Ecological Indicators*, 6(1), 104–13.

Potts, T.D. & Harrill, R. (1998). Enhancing communities for sustainability: a travel ecology approach, *Tourism Analysis*, 3, 133–42.

Potts, T.D. & Harrill, R. (2002). Travel ecology and developing naturally: making theory–practise connections. In R. Harris, T. Griffin, T. & P. Williams (eds.) *Sustainable Tourism: A Global Perspective* (pp. 45–57). New York: Butterworth & Hienemann

Ritchie, J.R.B. & Crouch, G.I. (2003). *The Competitive Destination: A Sustainable Tourism Perspective*. Cambridge, MA: CABI Publishing.

Seeliger, L. & Turok, I. (2013). Towards sustainable cities: extending resilience with insights from vulnerability and transition theory. *Sustainability*, 5, 2108–28.

Seidl, R. (2014). The shape of ecosystem management to come: anticipating risks and fostering resilience. *BioScience,* 16, 1159–69.

Simonsen, S.H., Biggs, R., Schluter, M., Schoon, M., Bohensky, E., Cundill, G., Dakos, V., Daw, T., Kotschy, K., Leitch, A., Quinlan, A., Peterson, G & Moberg, F. (2014). Applying resilience thinking: seven principles for building resilience in social-ecological systems. Stockholm: Stockholm University.

Smit, B. & Wandel, J. (2006). Adaptation, adaptive capacity and vulnerability. *Global Environmental Change-Human Policy Dimension,* 16, 282–92.

Stockholm Resilience Centre (2014) Applying resilience thinking: seven principles for building resilience in social-ecological systems. Stockholm: Stockholm University.

Tyler, D. & Dangerfield, J.M. (1999). Ecosystem tourism: a resource-based philosophy for ecotourism, *Journal of Sustainable Tourism*, 7(2), 146–58.

UNEP (2014). Bwindi Impenetrable National Park, Uganda. United Nations Environment Programme. Retrieved from http://www.eoearth.org/view/article/150830/

Walker, B.H. & Salt, D. (2006). *Resilience Thinking: Sustaining Ecosystems and People in a Changing World*, Washington, DC: Island Press.

Walker, B., Carpenter, S., Anderies, J., Abel, N., Cumming, G.S., Janssen, M., Lebel, L., Norberg, J., Peterson, G.D. & Pritchard, R., (2002). Resilience management in social–ecological systems: a working hypothesis for a participatory approach. *Conservation Ecology*, 6(1), 14. Retrieved from www.consecol.org/vol6/iss1/art14

Walker, B., Holling, C.S., Carpenter, S.R. & Kinzig, A. (2004). Resilience, adaptability and transformability in social-ecological systems. *Ecology and Society*, 9(2), 5.

3 The sustainable and resilient community

A new paradigm for community development

Alan A. Lew, Chin-cheng Ni, Tsung-chiung Wu and Pin T. Ng

The "holy grail"

Almost two decades ago, Graham Tobin (1999) referred to the integration of community sustainability and community resilience as the "holy grail" for natural hazards planning. In his conceptualization, sustainability was equated to a "mitigation" approach that sought to prevent natural disasters by reducing exposure, risk and vulnerability across the community. Resilience was associated with a "recovery" scenario that not only brought the community back to a state of normalcy but did so in an equitable manner with both short- and long-term policies that addressed the root causes of the disaster experience. He suggested that natural hazard/disaster planning could only be effective if sustainability and resilience were consistent and integrated, and equitably applied as part of an ongoing planning process that accounted for the changing social dynamics (geography, demography and social psychology) of a place. More recently, Weichselgartner and Kelman (2014) made a similar argument, suggesting that we need to recognize the limitations of sustainability and resilience, starting with the vague definitions that surround each concept.

Our study examined sustainability and resilience in rural tourism communities in Taiwan and resulted in conclusions that were very much in line with those of Tobin (1999) and Weichselgartner and Kelman (2014). Disasters, however, were not the primary focus of our research. Instead, we looked at how communities responded to the full range of changes impacting rural societies today, through the lenses of sustainability and resilience. Three of the eight communities that we studied, however, had experienced major natural disasters in recent memory. This was due to Taiwan's precarious geography on the Pacific Ring of Fire and directly within the Pacific typhoon path, and these disasters turned out to be a key factor in our research findings. We also differed from Tobin by adopting broader definitions of sustainability and resilience that extend beyond the natural hazards disaster context, which was his focus. As such, our conclusions also spread beyond recommendations for communities under threat from natural disasters. In addition, we argue that sustainability and resilience should be the development model for all communities for both short-term and long-term success in today's world.

An essential element in understanding Tobin's research findings, along with those of our own research and other similar studies, is how sustainability and resilience are defined. Based on the definitions that are adopted, the relationship between these two major concepts in contemporary community development can then be assessed, along with their application to real world contexts.

Definitions of resilience and sustainability

Numerous authors have attempted to define the relationship between sustainability and resilience. The result is that almost every possible configuration between the two concepts has been proposed (Table 3.1). Some view resilience as being the same as sustainability, but with a systems science perspective (Adger, 2003; Holling & Walker, 2003). Such a view seems common among sustainability enthusiasts in the general public as well. Others view resilience as being a subset of the broader concept of sustainability, either by providing indicators of sustainability (Walker & Salt 2006; Schianetz & Kavanagh, 2008; Magis, 2010), or as one way of implementing sustainability policies (Farrell & Twining-Ward, 2005; Anderies, et al., 2013). An opposite position to this places sustainability within the broader context of resilience (Levin, et al., 1998; Peirce, Budd & Lovrich, 2011; Strunz, 2012; Anderies, Folke, Walker & Ostrom 2013). This approach views strong resilience as a prerequisite for a system before a normative (ethical) sustainability approach can be successful.

Many of the definitions reviewed so far hint at how and why authors place the two concepts in opposition to one another, with sustainability being a mitigation and conservation policy and resilience being an adaptation and innovation approach (Tobin, 1999; Prasad, Ranghieri, Shah, Trohanis, Kessler & Sinha, 2009; Juech & Michelson, 2011; Peirce, Budd & Lovrich, 2011; Davoudi, 2012; McLellan, Zhang, Farzaneh, Utama & Ishihara, 2012; Jepson, 2016; Lew, 2014; Weichselgartner & Kelman, 2014; Lew, Ng, Ni & Wu, 2016). Others hold the same view, but are open to the possibility of overlap between the two (Brand & Jax, 2007; Derissen, Quaas & Baumgärtner, 2011). (Note that some authors seem to have expressed different opinions on these definitions over time, or perhaps in different contexts.)

With the exception of some of the definitions that see little or no distinction between sustainability and resilience, most authors are of the opinion that considerable differences exist. The idea that they are the same comes from the degree to which sustainability and resilience share some common goals and research perspectives when seen from a general point of view. One way to think of this broader view is to consider each concept as a metaphor (Carpenter et al., 2001; Strunz, 2012). As a metaphor, both sustainability and resilience can similarly apply to any action that sustains the current system and the current status quo indefinitely. System survival as a shared goal clearly applies to both concepts (Table 3.2) (Lew, Ng, Ni & Wu, 2016). Both approaches also hold the assumption that human societies and natural environments comprise a single ecosystem that has the potential to be in harmony through appropriate policy measures.

Our research into rural tourism communities in Taiwan began with the assumption that sustainability and resilience were two distinctly different approaches to

Table 3.1 Possible relationships between sustainability and resilience

	Relationship	Descriptions
	Sustainability = Resilience	Resilience as a systems science view of sustainability (Adger, 2003) A resilient socio-ecological system is synonymous with a region that is ecologically, economically, and socially sustainable (Holling & Walker, 2003, p. 1)
	Resilience ⊂ Sustainability	Sustainability is the broad social goal; resilience is how it can be implemented (Farrell & Twining-Ward; 2005; Fiksel, 2006; Anderies et al., 2013) Resilience is an indicator of sustainability (Walker & Salt, 2006; Schianetz & Kavanagh, 2008; Magis, 2010)
	Sustainability ⊂ Resilience	Resilience describes the overall condition of the system; sustainability provides a normative goal for system resilience (Derissen et al., 2011; Peirce, Budd & Lovrich, 2011; Strunz, 2012) Resilience is the base criterion for sustainable development and resources use (Adger, 1997; Perrings, 2006, p. 418)
	Sustainability ≠ Resilience	Sustainability mitigates change by maintaining resources above normative safe level; resilience adapts to change by building capacities to return to a desired state following a disruption (Tobin, 1999; Simmie & Martin, 2010; Davoudi, 2012)
	Sustainability ∩ Resilience	Pure resilience is descriptive and pure sustainability is normative, but with degrees of overlap when integrated (Brand & Jax, 2007; Derissen et al., 2011)

Source: Lew, Ng, Wu & Ni (2016)

community development. As the research developed, we shifted that view to one where they were still very different from one another, but with some degree of overlap. We view the overlap lying in areas where policy goals and actions are aligned with similarities defined in Table 3.2. This can happen when, for instance, we assessed the macro-level condition of the economies in our eight rural communities in Taiwan. Whether the economy is strong or weak has both sustainability and resilience implications, which cannot be clearly separated at that particular level of generalization.

Table 3.2 Similarities and differences in the assumptions and goals of sustainability and resilience

Similarities	Sustainability and Resilience
Assumptions	Human societies and natural environments are deeply linked in a single social-ecological system; harmony between the human and natural parts of this system is possible
Goals	System survivability; conservation of core resources ("slow controlling variables"); sense of place and belonging
Research and Practice	Community development; climate change policies and actions; education and learning as an implementation tool

Differences	Sustainability	Resilience
Assumptions	Stability and balance are the norm (or are at least possible)	Nonlinear and unpredictable change and chaos are the norm
Goals	Normative ideals (culture, environment and economic balance, efficiency and equity)	Adaptive management; diverse and redundant systems; learning institutions; social capital
Research Foci	Environmental and social impacts of economic development; over use of resources; carbon footprints	Natural and human crisis/disaster management; climate change impacts; adaptive cycles; social capital and networks
Methods and Practice	Conservation and mitigation against change; resource restoration; recycling and "greening"; education for behavior change	Reducing vulnerabilities and risk; increasing physical and social capacities for change (flexibility, redundancy and innovation); systems modeling
Criticisms	Radical and globalist agendas; highly contested politics and policies	Neoliberal agenda: "personal change" over "structural/governance change"; "adaptation" over "conservation"

Sources: Based on Adger, 1997; Fiksel, 2006; Brand & Jax, 2007; Prasad et al., 2009; Derissen et al., 2011; Magis, 2010; Redman, 2014; Lew, Ng, Ni & Wu, 2016

The two concepts differ primarily in their assumptions about the normal state and in the goals that result from those assumptions. Sustainability assumes that there is an ideal world which is moral and ethical in its relationships among humans and between humans and nature. This normative goal includes both intergenerational equity (for future generations) and intra-generational equity (which was defined as between rich and poor countries in the original *Brundtland Report*, WCED, 1987). Resilience, by most definitions (Brand & Jax, 2007) does not have an equity component. Instead, some of its earliest influences come out of chaos theory (Faulkner & Russell, 1997; McKercher, 1999) and disaster management (Schwab, 1998). In one of the earliest conceptualizations, Holling (1973, p. 14) defined the resilience of ecological systems, another major origin of the concept, as the descriptive and quantitative "measure of the persistence of systems and their ability to absorb change and disturbance and still maintain the same relationships between populations and state variables." However, it should be noted that critics of resilience contend that the assumption of resilience as being descriptive and non-normative actually hides a neoliberal agenda that is opposed to the equity goals of sustainability (MacKinnon & Derickson 2012; Weichselgartner & Kelman, 2014; Evans & Reid 2015).

Creating a stable and equitable society (sustainability) and creating an adaptable and innovative society that can bounce back from disturbances and change (resilience) are the most fundamental differences between the goals of sustainability and resilience. While these can be compatible, they can also be at odds with each other, especially when considering non-metaphorical scientific and policy definitions of sustainability and resilience. This has to do with the range of definitions that have been applied to these two important concepts over the years, which extend from the simple to the complex (Table 3.3). In its simplest definition, sustainability means to maintain the status quo indefinitely, which has a high degree of overlap with engineering resilience approaches. This is often what is meant when tourism destination marketing organizations, and similar tourism boosters, talk about the sustainability of the tourism economy (Butler, 1999). The next level of understanding of sustainability is a focus on the efficient use of carbon resources to reduce greenhouse gas emissions from an activity or a community. This has been the primary focus of most green certification programs that are common in the tourism sector. Some of these programs are now attempting to measure broader social impacts, which moves them into a more advanced definition of sustainability.

The intermediate definition of sustainability in Table 3 is the one that most lay-people consider when they hear the word "sustainability", which is to take actions that protect the natural environment (Peirce, Budd & Lovrich 2011). This was the primary focus of the *Brundtland Report* when it introduced the concept of sustainable development in 1987, although the report also had broader social goals (WCED, 1987). In general, only more knowledgeable individuals, including academics and other experts, are aware that cultural resources are also a major part of the concept of sustainability, which is the more advanced definition that the *Brundtland Report* was moving to address.

The comprehensive definition of sustainability is the least understood, and probably most perplexing to many, due to its complexity and all-inclusiveness. The United Nations Sustainable Development Goals for 2030 (UN, 2015) is an example of the comprehensive definition of sustainability. It includes 17 highly diverse objectives, including ending all poverty and hunger on the planet, protecting all life under water and on land, and eliminating gender and other social inequalities. Each of the 17 has more specific and quantifiable goals for the world to reach in 2030. Resilience is mentioned in several goals, but is specifically taken up in Goal 13 on climate change and it impacts, which mentions mitigation (sustainability; preventing climate change), but tends to focus more on adaptation (general resilience), and impact reduction and impact recovery (engineering resilience).

Table 3.3 The range of complexity in definitions of sustainability and resilience, from simple to comprehensive

Degree of Complexity	Simple	Moderate	Intermediate	Advanced	Comprehensive
Sustainability	Maintaining the status quo	Reducing carbon emissions and using recycling to mitigate climate change	Protecting the natural environment from deterioration and loss	Conserving and restoring natural and cultural resources	Creating fair and equitable well-being for all life on the planet
Resilience	Strength or toughness in adverse situations	Ability to bounce back from a disturbance or disaster	Capacity to adapt, innovate and transform in response to change and disasters	Adaptive cycles (temporal) and panarchy (spatial) responses to change	Adaptive management and learning institutions for nonlinear change and chaos
Forms of Resilience	Engineering resilience		General resilience	Ecological resilience	Evolutionary resilience

Source: Based on Lew, Ng, Wu & Ni (2016)

In terms of the different ways that people use the term resilience in a community context, the simplest is to mean toughness or persistence in some type of difficult situation, which can be viewed as very similar, if not identical, to the simplest definition of sustainability (Table 3.3). People who have not been exposed to resilience theory and resilience thinking are most likely to use this definition, which is almost always considered a good characteristic for a community or other system to have. A more moderate definition is the ability of a community to bounce back to a state of normalcy following a crisis or disaster. Crisis and disaster management is actually the most common type of published academic research related to community resilience (Meerow & Newell, 2015). Although some of that research falls under the engineering resilience approach that is most associated with disaster management, more of it focuses on the longer-term social recovery from the disaster, which is characteristic of the intermediate definition of resilience in Table 3.3. This also includes resilience to non-disaster change events, such as economic and cultural globalization, which aligns with the much sought after goal of measuring the overall general resilience of a community.

The advanced definition of resilience comprises the bulk of resilience theory literature that applies systems science modeling to understanding the social-ecological resilience of a place (system). Much of this comes out of the ecological sciences and generally tends to emphasize applications to natural ecosystems (Carpenter et al., 2001; Meerow & Newell, 2015), although social and economic systems have also been modeled (Adger, 1997; Baggio & Salnaghi, 2011; Ranjan, 2012). Strunz (2012) suggests that, due to the conceptual vagueness of resilience (which also applies to sustainability), it is better to think of all of the definitions of resilience as "resilience thinking," and limit the use of "resilience theory" to the narrow use of models to describe characteristics of a system (the advanced definition in Table 3.3). Strunz sees the vagueness in resilience thinking as allowing greater creativity, enabling interdisciplinary and transdisciplinary (outside of academia) communication, and offering more of a focus on problem solving instead of solving quantitative models (which is the focus of resilience theory). These characteristics have also been true for the concept of sustainability (Hunter 1997; Fennell, 2015).

Coming out of systems science, however, resilience also lends itself to more precise scientific and quantitative modeling to identify thresholds and regime shifts that signify the transformation of a system from one state to another (Carpenter et al., 2001; Fiksel, 2006; Strunz, 2012). This tends to be easier for environmental systems with minimal human impacts, although modeling similar economic shifts has been attempted (Weichselgartner & Kelman, 2014). However, such modeling can also be metaphorical, such as describing how the combined impacts of the Asian Economic Crisis of 1997–8 and the subsequent fall of President Suharto impacted international tourism to Indonesia in a manner that is reflected in the adaptive cycle model (Cochrane, 2010).

The comprehensive approach to resilience is the most complex, in part because evolutionary resilience is one of the newest concepts in resilience thinking (Simmie & Martin, 2010). The idea of constant change was suggested in the earliest definitions of resilience by Holling (1973), but was never advanced, taking a back seat to models that assumed various forms of stability and equilibria that are separated by threshold events. Ecological resilience rejects these concepts and attempts to build an understanding of a world that is in constant flux, if not in actual chaos (Davoudi, 2012). This is, however, a difficult concept to fully comprehend from a rational point of view and is still open to exploration and interpretation.

For our research on resilience and sustainability in rural Taiwan, we adopted the advanced definition of sustainability: the conservation or restoration of human and natural resources to a desired (normative) level or condition as a way of mitigating (preventing) undesirable change. For resilience, we adopted the intermediate definition: the capacity of a community for innovation and adaptation to change in response to both disaster events and slow to moderate environmental, social and economic change. These definitions are in line with Tobin (1999) and are supported by a numbers of other researchers, as noted above. This approach provides a clear conceptual foundation for comparing and operationalizing sustainability and resilience. It embraces the ethical and normative nature of sustainability, although it places it in a narrower context than the UN's 2030 Sustainable Development Goals. Our definition of resilience is metaphorical, but can be quantified to a degree, while also offering considerable qualitative research possibilities. The definitions we adopted of sustainability and resilience were well suited to clearly defining each concept and categorizing responses from our informants.

Rural Taiwan tourism communities

To empirically address the relationship between sustainability and resilience, we examined community development experiences in eight rural villages in Taiwan, with populations ranging from 500 to 1,000 residents. All of the villages had some form of tourism-related activities, although they ranged from the very small and nascent, to being major rural destinations. The villages were distributed across a range of geographic settings, including two each in high mountains (Alishan area), agricultural basins (western Taiwan), coastal wetlands (northeast and southwest Taiwan), and small islands (Pescadores/Penghu). Three of the villages had experienced one or both of the most significant natural disasters to ever hit Taiwan. These were the 921 Earthquake (September 21, 1999), which killed over 2,400 people in central Taiwan, and Typhoon Morakot in 2009, which brought record rainfalls, flooding and landslides, resulting in 461 deaths. The other five had not experienced major disasters within the memory of the people interviewed.

The data are based on 22 in-depth interviews with community informants in these village, ranging from 1.5 to 2.5 hours in length. The informants included: current and past elected village association leaders; non-governmental, non-profit, and social enterprise leaders; consultants working in the communities; and local entrepreneurs, farmers, tour guides and nature interpreters. Often one individual held more than one of these roles in the community. Our questionnaire contained specific sets of questions on local community budget expenditures, local knowledge and community well-being. This level of community organization in Taiwan usually has no regular budgetary allowance. Instead, they rely on grant proposals and other requests to the central government or county government to fund special, short-term projects.

To operationalize our definitions of sustainability and resilience, we adopted qualitative indicators that specifically point to their differences (Table 3.4). The differences are not in the subject categories (such as community education and economic development), but in the perspective and goals that the community seeks within each category. Based on our definitions (above), sustainability goals are conservation- and mitigation-oriented, while those of resilience are innovation- and adaptation-oriented. The orientation is reflected through public policies and programs. In addition, none of the options shown are new models or ways of doing things. Communities and other social groups have always had to make decisions along these lines. When a community development issue arises, the basic questions are: "What do we conserve or protect from change?" and "What do we innovate or allow to change?" "How" we conserve and innovate is, of course, also a crucial question in this context.

Table 3.4 Sustainability and resilience indicators

Indicator	Sustainability Goals	Resilience Goals
Economic Development	Improving economic efficiencies	Diversifying the economy
Community Education	Teaching traditional cultural knowledge	Teaching new and emerging skills
Environmental Education	Teaching traditional knowledge	Teaching science education
Environmental Management	Conservation and restoration	Adapting to environmental impacts
Infrastructure (roads, utilities)	Improving existing infrastructure	Creating diversified infrastructure options and redundancies (backups)
Health Care	Providing community-wide basic health care	Providing emergency or alternative health care

Source: Lew, Ng, Ni & Wu (2016)

Table 3.5 summarizes the sustainability and resilience policies and programs that we found in rural tourism communities in Taiwan. It compares them for the three communities that had experienced disasters in recent memory, and the five non-disaster experienced communities, using the indicator guideline from Table 3.4, although some of the categories are combined where appropriate. The results reflect both public policy options that the communities have taken and the outcomes of some of those options. Policy options are driven by a combination of the types of central government and county programs that are available to village communities, and the preferences within the villages as to which programs they will apply for, based on the limited resource that many of them have.

In terms of economic sustainability, both disaster and non-disaster experienced communities (hereafter referred to as "disaster communities" and "non-disaster communities", respectively) had very low rates of unemployment, with respondents often commenting that anyone who wants a job can likely find one locally. Poverty is also not an issue in any of the communities studied. However, the disaster communities clearly had stronger economies overall, both in tourism and in their more traditional agricultural activities. This contributed to lower rates of younger people migrating to larger cities for employment than was the case in non-disaster communities. Economic sustainability (defined as the efficient and non-wasteful use of resources: Lubin & Esty, 2010) was, therefore, more robust in the disaster communities studied. This situation is mostly likely due to the resilience measures taken by disaster communities in terms of economic development. These communities devote a much larger percentage of their local revenue to economic development, much of which is focused on developing tourism in different ways (accommodations, food, interpretation, recreation) to diversify their economic options. The high tourism employment percentage for these communities is a reflection of this resilience-building diversification.

The leisure agricultural programs of the Council of Agriculture (Executive Yuan, R.O.C.) are possibly the most important long-term resilience-building policies that was encountered in the study. These programs were in place in the two agricultural communities studied (one a disaster community; the other a non-disaster community). Taiwan's agricultural sector, not being able to compete with lower-wage countries, requires subsidies from the central government to enable farmers to survive. The Leisure Agricultural Area program, in particular, is helping areas that apply for funding under it to rebrand themselves as recreation and tourism destinations for Taiwan's urban populations. These areas are more likely to grow higher-value organic and specialty crops and visitors are more able to meet the farmers and buy directly from them, both in-person and online. The program has a strong educational component to teach farmers how to brand themselves in these ways and how to market their products online.

In interpreting the budgetary spending rates in Table 3.5, it is important to recognize that average annual budgets for disaster communities were five times higher than that of non-disaster communities (US$152,000 compared to US$29,000). Spending rates shown are the percentage of total budget spent on each category. Because local village budgets are based mainly on successful grant applications to specific programs, they vary considerably from one year to the next. The averages shown here were based on the current and previous one to two years for communities in each grouping. The significance of this difference is discussed in greater detail below.

In terms of local knowledge, sustainability approaches are very strong in both disaster and non-disaster communities, as reflected in comparably high rates of environmental knowledge and fairly high use of natural resources obtained from the environment (excluding gardening and farming). Traditional religious practices were extremely high in all of the communities interviewed, which reflects the strength of traditional Chinese religion (a combination of Taoism and Buddhism) across Taiwan (Hu & Yang, 2014). From a resilience perspective, however, disaster communities were much more likely to engage their residents in different forms of formal education than were non-disaster communities. Education is a key component in building resilience because it fosters problem solving, creativity and innovation in adapting or transforming to changing conditions (Walker et al., 2004; Berkes, 2007; Strunz, 2012).

For environmental sustainability, non-disaster communities devote twice as much of their budgets to environment conservation and environmental education efforts than do disaster communities. However, when the much higher total budgets of disaster communities are taken into account, disaster communities are seen to spend twice as many dollars on environmental sustainability initiatives. Thus, it is generally safe to suggest that environmental sustainability is a significant value shared across all of the communities in rural Taiwan. From a resilience perspective, all of the communities in which agriculture is a major practice (which excluded the wetland and island villages) have experienced some changes in the

Table 3.5 Sustainability and resilience indicators for disaster- and non-disaster-experienced tourism communities in rural Taiwan

Indicator	Sustainability Goals	Resilience Goals
Economic Development		
	Improving Economic Efficiencies	*Diversifying the Economy*
Disaster-experienced Communities[1]	**Very low** unemployment rates (3%), poverty and income inequality; **Lower** rates of employment migration (24%); **Stronger** and better integrated economies in both the tourism and agriculture sectors[2]	**Much higher** rates of tourism employment (23%); **Higher** spending rates for economic development (11%); **Leisure Agricultural Area** programs in agricultural communities
Non-Disaster-experienced Communities[1]	**Very low** unemployment rates (4%), poverty and income inequality; **Higher** rates of employment migration (36%); **Somewhat weaker** tourism and agricultural sectors[2]	**Lower** rates of tourism employment (5%); **Lower** spending rates for economic development (7%); **Leisure Agricultural Area** programs in agricultural communities
Community Education and Local Knowledge		
	Teaching Traditional Cultural and Environmental Knowledge	*Teaching New and Emerging Skills, and Science Education*
Disaster-experienced Communities	**High** rates of traditional environmental knowledge (70%) and use of local environmental resources (31%); **High** rates of religious participation (99%)	**Much higher** rates of formal environmental (41%), agriculture (39%) and disaster (45%) education
Non-Disaster-experienced Communities	**High** rates of traditional environmental knowledge (63%) and use of local environmental resources (37%); **High** rates of religious participation (99%)	**Lower** rates of formal environmental (12%), agriculture (19%) and disaster (13%) education
Environmental Management		
	Conservation and Restoration	*Adapting to Environmental Change*
Disaster-experienced Communities	**Lower** spending rates for environmental protection (8%) and environmental education (6%), but **higher** total dollar amounts[3]	**Some** changes in crops cultivated in agricultural areas to adapt to climate, domestic tourism and global market changes
Non-Disaster Experienced Communities	**Higher** spending rates for environmental protection (16%) and environmental education (14%), but **lower** total dollar amounts[3]	**Some** changes in crops cultivated in agricultural areas to adapt to climate, domestic tourism and global market changes

Infrastructure (roads, utilities) and Health Care

	Improving Existing infrastructure; Providing Basic Health Care	*Creating Diversified Infrastructure Options and Redundancies (backups); Providing Emergency or Alternative Health Care*
Disaster-experienced Communities	**Higher** spending rates for infrastructure (60%) for strengthening existing infrastructure; **Lower** spending rates for community health (5%)[3]	**Higher** spending rates for infrastructure (60%) – some of this is for developing alternative road options
Non-Disaster-experienced Communities	**Lower** spending rates for infrastructure (35%); **Higher** spending rates for community health (10%)[3]	**Some** alternative health care programs in one community

Notes:
1. Natural disaster experienced communities = 3; Non-disaster experienced communities = 5; percentages are for either resident or enrolled population, unless spending is mentioned in the description.
2. Although listed in the sustainability column, the assessment of overall economic strength is an overlapping variable that has implications for both sustainability and resilience, as discussed in the text.
3. Spending rates and total dollar amounts refer to village funds spent. Disaster experienced communities had average annual budgets that were five times higher than that of non-disaster experienced communities (US$152,000 compared to US$29,000). Spending rates shown are the percentage of total budget spent on each category.

Source: Authors

types of crop that they plant, which reflects slow change in both environment conditions (climate change), in the domestic tourism market and in the global agricultural economy. Overall, we did not encounter any major differences between disaster and non-disaster communities in terms of environmental sustainability and resilience.

This lack of difference between the two community groups was also true for health care when total budget dollars are taken into account. Health care is provided mostly through counties in Taiwan and supported by easily obtained grants from county governments. On the other hand, spending on infrastructure was significantly different, with disaster communities devoting a much higher percentage of the budgets and an even higher total dollar amount to the upgrading and maintenance of roads and utilities, which is a sustainability approach to infrastructure. The much higher budgets in disaster communities is a reflection of their greater infrastructure needs due to the more precarious geographies that makes them more susceptible to disaster in the first place.

Infrastructure strengthening is referred to as "engineering resilience" in the resilience literature (Holling 1973; Pimm 1984) and is conceptually much more narrowly defined than are ideas related to social-ecological resilience. Based on our indicators (Table 3.4), engineering resilience is a form of sustainability when the focus is on strengthening existing infrastructure. It would be a resilience indicator if the focus were instead on building redundancies and options into the system. Based on this, the disaster communities appear to have more sustainable infrastructure systems, although this can only be measured by how well it responds to the next disaster.

In summary, sustainability policies and outcomes were found to be mostly comparable across communities whether or not they have experienced natural disasters. Resilience policies and practices were much stronger in the natural disaster communities, mostly in the areas of economic development and formal community education. It may be that the other areas (environmental management, infrastructure and health care) are not good indicators of community resilience because, at least in the Taiwan cases, they are either too dependent on the distinct geographies of a place or are too uniform across the political landscape due to central government policies.

Not reflected in the numbers, but evident in interviews, was a much higher degree of collaboration and cooperation in disaster communities, as well as higher levels of entrepreneurship and forward vision. These less definable characteristics appear to be a direct result of the disaster experience which all of the interviewees in the three communities considered a turning point in their development. While such an appraisal is unlikely in every post-disaster instance, these specific cases experienced positive outcomes due to:

- special post-disaster recovery funding from the central government for 20 months, which provided better knowledge of government funding programs and the application processes after the disaster;
- higher levels of community cooperation, shared vision and willingness to innovate after the disaster experience;
- greater general awareness and identification with communities across Taiwan and abroad that have had similar disaster and recovery experiences; and
- new opportunities for some entrepreneurs following the destruction of their buildings and businesses.

Tourism discussion

Although not a specific focus of the research cited above, tourism was a significant development concern in all the communities that were interviewed, and the results provided insight into potential relationships between tourism development and community resilience. In rural Taiwan, as in many other parts of the world, rural landscapes are being transformed by a new "Romanticism" among the ever increasing numbers of urban residents who see them as a domain of escape and rejuvenation (Urry, 1995; Garrod, Wornell & Youell, 2006; Knudsen & Greer, 2011). Both natural disaster and non-disaster experienced communities in Taiwan have made tourism a key component in their local economic revitalization plans, encouraged by the central government and in response to growing domestic and international tourist arrivals. In our sample, the natural disaster communities clearly had greater success in developing tourism.

The disaster experience offered opportunities for individuals and communities to innovate as they were less bound by the legacies of the past path-dependencies. For example, Ruili is a disaster community in the Alishan mountain scenic area that was hit by both the 921 Earthquake in 1999 and massive flooding and landslides from Typhoon Morakot in 2009. Its traditional economy is based on tea production, which continues to be a significant source of wealth for many of its residents. Tourism is also prevalent in Ruili, primarily through the 39 bed and breakfast establishments that operate there. In general, the more a business person is involved in tourism, the more they appear to have a desire to see the community work together as a whole. Those who operate solely in the tea sector were more focused on their individual work activities and sometimes expressed annoyance about visiting tourists. Tourism entrepreneurs (many of whom also grow tea) were focused on the special sense of place that tea created for Ruili, and initiated training in the traditional tea ceremony for local residents as a way to strengthen that sense of place. At the same time, there was also more awareness of disaster issues because their sector was more susceptible to declines in tourist arrivals due to damaged infrastructure. Tea plantations, on the other hand, have high physical resilience to both typhoon flooding and earthquakes.

Communities such as Ruili exhibited many of the characteristics identified as indicators of both community resilience and sustainability. Sustainability is seen in their care for the traditional tea economy and recovery of traditional tea culture (which had been largely lost for most residents). Resilience is seen in their innovation and diversification, including various educational initiatives that are particularly strong within the tourism sector (Tsai et al., 2016). This is complemented by government programs to support environmental sustainability. The natural environment, which is also seen as a tourism resource, includes hiking trails and the nighttime viewing of fireflies and flying squirrels. In this way, the most successful rural communities in our study had indicators showing both a high degree of resilience and high degree of sustainability (Peirce, Budd & Lovrich, 2011).

Conclusions: sustainable and resilient communities

Sustainability (mitigating against change) and resilience (embracing change) are not new ideas and are at the core of policy decisions that people and communities make all the time. Sustainability has become the major development paradigm throughout the world since it was introduced in 1987. For most of that time, the only alternative to sustainability was "unsustainability", which was usually attributed to neoliberal (free market) economic policies that overexploited limited human and natural resources. The growing interest in resilience as an alternative approach to understanding contemporary development issues is based on the belief that it is different from sustainability and that it does not carry perceptions of being anti-business. Resilience is thought to be able to achieve roughly similar results to sustainable development at a time when climate change appears to be overwhelming sustainability efforts to control it. This may be true, but unfortunately, because many are not aware of the subtle differences in goals and impacts of these two approaches, the terms "sustainability" and "resilience" are often misunderstood and misused.

In their most simplistic definitions, sustainability and resilience are almost identical in meaning (Table 3.3). On the other hand, their most complex definitions can be meaningless in their complexity and all-inclusiveness. We adopted moderate definitions that point out how very different sustainability and resilience approaches can be in terms of their assumptions, goals, research focus and implementation methods. What they share in common is a desire to improve the quality of life of communities. In essence, they provide two different mechanisms for what are often perceived as two very different problems: intermittent fast change disasters, and persistent slow changes. The comparison of disaster experienced communities and non-disaster experienced communities shows that sustainability is a component of both types of community, partly driven by their need to respond to the common slow changes that affect all rural communities (Lew, 2014). This was seen in the non-disaster experienced communities spending higher percentages of their local budget funds for environmental protection and environmental education, and community health (Table 3.5). It may also be implied by their lower levels in characteristics that were more resilience-oriented.

Sustainability, therefore, is seen to be the development path of choice for addressing slow variable changes, such as economic globalization and climate change. Sustainability responses are also partly driven by the larger institutional context that operate within and make up government policies and programs. This finding (that policies and actions to address slow change are primarily aligned with the sustainability approach) has not been addressed in a significant way in the resilience literature. Tobin (1999) referred to this association by equating most disaster planning actions prior to a major event with sustainability. Similarly, Redman (2014) defined sustainability as "adaptation" and resilience as "transformation." Within the sustainability adaptation approach, he included incremental change, maintaining previous order and building adaptive capacity as key measures. This approach recognizes that sustainability, as pure mitigation and conservation, is essentially impossible in a world that is undergoing continuous non-linear change. However, it does not recognize the political differences between policies of conservation and those of adaptation which could make an adaptation approach to sustainability untenable. We contend that sustainability offers one approach to addressing slow change issues and resilience offers another.

As evidenced in our fieldwork in rural Taiwan, what made disaster experienced communities stand out was the strength of their resilience orientation, developed and enhanced as a result of their disaster experience. Based on the indicators in Table 3.4, the disaster experienced communities had higher levels of the following resilience characteristics (Table 3.5):

- They had actively built more diversified economies, mostly through the addition of a tourism economy to their traditional agricultural economy, which enabled more local youths to stay and work within the rural villages.

- While their overall rates of natural resource knowledge and use were no different, the disaster experienced communities had considerably higher rates of formal environmental, agricultural and disaster education.
- They had higher total budgets and higher rates of spending and total spending amounts on infrastructure, although these could, to some degree, be due to their more precarious natural hazard locations.

The disaster experienced communities have learned through experience that building resilience is an effective approach to addressing fast-change events. The question that remains is how to assist non-disaster communities to both maintain their sustainability orientation and strengthen their resilience. Many government and non-governmental programs already support sustainability initiatives due to its dominance in global development debates (Lubin & Etsy, 2010). Having more government programs that incentivize resilience planning (such as Taiwan's leisure agricultural programs) can help bring more communities to the preferred state of being both resilient and sustainable. But that can only occur if adequate funding and political leadership encourages resilience in the same way that sustainability has been supported. Only in this way can communities be best prepared for the complexity of change in the world today.

Acknowledgement

This research was funded by the Chiang Ching-Kuo Foundation (RG014-A-13) and the Taiwan Ministry of Foreign Affairs, Taiwan Fellowship Program.

References

Adger, W.N. (1997). *Sustainability and social resilience in coastal resource use.* CSERGE Working Paper GEC 97–23. University of East Anglia, UK: Center for Social and Economic Research on the Global Environment.

Adger, W. N. (2003). Building resilience to promote sustainability. IHDP (International Human Dimensions Program on Global Environmental Change) *Update* 2: 1–3.

Anderies, J. M., Folke, C., Walker, B. & Ostrom, E. (2013). Aligning key concepts for global change policy: robustness, resilience, and sustainability. *Ecology and Society*, 18(2): 8. http://dx.doi.org/10.5751/ES-05178-180208

Baggio, R. & Salnaghi, R. (2011) Complex and chaotic tourism systems: toward a quantitative approach. *International Journal of Contemporary Hospitality Management.* 23(6): 840–61.

Brand, F.S. & Jax, K. (2007). Focusing the meaning(s) of resilience: resilience as a descriptive concept and a boundary object. *Ecology and Society*, 12(1): 23. Retrieved from www.ecologyandsociety.org/vol12/iss1/art23

Berkes, F. (2007). Understanding uncertainty and reducing vulnerability: lessons from resilience thinking. *Natural Hazards* 41: 283–95.

Butler, R. (1999). Sustainable tourism: a state of the art review. *Tourism Geographies* 1(1): 7–25.

Carpenter, S., Walker, B., Anderies, J.M. & Abel, N. (2001). From Metaphor to Measurement: Resilience of What to What? *Ecosystems* 4: 765–81. DOI: 10.1007/s10021-001-0045-9

Cochrane, J. (2010). The Sphere of Tourism Resilience. *Tourism Recreation Research* 35(2): 173–85.

Davoudi, S. (2012). Resilience: A bridging concept of a dead end? *Planning Theory and Practice*, 13(2): 299–333, http://dx.doi.org/10.1080/14649357.2012.677124

Derissen, S., Quaas, M.F. & Baumgärtner, S. (2011). The relationship between resilience and sustainability of ecological-economic systems. *Ecological Economics*, 70: 1121–28.

Evans, B. & Reid, J. (2015). Exhausted by resilience: response to the commentaries. *Resilience: International Policies, Practices and Discourses*, 3(2): 154–59. DOI:10.10 80/21693293.2015.1022991

Farrell, B.H. & Twining-Ward, L. (2004). Reconceptualizing tourism. *Annals of Tourism Research*, 31(2): 274–95.

Farrell, B. H. & Twining-Ward, L. (2005). Seven steps towards sustainability: Tourism in the context of new knowledge. *Journal of Sustainable Tourism*, 13(2): 109–22.

Faulkner, B. & R. Russell. (1997). Chaos and Complexity in Tourism: In Search of a New Perspective. *Pacific Tourism Review* 1:93–102.

Fennell, D. (2015). Tourism and the precautionary principle in theory and practice. In C.M. Hall, S. Gössling & D. Scott, eds., *The Routledge Handbook of Tourism and Sustainability*. London: Routledge.

Fiksel, J. (2006). Sustainability and resilience: toward a systems approach. *Sustainability: Science, Practice, & Policy*, 2(2): 14–21.

Garrod, B., Wornell, R. & Youell, R. (2006). Re-conceptualising rural resources as countryside capital: The case of rural tourism. *Journal of Rural Studies* 22: 117–28.

Hamzah, A. & Hampton, M.P. (2012). Resilience and Non-Linear Change in Island Tourism. *Tourism Geographies* 15(1): 43–67. DOI: 10.1080/14616688.2012.675582

Holling, C. S. (1973). Resilience and stability of ecological systems. *Annual Review of Ecology and Systematics* 4: 1–23.

Holling, C. S. & Walker, B. (2003). Resilience Defined, prepared for the *Internet Encyclopedia for Ecological Economics*. The International Society for Ecological Economics. Online at: http://isecoeco.org/pdf/resilience.pdf

Hu, A. & Yang, F. (2014). Trajectories of Folk Religion in Deregulated Taiwan. *Chinese Sociological Review* 46(3): 80–100.

Hunter, C. (1997). Sustainable tourism as an adaptive paradigm. *Annals of Tourism Research* 24(4): 850–67.

Jepson, E. J. (2016). "A new metric for community resilience". Planetizen.com (23 January 2016). Retrieved from www.planetizen.com/node/83342/new-metric-community-resilience.

Juech, C. & Michelson, E. S. (2011). Rethinking the future of sustainability: From silos to systemic resilience. *Development*, 54(2): 199–201.

Knudsen, D., Greer, C. (2011). Tourism and nostalgia for the pastoral on the island of Fyn, Denmark. *Journal of Heritage Tourism* 6(2): 87–98.

Levin, S.A., Barrett, S., Anyar, S., Baumol, W., Bliss, C., Bolin, B., Dasgupta, P., Ehrlich, P., Folke, C., Gren, I.M., Holling, C.S., Jansson, A., Jansson, B.-O., Mäler, K.-G., Martin, D., Perrings, C., Sheshinski, E., 1998. Resilience in natural and socioeconomic systems. *Environment and Development Economics* 3: 221–35.

Lew, A.A. (2014). Scale, change and resilience in community tourism planning. *Tourism Geographies*, 16(1): 14–22. DOI:10.1080/14616688.2013.864325.

Lew, A.A., Ng, P. T., Ni, C-C. & Wu, T-C. (2016). Community sustainability and resilience: similarities, differences and indicators. *Tourism Geographies*, 18(1): 18–27. DOI:10.1080/14616688.2015.1122664

Lew, A.A., Ng, P.T., Wu, T-C. & Ni, C-C. (2016). "Some New Resilience Figures and Diagrams. Collaborative for Sustainable Tourism and Resilient Communities Blog" (30 September). Retrieved from www.tourismcommunities.com/blog/some-new-resilience-figures-and-diagrams.

Lubin, D.A. & Esty, D.C. (2010). The sustainability imperative. *Harvard Business Review* 88(5):42–50.

MacKinnon, D. & Derickson, K.D. (2012). From resilience to resourcefulness: A critique of resilience policy and activism. *Progress in Human Geography*, 37(2): 253–70.

McKercher, B. (1999). A Chaos Approach to Tourism. *Tourism Management* 20:425–34.

McLellan, B., Zhang, Q., Farzaneh, H., Utama, N.A. & Ishihara, K. N. (2012). Resilience, sustainability and risk management: A focus on energy. *Challenges*, 3: 153–82. doi:10.3390/challe3020153

Magis, K. (2010). Community resilience: An indicator of social sustainability. *Society & Natural Resources,* 23: 401–16.

Meerow, S. & Newell, J. P. (2015). Resilience and complexity: A bibliometric review and prospects for industrial ecology. *Journal of Industrial Ecology*, 19(2): 236–51.

Peirce, J. C., Budd, W. W. & Lovrich, N.P. (2011). Resilience and sustainability in US urban areas. *Environmental Politics*, 20(4): 566–84

Pimm, S. L. (1984) The complexity and stability of ecosystems. *Nature*, 307: 321–26.

Prasad, N., Ranghieri, F., Shah, F., Trohanis, Z., Kessler, E. & Sinha, R. (2009). *Climate Resilient Cities: A Primer on Reducing Vulnerabilities to Disasters*. Washington, DC: The World Bank.

Ranjan, R. (2012). Natural resource sustainability versus livelihood resilience: Model of groundwater exploitation strategies in developing regions. *Journal of Water Resources Planning and Management* 138(5): 512–22.

Redman, C. L. (2014). Should sustainability and resilience be combined or remain distinct pursuits? *Ecology and Society*, 19(2): 37. http://dx.doi.org/10.5751/ES-06390-190237

Schianetz, K. & Kavanagh, L. (2008). Sustainability indicators for tourism destinations: A complex adaptive systems approach using systemic indicator systems. *Journal of Sustainable Tourism* 16: 601–28.

Schwab, J.C. (1998). *Planning for Post-Disaster Recovery and Reconstruction*. Planning Advisory Service Report #483/484 (December). Chicago: American Planning Association.

Simmie, J. and Martin, R. (2010) The economic resilience of regions: towards and evolutionary approach. *Cambridge Journal of Regions, Economy and Society* 3: 27–43. doi:10.1093/cjres/rsp029

Strunz, S. (2012). Is conceptual vagueness an asset? Arguments from philosophy of science applied to the concept of resilience. *Ecological Economics* 76: 112–18.

Tobin, G.A. (1999) Sustainability and community resilience: the holy grail of hazards planning? *Environmental Hazards* 1: 13–25.

Tsai, C-H., Wu,T-C., Wall, G. and Linliu, S-C. (2016). Perceptions of tourism impacts and community resilience to natural disasters. *Tourism Geographies* 18(2): 152–73. DOI: 10.1080/14616688.2016.1149875

UN (2015). "Sustainable Development Goals". Sustainable Development Knowledge Platform. Retrieved 8 June 2016 from https://sustainabledevelopment.un.org/topics/sustainabledevelopmentgoals

Urry, J. (1995). A middle-class countryside? In J. Urry, ed., *Consuming Places* (pp. 211–29). London: Routledge.

Walker, B. & Salt, D. (2006). *Resilience Thinking: Sustaining People and Ecosystems in a Changing World*. Washington DC: Island Press

Walker, B.H., Holling, C.S., Carpenter, S.R. & Kinzig, A. (2004). Resilience, adaptability and transformability in social–ecological systems. *Ecology and Society* 9(2):5. Retrieved from www.ecologyandsociety.org/vol9/iss2/art5/

Weichselgartner, J. & Kelman, I. (2014) Geographies of resilience: Challenges and opportunities of a descriptive concept. *Progress in Human Geography* 39(3): 249–67.

WCED (1987). *Our Common Future. The Brundtland Report*, World Commission for Environment and Development. Oxford: Oxford University Press.

Part II
Nature-based tourism and climate change

4 Searching for resilience

Seal-watching tourism as a
resource for community
development in Iceland

Georgette Leah Burns

Introduction

Icelandic culture, from its modern complexities to the very basics of continued existence of people on this northern island country, epitomizes resilience. Iceland was settled in approximately AD 870 (Dugmore, Newton, Larsen & Cook, 2000) and over the last thousand years has represented "a case of an apparent near miss" (Streeter, Dugmore & Vésteinsson, 2012, p. 3664) in terms of human survival. Despite adversities wrought by volcanic eruptions, severe weather events, famine and plagues, Icelandic society has endured.

Based on ethnographic research, this chapter presents a case study of wildlife tourism in a rural region in northwest Iceland. Human and seal residents of Húnaþing vestra have a long history of interactions in which their shared social-ecological system has demonstrated resilience. These interactions have involved humans hunting seals as a resource for subsistence and commercial gain, and seal prominence in Icelandic folklore. Exploring the history of these interactions demonstrates both change and adaptation over time.

More recently, establishment of the Icelandic Seal Center in 2005 served to provide Húnaþing vestra with a unique tourism brand, at a time when rural communities across Iceland were experiencing declining populations and competing to attract the tourist dollar. The addition of wildlife tourism has added a third actor, the tourist, to the existing system, and, as a consequence, interactions with, and attitudes toward, seals as a resource have changed again. Despite lack of formal planning for tourism in Húnaþing vestra, the residents, the tourists and the seals must find a way forward. This chapter demonstrates how examination of the literature and case studies of ethics and responsibility in wildlife tourism can help create a path to move beyond sustainable tourism and toward resilience for this community of people and wildlife.

Location and methods

Húnaþing vestra, a municipal area in the northwest of Iceland covering 3,019km^2 (Figure 4.1), contains a human population of 1,173 (Karlsdóttir, 2016). Hvammstangi, situated on the western side of the Vatnsnes peninsula facing Miðfjörður, is the service and government administrative centre for the

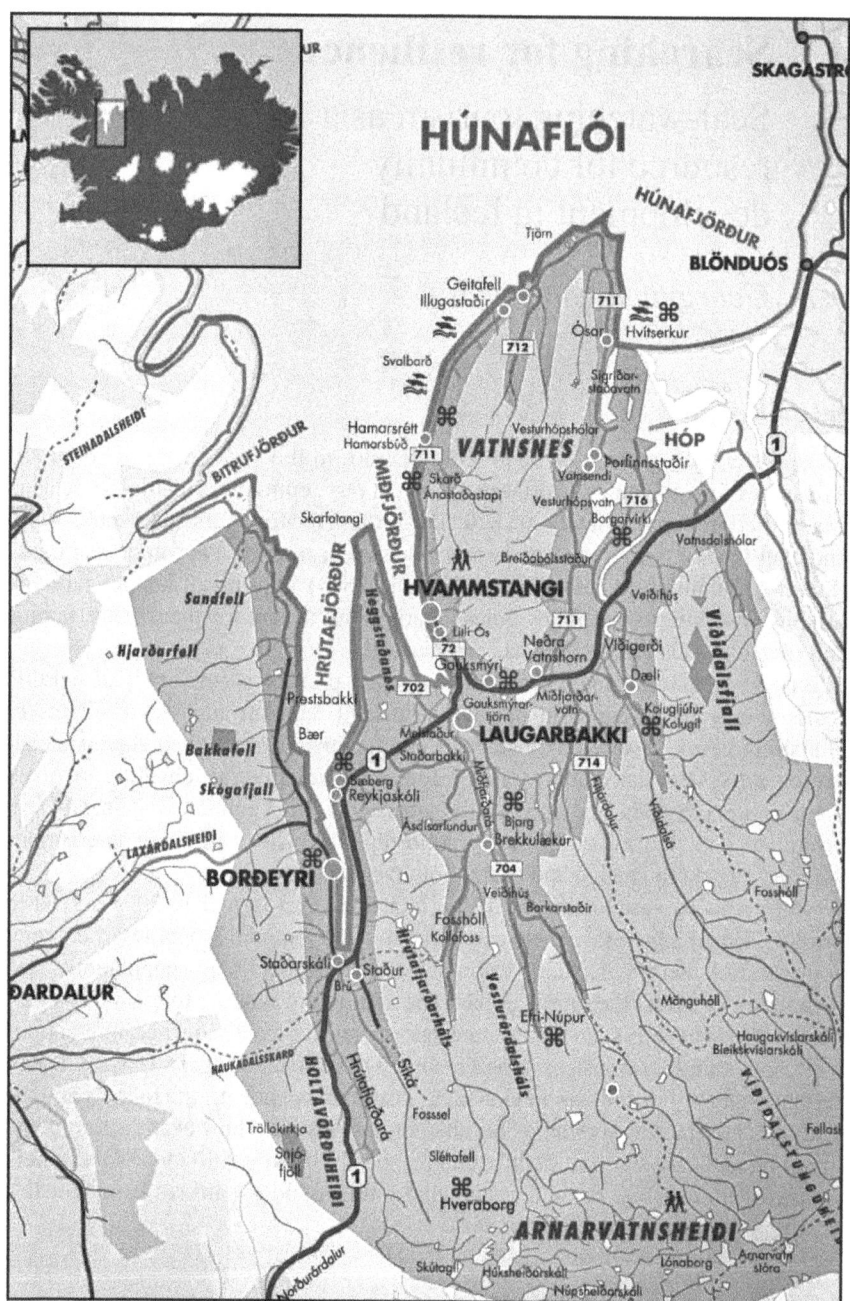

Figure 4.1 Map of Húnaþing vestra
Source: www.visithunathing.is/is/hunathing-vestra/kort-af-hunathingi-vestra

municipality. Located 197km from the capital city of Reykjavik, Hvammstangi is the most densely populated area in the municipality with a 2014 population of 558 (Ragnarsson, 2015, p. 10). Hvammstangi's businesses, which provide employment for local residents, include a school, which educates students from grades one to ten, a harbour, grocery store, hardware store, restaurant, swimming pool and sports centre, medical centre, wool factory, craft store, museum, two mechanics' workshops and a variety of accommodation facilities. People residing outside the town are most likely to earn a living as sheep farmers, though there is also some farming of cattle, chicken and eider ducks. Cultural activities in Hvammstangi include a church, several choirs, a dramatic society, music school, sports society and a community centre for young people.

Húnaþing vestra and Hvammstangi share a storyline similar to many peripheral communities around the world (George, Mair & Reid, 2009). It is the story of a community that has never been large but has maintained its existence through access to natural resources, and has adapted to changes over time, demonstrating its resilience. In recent years, it is also a story of population decline and attempts to retain residents in rural areas by using tourism as a tool for rural development. The empirical data presented in this chapter is based on 32 months of ethnographic research, primarily using participant observation, while living in Hvammstangi. During this time, my role in the town was as a tourism researcher while I was employed at the Icelandic Seal Center, which is a key player in this story.

Resilience

The distinction between resilience and sustainability in this context is important to make as they are related, but different, concepts. The concept of resilience, like sustainability, hails from ecological science in the 1970s (Brand & Jax, 2007) but is now widely used beyond ecology (Lew, 2014) with more recent adoption into analysis of social-ecological systems (Folke, 2006; Folke et al., 2002; Lebel et al., 2006) and theoretical musings on community development (Lew, Ng, Ni & Wu, 2016).

Resilience can be defined as "the capacity of a system to deal with change and continue to develop," where that system can be a landscape, a city or a coastal region (Stockholm Resilience Centre, 2014, p. 3), and focus is on building capacity and adapting to change. As such, it is commonly referred to as a descriptive concept of system dynamics (Derissen, Quaas & Baumgärtner, 2011). Sustainability, by contrast, focuses on mitigating or preventing change, and is considered a normative concept (Derissen et al., 2011). The resilience approach operates in a holistic framework that locates humans as part of the biosphere, thus moving beyond a view of them as "external drivers of ecosystem dynamics" (Stockholm Resilience Centre, 2014, p. 3). In this chapter, the system investigated is the municipality of Húnaþing vestra and the holistic framework includes the roles of residents, seals and visitors to both establish and maintain resilience.

Folke et al. (2002) determine resilience as related to:

- the magnitude of shock that a system can absorb and still remain within a given state;
- the degree a system is capable of self-organization; and
- the degree to which the system can build capacity for learning and adaptation.

Resilience thus equates to the level of change a system can endure while maintaining control of internal structure and function (Lebel et al., 2006) and many understandings contribute to a system's innate resilience and to the building of resilient systems (Folke et al., 2002; Lebel et al., 2006).

In the context of tourism discourse, focus has been largely upon economic resilience, with the most common perspective being the recovery of tourism industries and arrival numbers following fast variable changes (Lew, 2014). The situation described here joins the body of growing literature on case studies of tourism and resilience (e.g., Biggs, Hall & Stoeckl, 2012; Hall, Timothy & Duval, 2013; Orchiston, 2013) while expanding to community issues beyond the narrow confines of just economics; thus, maintaining a complex systems approach which is at the core of resilience thinking (Lew, 2014). It also moves away from the tendency in the literature to focus on resilience in the context of external drivers and fast variable changes to also consider slow variable changes.

Enhancing the ability of a system (here, a rural community) to build adaptive capacity to tourism impacts is vital to the effectiveness of achieving strong resilient communities able to adapt to the changes tourism brings. Capacity in this context is the resource and strength held by the community that lessens the risk or consequences of adversity that may arise from increased tourism in remote and small-scale locations, such as the one described in this chapter. The community is less vulnerable when it has built effective adaptive capacity (Smit & Wandel, 2006), as can be seen in the case of Iceland nationally and Húnaþing vestra locally.

Iceland and resilience

> In debates on societal collapse, Iceland occupies a position of precarious survival, defined by not becoming extinct, like Norse Greenland, but having endured, sometimes by the narrowest of margins
>
> (Streeter et al., 2012, p. 3664)

In the early years of Icelandic settlement, the "near miss" of human survival was largely due to environmental hardship, some of which was caused by human occupation and maladaptive use of the landscape (Dugmore et al., 2000). For example, from the time of settlement through to the 1300s, human activities led to the clearing of much of the forested regions in the country (Mairs, Church, Dugmore & Sveinbjarnardóttir, 2006). This resulted in heavy reliance on the limited supplies of driftwood (Brunon, 2013) for building houses, boats and bridges, as well as for farm implements and household utensils (Kristjánsson, 1980, p. 446).

During the "Age of Settlement," from AD 874 to 930, between 12,000 and 15,000 settlers arrived in Iceland from Norway and the British Isles (Kristjánsson, 1980, p. 441). This was closely followed by the volcanic eruption of Eldgjá from 934 to 940 (Vasey, 1996) and the first recorded famine in 980, resulting in extensive loss of life (Kristjánsson, 1980, p. 441). In the earliest centuries of settlement pigs were highly valued livestock in the sedentary farming communities, but their popularity declined in favour of sheep, cattle, goats and horses (McGovern et al., 2007). Farming was supplemented by hunting birds and marine mammals, and fishing in oceans, lakes and streams. Birds and fish remain important to the Icelandic diet.

Geothermal activity from the many volcanic systems assisted human survival by supplying hot water for heating and cooking in an otherwise extremely cold climate, but volcanoes were also the cause of extensive damage to farming land and infrastructure, and disruption to air quality and climate, when they frequently erupted: 205 eruptive events were recorded between settlement and 2007 (Thordarson & Larsen, 2007, p. 118), averaging 20–25 events per 100 years, and 30 different volcanic systems have been identified (ibid., p. 124). While not all of these eruptions have been catastrophic, many have been large enough to cause significant disruption to life and livelihood; for example, Hekla in 1693 (Thórarinsson, 1967) and Laki in 1783 (Thordarson & Self, 2003). Iceland experienced substantial climatic changes in the period 1300–1700 (Streeter et al., 2012, p. 3668), and in the fifteenth century two plagues (1402–4 and 1494–5) each caused the death of approximately half the population (Karlsson, 1996).

External drivers of a system, such as those described above, result in changes to slow variables which, as they approach threshold levels, influence the faster moving variables in the system. The system fluctuates more and together these variables can push the system across a threshold into an alternative regime of (in) stability (Walker, Carpenter, Rockstrom, Crépin & Peterson, 2012, p. 30). That rapid demographic downturns, volcanic disruptions and climatic changes did not lead to societal collapse strongly suggests that the social-ecological system was resilient. The response by the population seems to have been to scale back farming practices in times of rapid change, but keep the methods the same (Streeter et al., 2012, p. 3669), thus revealing an example of a slow controlling variable, and a resilience that was both ideological and functional.

Set against this historical backdrop, of survival in the face of extreme adversity, Iceland moved into the modern era, where the climate has not become any gentler and volcanic eruptions persist, as with, e.g., Eldfell (1973), Eyjafjallajökull (2010) and Bárðarbunga (2014–15). A turning point for rural communities has been the ability to harness forms of renewable energies and Hvammstangi has used geothermal sources for heating since 1973 (Yuanyuan, 2010, p. 297).

More recently, Iceland suffered financial hardship following the global financial crisis (GFC) in 2008. During the last decades of the nineteenth century, before commercial fishing emerged, Iceland was one of the poorest countries in Europe (Ólafsson, 2005). Following mechanisation of the primary industries of fisheries and agriculture during the latter half of the twentieth century, however, Iceland prospered, topping the United Nations Human Development Index in 2008

(UNDP, 2008, cited in Jóhannesson & Huijbens, 2010, p. 424). Prior to the GFC, Iceland, along with several other European countries and the US, experienced an "asset bubble" in the housing and commercial property market (Dubrowski, 2010, p. 39). When this bubble burst, Icelandic banks and their international depositors were left in significant financial trouble. Some 85 per cent of the Icelandic bank sector collapsed, the value of the Icelandic Krona plummeted (Matthiasson, 2008) and unemployment skyrocketed (Vinnumálastofnun, 2009, cited in Jóhannesson & Huijbens, 2010, p. 424) as did public sector debt (Fjármáláráðuneytið, 2009). The economic and social consequences felt across the country resulted in "abrupt social change" (Ragnarsdóttir, Bernburg & Ólafsdóttir, 2013, p.755). Iceland was forced to request IMF rescue, co-sponsored by the EU (Dubrowski, 2010), and searched for new paths to a resilient future. It turned to tourism.

The promotion of tourism to Iceland post-GFC was strong, deliberate and effective. Although Iceland had a long history as a destination for travellers and explorers, overall numbers were very low and tourism was almost non-existent right up to the mid-twentieth century (Jóhannesson & Huijbens, 2010). Numbers began steadily increasing in the mid 1960s, and rose from 200,000 to 500,000 registered arrivals in the ten-year period from 1997 in 2007 (Ferðamálastofa, 2009, cited in Jóhannesson & Huijbens, 2010, p. 426). The greatest growth, however, has occurred since the GFC, with foreign arrival numbers doubling between 2010 and 2015, and averaging an increase of 21.6 per cent per year (Óladóttir, 2016), well above the global average of 4 per cent (UNWTO 2016). As a consequence of this adaptive response, through increased tourism marketing, to a time of financial hardship, tourism now surpasses the fishing industry and aluminium production as the main source of foreign exchange income for Iceland (Óladóttir, 2016).

Seals and Icelandic culture

Waters surrounding the Vatnsnes peninsula in Húnaþing vestra contain some of the largest populations of harbour seals found in Iceland and human residents of this region have a long relationship with seals as a resource (Figure 4.2). Seals are likely to have played an important role in the settlement of Iceland by humans. Abundant colonies of harbour and grey seals provided access to products such as meat, blubber and skins (Hauksson & Einarsson, 2010, p. 341).

Seal hunting (Hauksson & Einarsson, 2010) is an important part of Iceland's history and continues in contemporary times, though it is no longer done for commercial purposes and seldom for subsistence. In the seventeenth century, 364 seal hunting farms were registered. This number dropped to 215 in the eighteenth century, with 264 registered in 1932 (Kristjánsson 1980, p. 447). Hunting seals was traditionally a legal right for farmers whose land abutted the waters in which seals were found. A licence was not required to shoot seals and there was no quota system to limit the number taken (Hauksson & Einarsson, 2010). Seal hunting rights were considered extremely profitable (Kristjánsson, 1980, p. 447). "Seal hunting was regarded as an important supplement to other economic resources, and in certain regions it at times provided the only means of subsistence" (Kristjánsson, 1980, p. 448). Seal skins

Figure 4.2 Harbour seals on the Vatnsnes peninsula, Iceland
Source: Andreas Muhar, used with permission.

were in high demand for making shoes, outer garments, bags and ropes. They were also used for binding manuscripts, and as containers and decorations (Kristjánsson, 1980, p. 448), but skins were not the only valuable product from seal hunting. Blubber was used for oil, cooking, making paint, as food for both humans and livestock, and illumination before kerosene (Kristjánsson, 1980, p. 448). It was also exported, as were skins. Seal meat was eaten fresh, salted and smoked, and was also used as bait (Kristjánsson, 1980, p. 448).

Given this history, it is not surprising that seals hold a prominent place in Icelandic folklore (Puhvel, 1963). Their cultural importance is also evident in the naming of places, many of which probably date back to the earliest settlements in Iceland (Kristjánsson, 1980, p. 446). For example, the town of Selfoss in south Iceland is a combination of the Icelandic words for seal (selur) and waterfall (foss).

The global anti-seal hunting campaign in the 1970s and a reduction in the reliance on seal meat within Iceland (Granquist and Hauksson 2016), coupled with a European Union ban on importing and commercializing seal products (Wegge 2013), led to a significant reduction in seal hunting. In Iceland no reporting is required of the numbers shot annually, so accurate records of seal populations are difficult to keep. However, seals are still hunted around estuaries to decrease suspected predation on migrating salmon, and annual counting by the Icelandic Seal Center indicates a recent and severe decrease in the harbour seal population (Granquist, Hauksson & Stefánsson, 2015).

Tourism in Húnaþing vestra

The Icelandic Seal Center (Figure 4.3) was established in Hvammstangi in 2005 as a non-profit organization. It is a museum, research centre and gift/souvenir store and also serves as the tourist information centre for the region. It has established three seal watching sites, in conjunction with landowners, on the Vatnsnes peninsula, and is funded through profit from museum entry and sales, research grants and some government support.

Destinations that are far from the main hubs of tourism activity need to offer something special or unique to attract tourists (Sharpley, 2007; Prideaux, 2002). Impetus for the creation of the Icelandic Seal Center came from the local community, who wanted to attract tourism to the town, located 6km off the main road that circles the island, which had not yet gained a reputation as a popular tourist destination. Benefits of increased visitor numbers were anticipated to include direct and indirect financial gains for the community.

Rural population decline is common across Iceland, raising concerns about the long-term viability of some of these communities. The population of Húnaþing vestra dropped from 1,412 in 1998 to 1,107 in 2012, while the population of northwest Iceland as a whole dropped from 8,252 to 7,299 in that same 14-year period (Ragnarsson, 2015, p. 9). For Hvammstangi, the population decreased from 698 in 2004 (Gunnarsdóttir & Gissurarson, 2008) to 558 in 2014 (Ragnarsson, 2015, p. 10). By attracting more tourism to the area, it was hoped that this trend of declining resident numbers could be reversed and a resilient community maintained.

Figure 4.3 The Icelandic Seal Center, beside the harbour in Hvammstangi, Iceland
Source: Icelandic Seal Center, used with permission

The aim to attract more visitors was successful, reflecting the nationwide increase in visitor numbers. The deliberate, targeted, strategy to attract visitors yielded a ten-fold increase over ten years, with the recorded visitor numbers in Hvammstangi increasing from 2,200 in 2005 to 27,150 in 2015 (Figure 4.4). This rate of increase shows no signs of abating. In the first eight months of 2016, 35,000 visitors were counted at the Icelandic Seal Center (2016). The influx of this many visitors has potential to challenge the functioning capacity of the small receiving community, though in different and perhaps more gradual ways than the natural disasters of the past to which Icelanders are accustomed. Careful management is required to ensure the capacity for resilience already demonstrated by this culture is harnessed in such a way as to meet the challenges associated with tourism as an external driver of change.

The rapid increase in tourism is a consequence of Hvammstangi's success in establishing itself as a recognized tourism destination in Iceland. Seals were identified as the largest pull-factor in the region and became the destination's key resource for development and marketing (Ram, Björk & Weidenfeld, 2016). The resultant tourism product, seal watching, fits with Hu and Wall's (2005, p. 619) definition of a tourism attraction as a "permanent resource, either natural or human-made, which is developed and managed for the primary purpose of attracting visitors." The Icelandic Seal Center was established and then attracted other related businesses to the area. In this way, the Center has contributed "to the development of critical mass in the destination offering" becoming "only one part of a complex network of tourism service providers" (Leask, 2008, p. 11).

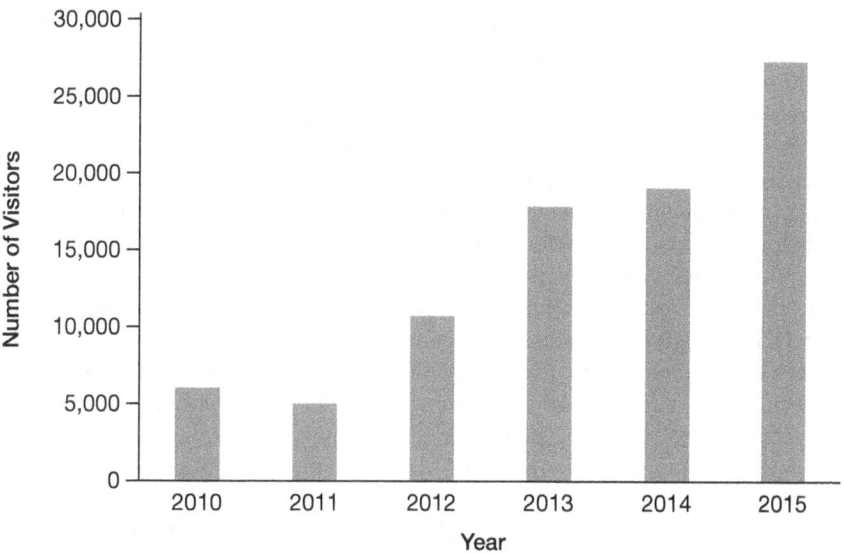

Figure 4.4 Visitor numbers to the Icelandic Seal Center in Hvammstangi, 2010–15
Source: Author (based on data from Þórisson, 2016)

How can the concept of resilience help us to understand this situation? Increasing tourism has changed this community's relationship with seals as a resource. Seals were originally valued as a means of subsistence, but are now significant objects of tourism (Burns, 2015a). The community has already demonstrated resilience to environmental and economic conditions, reinventing itself as the "land of seals" to successfully bring tourists and related businesses into the town. The current pressing question is, though, how to ensure this continues and social-ecological resilience is also maintained in the face of negative impacts from tourism?

The challenges faced by communities using tourism as a tool for development are mostly slow and gradual, emerging over time as number of visitors increase and both social and environmental impacts become apparent (George et al., 2009; Lew, 2014, p.18). The challenges can be experienced at different levels throughout the community, from individuals and single businesses to community collectives, public commons and local government (Lew, 2014, p18), and thus require managing both at, and across, these multiple levels.

In Iceland, weather conditions ensure very stratified seasonal restrictions to tourism activities. In Húnaþing vestra, peak tourism and visitation occurs over the three months of summer, June, July and August. Increased marketing of the "shoulder season," either side of summer, is gradually attracting more visitors during these times but summer remains, and is likely to continue to remain, the peak season. Consequently, the community is annually given time to build capacity over the winter months—reflecting on, and readjusting to, the shock to the system over summer and being able to self-organize through constant learning and adaption. Of course, not all members of the community engage in this practice. For those businesses engaged in providing everyday services, such as vehicle repair businesses, tourism has little impact on their daily routines regardless of the time of year. For this reason, internal tourism brokers in the community take the very conscious, and sometimes paid, roles of analysing the success of the previous season and planning for the following one. This takes the form of not just aiming to increase numbers and thus profit from sales, tours and accommodation, but also liaising with community members, such as farmers living on the peninsula, for whom tourists are not crucial for their livelihood but can nevertheless have indirect positive and negative influences on their lives.

In Húnaþing vestra, positively perceived impacts of tourism include direct financial benefits for those individuals and business operators who sell goods and services directly to tourists; for example, the craft store, museum, wool factory, seal watching boat tours and the camping ground. Over the past five years several new accommodation options have opened in the municipality, including bed and breakfast accommodations in residents' homes and the listing of properties online through sites such as airbnb and booking.com. Significant indirect benefits are also apparent. For example, with several hundred visitors per day passing through Hvammstangi in summer, the general store is able to stock a larger variety and volume of consumable products than might be otherwise possible for a community of under 600 people. In March 2015 a restaurant opened in the town, catering to both visitors and residents all year round and also doubling as the kitchen that makes school lunches during school terms.

Negatively perceived impacts focus generally on intrusion on local lifestyle, with the change from a small quiet town over winter to one dominated by tourists and their activities over summer. This includes safety concerns about the increased traffic both through town and on the peninsula road, particularly during peak visitation season when tourists arrive in vehicles ranging from small hire cars to large campervans and buses. Increased litter, both in town and along the coastline, is also a concern for the local community as is the potential influence from demonstrations of alternative cultural behaviours and practices to the town's residents. Researchers from the Icelandic Seal Center have documented the potential impact on the seal population of increasing tourism (Granquist & Sigurjonsdóttir, 2014) and intrusion on local lifestyle, and monitoring of these continues.

Adaptive capacity in Húnaþing vestra

Adaptive capacity, a key concept in resilience literature, varies in different countries, communities, within social groups and in relation to individuals (Folke et al., 2002; Smit & Wandel, 2006). It also shifts over time and is determined by specific contextual issues (Smit & Wandel, 2006). However, key determinants for adaptive capacity are likely to include:

- human capital;
- localized adaptive capacity;
- resource distribution;
- kinship networks;
- the institutional environment in which adaptations occur; and
- access to resources (information, financial, technological) (Smit & Wandel, 2006, p. 287).

The close knit communities of Hvammstangi and Húnaþing vestra have strong, longstanding, kinship networks in the area of increased tourism. They also have a long history of building strong social and human capital and have demonstrated the capacity to work effectively together over time to maintain their lifestyle in this remote region of northern Iceland. The new influx of tourism has the potential to challenge and change this lifestyle; however, a localized adaptive capacity has already been demonstrated, as is a supportive institutional environment in the form of the Icelandic Seal Center and associated businesses in which adaptation can occur. The Seal Center also has access to resources that can inform adaptation and mitigate negative changes. These resources include the skill set amongst staff able to consider both social and biological aspects of humans interacting with seals, as well as those with experience and skills in running a successful tourism business. Without the presence of this Center in the town, seal watching is less likely to be a social and ecological success. Importantly too, as access to employment in the local tourism industry grows, this may, in the long term, serve to boost resident numbers in the region and stem the flow of migration to larger, urban areas. The return of residents to the region, and the retention of those who might otherwise have left, is likely to further enhance the social and human capital necessary for building capacity for resilience.

Ethics and resilience

A recent trend in tourism literature has been a call for more ethically responsible tourism (e.g., Burns, 2015b, pp. 49–51; Fennell, 2012; Fennell, 2014; Lovelock & Lovelock, 2013) in the quest to both understand and mitigate the potential negative impacts of tourism. If the concept of resilience is to engage meaningfully with tourism practice, consideration of ethics is required. To date, the topic of ethics in resilience has been largely confined to debate over whether resilience is normative or not. Most commonly, resilience is perceived as a descriptive concept and sustainability as a normative concept (Derissen et al., 2011) and thus combination of the two is deemed necessary for an ethical perspective to arise.

In the context of wildlife tourism, Burns, Macbeth & Moore (2011) proposed a management framework based on a set of ecocentric ethical principles, and these principles have been applied to the case of seal watching in Iceland (Burns, 2015a). The framework has obvious synergy with the concept of resilience in that both strive for a holistic approach to, and recognition of, a complex system. They also both clearly situate humans and their role as part of the ecosystem; however, divergence is apparent when comparing the framework with the notion of resilience linked with environmental economics and the ecosystems services literature.

The theme of anthropocentrism is pervasive throughout the environmental economics classification of ecosystems services which sees these services as only created on an anthropocentric utilitarian basis (McCauley, 2006; Redford & Adams, 2009; Sagoff, 2008). The classification provides no basis for intrinsic valuation of the environment because such a valuation is considered of little use in decision making (Schellnhuber et al., 2001). Anthropocentrism promotes an exploitative human–environment relationship in which humans, as consumers, are increasingly separated and alienated from the environment (Robertson, 2012). More holistic views of nature that may be held by long-term residents are excluded from, or even contradicted by, this relationship (Fairhead, Leach & Scoones, 2012). Despite claims of holism and recognition of a complex system, anthropocentrism remains the dominant perspective in resilience discourse. Where ethics are considered in literature that combines tourism and resilience, normative questions are based on considerations of what leads to the best outcomes for humans (Biggs, Ban & Hall, 2012; Calgaro, 2011; Mahrouse, 2011).

In contrast, the framework proposed by Burns et al. (2011) argues for recognition of intrinsic value and a more ecocentric approach to how humans view, and manage, their relationships with nature and wildlife in the context of tourism. A central tenant of this framework is the Precautionary Principle. The question of how resilient or vulnerable social-ecological systems are underlies interpretations of the Precautionary Principle (O'Riordan & Jordon 1995). In the context of wildlife tourism, this principle claims that "If a wildlife tourism action has a suspected risk of causing harm to animals or their habitat, in the absence of scientific consensus that the action is harmful, then the burden of proof that it is not harmful falls on those proposing the action" (Burns, 2015a, p. 53). This principle is especially pertinent to seal watching areas in Iceland where current census data suggest that harbour seal populations across the country are in decline, and the reasons for this trend are unknown (Granquist & Hauksson 2016).

Employing the Precautionary Principle has potential to minimize harm to the seals and their habitat, and enhance resilience of communities dependent on this species as a tourist attraction. Precautionary planning should include "reducing the vulnerability of key social and environmental systems, building capacity to respond to anticipated needs" (Lew, 2014, p. 18). In the context of tourism, the sector can assist by "supporting public education and awareness of known vulnerabilities through sites, museums and events" (Lew, 2014, p. 18). In Húnaþing vestra, the Iceland Seal Centre currently offers this support, and has plans to expand this education and awareness of seals and tourism. For example, a new design for the museum includes an exhibit that focuses on the effects of human interaction with seals, interactively demonstrating that if visitors are too close or too noisy then the seals will move away. In this way, the Seal Centre has adopted a keen role in assuring resilience by building community support and educating both residents and visitors. The goal of Burns et al. (2011) is to provide a more sustainable way of managing wildlife, and resilience is not a focus of this discussion. However, by moving this framework into the field of adaptation rather than just mitigation, it, and the ecocentric ethical principles it encompasses, can also be used as a tool to aid and build capacity for resilience.

Conclusion

This may seem like a small story from a small town, but it is in fact a big story from one of the world's current hot spots in tourism. Its circumstances parallel others in Iceland and around the world where tourism is used for rural development. Consequently, it is significant for the tourism industry in Iceland as a whole, provides an important example for other small towns using tourism as a way to develop, and offers insight into understanding the concept and application of resilience in tourism.

From the early history of Iceland we can piece together a picture of a people who have survived by choosing continuity over change, with society members maintaining "confidence in the basic soundness of their system and the belief it would recover" (Streeter et al., 2012, p. 3669) in the face of natural, and human-induced, disasters. A side effect of this choice is likely to have been "reduced dynamism and further entrenchment of the established order" (Streeter et al, 2012, p. 3669). Nevertheless, this choice has worked. On the surface, this continuity and reduced dynamism may seem an antithesis to the framework of resilience that stresses the importance of adaptation and change. But for Húnaþing vestra the residents have built on established strong kinship networks and human capacity for working together to create a framework in which adaptations can, and do, occur.

It is vital that Húnaþing vestra plans for long-term management of the persistent changes that seem inevitable given the type of tourism currently being attracted to the area and the demonstration of its growth. The plan would do well to take into account the long-term resilience of the community and consider how to continue to strengthen it in the context of future tourism. By continuing to maintain the strengths already demonstrated for adaptive capacity and for long-term resilience, and coupling these with new knowledge about ethics and responsibility, and indeed the ongoing research at the Seal Centre into ways of enhancing them, seal watching in Húnaþing vestra will provide a positive example of how tourism and rural community development can combine in the search for resilience.

Moving away from the focus on economic resilience common in tourism discourse, this case study of wildlife tourism has highlighted community issues to further illustrate the complex systems approach at the core of resilience thinking. The argument has been made that resilience, particularly as it is conceptualized through the ecosystems services literature, is fundamentally anthropocentric. By exploring a more ecocentric ethical approach, and applying the precautionary principle to tourism management, an opportunity exists to embrace intrinsic values as a component of the holism in resilience and expand the practical application of resilience theory.

References

Biggs, D., Ban, N. C. & Hall, C. M. (2012). Lifestyle values, resilience, and nature-based tourism's contribution to conservation on Australia's Great Barrier Reef. *Environmental Conservation,* 12(4), 370–79.

Biggs, D., Hall, C. M. & Stoeckl, N. (2012). The resilience of formal and informal tourism responses to disaster: Reef tourism in Phuket, Thailand. *Journal of Sustainable Tourism*, 20, 645–65.

Brand, F. S. & Jax, K. (2007). Focusing the meaning(s) of resilience: resilience as a descriptive concept and a boundary object. *Ecology and Society,* 12(1), 23. Retrieved from www.ecologyandsociety.org/vol12/iss1/art23/

Burns, G. L. (2015a). Animals as tourism objects: ethically refocusing relationships between tourists and wildlife. In K. Markwell (Ed.), *Animals and Tourism: Understanding Diverse Relationships* (pp. 44–59). Bristol: Channel View Publications.

Burns, G. L. (2015b). Ethics in tourism. In C. M. Hall, S. Gossling & D. Scott (Eds.), *The Routledge Handbook of Tourism and Sustainability* (pp. 117–26). London: Routledge.

Burns, G. L., Macbeth, J. & Moore, S. (2011). Should dingoes die? Principles for engaging ecocentric ethics in wildlife tourism management. *Journal of Ecotourism*, 10(3), 179–96.

Brunon, H. (2013). Forest Lost and Paradise Regained. In P. Boschiero, L. Latini & D. Luciani (Eds.), *The XXIV International Carlo Scarpa Prize for Gardens* (pp. 150–57). Trevise: Fondazione Beneton Studi Ricerche.

Calgaro, E.L. (2011). Building resilient tourism destination futures in a world of uncertainty: assessing destination vulnerability in Khao Lak, Patong and Phi Phi Don, Thailand to the 2004 Tsunami. PhD thesis. Macquarie University, Australia.

Derissen, S., Quaas, M. F. & Baumgärtner, S. (2011). The relationship between resilience and sustainability of ecological-economic systems. *Ecological Economics*, 70, 1121–28.

Dubrowski, M. (2010). The global financial crisis: lessons for European integration. *Economic Systems,* 34, 38–54.

Dugmore, A. J., Newton, A., Larsen, G. & Cook, G. (2000). Tephrochronology environmental change and the Norse settlement of Iceland. *Environmental Archaeology*, 5, 21–34.

Fairhead, J., Leach, M. & Scoones, I. (2012). Green grabbing: a new appropriation of nature? *Journal of Peasant Studies*, 39(2), 237–61.

Fennell, D. A. (2012). *Tourism and Animal Ethics*. London: Routledge.

Fennell, D. A. (2014). Exploring the boundaries of a new moral order for tourism's global code of ethics: an opinion piece on the position of animals in the tourism industry. *Journal of Sustainable Tourism*, 22, 983–96.

Fjármálaráðuneytið. (2009). *Islenskur þjóðarbúskapur og ríkisfjármál* [Icelandic economy and national treasury]. Retrieved from www.fjarmalaraduneyti.is/media/frettir/Islenskur_tjodarbuskapur_og_riksifjarmal.pdf

Folke, C. (2006). Resilience: the emergence of a perspective for social-ecological system analyses. *Global Environmental Change*, 16(3), 253–67.

Folke, C, Carpenter, S., Elmqvist, T., Gunderson, L., Holling, C.S. & Walker, B. (2002). Resilience and sustainable development: building adaptive capacity in a world of transformations. *Ambio*, 31, 437–40.

George, E. W., Mair, H. & Reid, D. G. (2009). *Rural Tourism Development: Localism and Cultural Change*. Bristol: Channel View Publications.

Granquist, S. M. & Hauksson, E. (2016). *Management and Status of Harbour Seal Population in Iceland 2016: Catches, Population Assessments and Current Knowledge*. Reykjavik: Veiðimálastofnun [Institute for Freshwater Fisheries].

Granquist, S. M. & Sigurjonsdóttir, H. (2014). The effect of land-based seal watching tourism on the haul-out behaviour of harbour seals (Phoca vitulina) in Iceland. *Applied Animal Behaviour Science*, 156, 85–93.

Granquist, S. M., Hauksson, E. & Stefánsson, T. (2015). *Landselatalning ári. 2014: Notkun Cessna yfir.ekju flugvélar, yrilvængju og ómanna.s loftfars (flygildi) vi. talningu landsela úr lofti* [Harbour seal population assessment 2014: An aerial survey using Cessna airplane, helicopter and drone to count harbour seals]. Reykjavik: Veiðimálastofnun [Institute for Freshwater Fisheries].

Gunnarsdóttir, M. J. & Gissurarson, L. R. (2008). HACCP and water safety plans in Icelandic water supply: preliminary evaluation of experience. *Journal of Water and Health*, 6, 377–82.

Hall, C. M., Timothy, D. J. & Duval, D. T. (Eds.) (2013). *Safety and Security in Tourism: Relationships, management, and marketing*. London: Routledge.

Hauksson, E. & Einarsson, S. T. (2010). Review on utilization and research on harbor seal (*Phoca vitulina*) in Iceland. *NAMMCO Scientific Publications*, 8, 314–53.

Hu, W. & Wall, G. (2005). Environmental management, environmental image and the competitive tourist attraction. *Journal of Sustainable Tourism*, 13, 617–35.

Icelandic Seal Center (2016). Facebook post, September 11, 2016. www.facebook.com/IcelandicSealcenter

Jóhannesson, G. T. & Huijbens, E. H. (2010). Tourism in times of crisis: exploring the discourse of tourism development in Iceland. *Current Issues in Tourism*, 13, 419–34.

Karlsdóttir, G. H. (2016). *Húnaþing vestra*. Retrieved from www.hunathing.is

Karlsson, G. (1996). Plague without rats: The case of 15th-century Iceland. *Journal of Medieval History*, 22, 263–84.

Kristjánsson, L. (1980). Íslenzkir Sjávarhættir [Icelandic Fisheries]. Reykjavik: Bókaútgáfa.

Leask, A. (2008). The nature and role of visitor attractions. In A. Fyall, B., Garrod, A. Leask & S. Wanhill (Eds.), *Managing Visitor Attractions* 2nd ed. (pp. 3–15). Oxford: Butterworth Heinemann.

Lebel, L., Andreies, J., Campbell, B., Folke, C., Hatfield-Dodds, S., Hughes, T. & Wilson, J. (2006). Governance and the capacity to manage resilience in regional social-ecological systems. *Ecology and Society*, 11(1), 1–19.

Lew, A. A. (2014). Scale, change and resilience in community tourism planning. *Tourism Geographies*, 16, 14–22.

Lew, A. A., Ng., P. T., Ni, C-C. & Wu, T-C. (2016). Community sustainability and resilience: similarities, differences and indicators. *Tourism Geographies*, 18, 18–27.

Lovelock, B. & Lovelock, K. M. (2013). *The Ethics of Tourism: Critical and Applied Perspectives*. New York: Routledge.

McCauley, D. J. (2006). Selling out on nature. *Nature*, 443, 27–8.

McGovern, T. H., Vésteinsson, O., Friðriksson, A., Church, M., Lawson, I., Simpson, I., ... Dunbar, E. (2007) Landscapes of settlement in northern Iceland: historical ecology of human impact and climate fluctuation on the millennial scale. *American Anthropologist*, 109, 27–51.

Mahrouse, G. (2011). Feel-good tourism: an ethical option for socially-conscious westerners? *ACME*, 10(3), 372–91.

Mairs, K. A., Church, M. J., Dugmore, A. J. & Sveinbjarnardóttir, G. (2006). Degrees of success: evaluating the environmental impacts of long term settlement in South Iceland. In J. Arneborg & B. Grønnow (Eds.), *The Dynamics of Northern Societies* (pp. 365–73). Copenhagen: Publications from the National Museum (PNM).

Matthiasson, T. (2008). Spinning out of control: Iceland in crisis. *Nordic Journal of Political Economy*, 34, 1–19.

Óladóttir, O. (2016). *Tourism in Iceland in Figures: May 2016*. Reykjavik: Icelandic Tourist Board.

Ólafsson, S. (2005). Normative foundations of the Icelandic welfare state: on the gradual erosion of citizenship-based welfare rights. In N. Kildal & S. Kuhnle (Eds.), *Normative Foundations of the Welfare State: The Nordic Experience* (pp. 214–36). London: Routledge.

Orchiston, C. (2013). Tourism business preparedness, resilience and disaster planning in a region of high seismic risk: the case of the Southern Alps, New Zealand. *Current Issues in Tourism*, 16, 477–94.

O'Riordan, T. & Jordon, A. (1995). The precautionary principle in contemporary environmental politics. *Environmental Values*, 4(3), 191–212.

Prideaux, B. (2002). Building visitor attractions in peripheral areas: can uniqueness overcome isolation to produce viability? *International Journal of Tourism Research*, 4, 379–89.

Puhvel, M. (1963). The seal in the folklore of northern Europe. *Folklore*, 74, 326–33.

Ragnarsdóttir, B. H., Bernburg, J. G. & Ólafsdóttir, S. (2013). The global financial crisis and individual distress: the role of subjective comparisons after the collapse of the Icelandic economy. *Sociology*, 47, 755–75.

Ragnarsson, A. (2015). *Norðurland Vestra, Stöðugreining 2014* [Northwest Status Analysis 2014]. Iceland: Byggðastofnun.

Ram, Y., Björk, P. & Weidenfeld, A. (2016). Authenticity and place attachment of major visitor attractions. *Tourism Management*, 53, 110–22.

Redford, K. H. & Adams, W. M. (2009). Payment for ecosystem services and the challenge of saving bature. *Conservation Biology*, 23, 785–87.

Robertson, M. (2012). Measurement and alienation: making a world of ecosystem services. *Transactions of the Institute of British Geographers*, 37, 386–401.

Sagoff, M. (2008). On the economic value of ecosystem services. *Environmental Values*, 17, 239–57.

Schellnhuber, H. J., Kokott, J. Beese, F.O., Fraedrich, K., Klemmer, P., Kruse-Graumann, L., … Zimmermann, H. (2001). *World in Transition: Conservation and Sustainable Use of the Biosphere.* London: Earthscan Publications.

Sharpley, R. (2007). Flagship attractions and sustainable rural tourism development: the case of Alnwick Garden, England. *Journal of Sustainable Tourism,* 15, 125–43.

Smit, B. & Wandel, J. (2006). Adaptation, adaptive capacity and vulnerability. *Global Environmental Change,* 16, 282–92.

Streeter, R., Dugmore, A. J. & Vésteinsson, O. (2012). Plague and landscape resilience in premodern Iceland. *Proceedings of the Natural Academy of Sciences of the United States (PNAS),* 109, 3664–9.

Stockholm Resilience Centre (2014). *Applying Resilience Thinking: Seven Principles for Building Resilience in Social-Ecological Systems.* Stockholm: Stockholm University.

Thordarson, T. & Larsen, G. (2007). Volcanism in Iceland in historical time: volcano types, eruption styles and eruptive history. *Journal of Geodynamics,* 43, 118–52.

Thordarson, T. & Self, S. (2003). Atmospheric and environmental effects of the 1783–84 Laki eruption: A review and reassessment. *Journal of Geophysical Research Atmospheres,* 108, 4011.

Thórarinsson, S. (1967). *The eruptions of Hekla in historical times.* Reykjavik: Leiftur.

UNWTO (2016). *UNWTO Annual Report 2015.* Madrid: World Tourism Organisation.

Vasey, D. E. (1996). Population regulation, ecology, and political economy in preindustrial Iceland. *American Ethnology,* 23, 366–92.

Walker, B. H., Carpenter, S. R., Rockstrom, J., Crépin, A-S. & Peterson, G. D. (2012). Drivers, "slow" variables, "fast" variables, shocks, and resilience. *Ecology and Society,* 17(3): 30.

Wegge, N. (2013). Politics between science, law and sentiments: explaining the European Union's ban on trade in seal products. *Environmental Politics,* 22, 225–73.

Yuanyuan, L. (2010). *Chemical characteristics and the formation conditions of geothermal fluids in Reykir at Reykjabraut, N-Iceland.* Geothermal Training Programme (Report No. 17). Reykjavík: Orkustofnun.

Þórisson, S. L. (2016). *Selasetur Íslands Annual Report 2015.* Hvammstangi: Icelandic Seal Center.

5   Tourism development and resilience in small oceanic islands in Australia and Brazil

Leonardo Nogueira de Moraes

Introduction

The employment of the concept of resilience to tourism studies is a promising avenue to research on the sustainability of tourism development by emphasising its dynamic aspects and the importance of adaptation and mitigation for long-term ecological conservation and social well-being (see Allison, Moore & Strickland-Munro, 2010; Hamzah & Hampton, 2012; Nogueira de Moraes, 2014; Scheyvens & Momsen, 2008). Understanding the drivers and inhibitors underlying the processes leading to its manifestation is as important as discussing the applicability of resilience as a societal and ecological goal (Walker, Carpenter, Rockstrom, Crépin & Peterson, 2012). Hence, if resilience is to be understood as the 'tip of the iceberg', the question that arises is: what lies underneath?

This chapter investigates the relationships between the contextualised design of tourism development and the resilience of social-ecological complex adaptive systems (SECASs) in two small oceanic island case studies, namely Fernando de Noronha Archipelago, Brazil, and Lord Howe Island, Australia, with the intenttion of shedding light on the underlying processes leading to different degrees of resilience. The findings articulated stem from qualitative research framed by a conceptual model developed from complex adaptive systems theory (Mitchell, 2009; Norberg & Cumming, 2008), network theory (Newman, 2010) and social-ecological systems and resilience theory (Lew, Ng, Ni & Wu, 2016; SRC, 2014). This conceptual model recognises a 'tourism development complex adaptive system' operating within a 'contextualised global–local social-ecological complex adaptive system', accounting for 'local tourism development' and 'global tourism' as complex adaptive properties emerging from processes of self-organisation that occur in these same systems.

Data collected were organised to allow understanding of causality respecting the characteristics of complex adaptive systems theory: global, national and local *contexts* are depicted as the background for the case studies; environmental and demographic features are explained as *intervening conditions* acting upon the *phenomenon* of self-organisation by means of competition and cooperation; specific historical events, activities and regulations are portrayed as *causal conditions* for the studied *phenomenon*, whose *consequences* are understood here as specific sustainability outcomes and degree of resilience. The processes of *tourism development* and *localised conservation* in both case studies are explained as *(inter)action strategies* acting upon the *phenomenon* or as system-wide emergent properties.

Specific sustainability outcomes related to local empowerment and local social cohesion are identified as consequences of the studied phenomenon and as evidence of the degree of resilience experienced by both case studies. The study of the phenomenon of self-organisation by means of competition and cooperation is carried out by categorising these relationships against a global, global–local and local continuum, allowing understanding of how specific combinations of different degrees of three types of competition and cooperation seem to influence the level of local empowerment and social cohesion leading to different degrees of resilience. Finally, the role of tourism development and of localised conservation as influencers of self-organisation is analysed.

Research conceptual framework

Before discussing tourism development from the viewpoint of resilience, reviewing the rise of sustainability as a widely employed concept by academics, governments, citizens and enterprises is adopted, framing a concept of sustainability that informed research design and practice based on seven cornerstones:

1 the development of ecology and environmental conservation;
2 the growing understanding of human–environmental relations;
3 the construct of Earth's limits and resulting understanding of limits to growth and the need to address inequalities in access to and use of resources;
4 the emergence of the concept of sustainable development in the international arena;
5 the debate on priorities, degrees and dimensions of sustainability;
6 the emergence of sustainability science and complex adaptive systems theory; and
7 the plurality of (sometimes divergent) concepts of sustainability, leading to contradictory implementation strategies.

Resilience theory (Lew et al., 2016; SRC, 2014) is identified as being transversal to many of these cornerstones, especially to ecology and environmental conservation, to the understanding of human–environmental relations, to limits to growth and to complexity theory. Despite being connected, sustainability and resilience are different concepts that can be understood as being complementary but sometimes also in conflict due to the breadth of meanings attributed by those with different interests (see Lew et al., 2016). Overall, conflicting concepts arise from the distorted view of sustainability as only seeking to promote the continuity of the status quo and the restricted understanding of resilience as being opposed to conservation by understanding that it only promotes adaptation as key to the survival of the human species, therefore relieving the need to mitigate human impacts on the environment.

Conversely, reviewing the evolution of the context of sustainability led to the understanding that sustainability is an anthropocentric concept related to the human species capacity to survive by undergoing adaptation and undertaking paths of human development that redesign human and environmental systems, respecting existing and future thresholds, while catering for social-ecological systems that are more resilient. The conceptual framework employed by this research suggests that the sustainability of human (and tourism) development is dependent on how it affects the resilience of the social-ecological systems in which it occurs.

When analysed from the perspective of complex adaptive systems theory, sustainability and resilience are relative (and not absolute) concepts as, through feedback processes, thresholds are dynamic, thus constantly changing. Sustainability is always a future condition, and human development, through the redesigning of environmental and human systems, can be understood as undertaking pathways leading to more or less sustainable futures. However, achieving a purely sustainable (or unsustainable) current condition will never occur, because the current situation is complex and always relative to an even more sustainable future. Therefore, this chapter prefers to employ the terms sustainability and resilience, instead of sustainable and resilient.

Construction of the conceptual framework also involved reviewing different tourism (development) concepts and definitions from a complex adaptive systems perspective and in contrast to the concept of sustainability. The upshot from this review suggests that human needs that cannot be satisfied in tourist residency areas motivate tourists to seek destinations that will cater for those needs. As a consequence of this movement of people around the globe, tourism-oriented redesigning of human and environmental systems and resources takes place, leading to feedbacks that reverberate not only in residency, transit and destination areas, but also on other areas of the planet and on the planet as a whole – carbon emissions being a practical example to illustrate this point. This way, tourism development promotes redesigning of resources and systems for tourism purposes, on top of existing redesigning for human development purposes.

The concept also suggests that the sum of resources and systems in their natural state with those that have been redesigned for purposes other than tourism would result in the destination's endogenous attributes or its degree of authenticity, whereas those resources and systems redesigned for tourism purposes would represent its exogenous attributes or its degree of tourism orientation. This understanding is aligned with Plog's (2001) psychocentric theory on tourism development which proposes that tourism could become the seed of its own destruction by excessive development of infrastructure leading to loss of authenticity and decreased capacity to attract tourist flows. As a result, the suggested concept also identifies three dimensions of sustainability associated with tourism development:

- local (affecting the capacity of local communities to survive);
- global (affecting the capacity of the human species to survive); and
- sectoral (affecting the capacity of tourism to continue to exist).

Once base definitions for sustainability and for tourism development in the context of sustainability were developed, a theoretical framework of analysis for researching the dynamics of sustainability and tourism development was also constructed with the aid of concepts and theories related to networks, (social) structure and agency, social-ecological systems, complexity, resilience, competition and cooperation, and tourism clusters. As a result, a conceptual model of a local–global social-ecological complex adaptive system portraying tourism development and nature conservation was also constructed.

This concept suggests that agents are open systems that interact with one another with the support of and being influenced by institutional and relational structures (López & Scott, 2000) while also contributing to their redesigning through individual agency. It also implies that agents have limited contact with these structures and therefore incorporate only part of them, what is incorporated becoming embodied structures (López & Scott, 2000) they use to inform decision-making and actions. The concept also implies that agents are subject to environmental conditions as well as relational and institutional structures. By taking into account complex adaptive systems being hierarchical systems (Chu, Strand & Fjelland, 2003), the concept suggests that local social-ecological complex adaptive systems are contained within global systems, allowing a constant transfer and process of global and local stock of matter, energy, information and agents to take place between the two systems. In that respect, tourists and temporary migrant workers could be examples of agents that leave the global stock and become part of the destination's local stock for a limited time, returning home with experiences that may have transformed them as individuals. Within the organisational network that is established among local and global agents it is possible to identify one where stakeholders are involved with tourism and another with nature conservation. As a result of the relationships established within these networks, one can observe the emergence of the following complex adaptive properties that reinforce and regulate those same relationships:

- a global nature conservation movement;
- global tourism;
- local nature conservation practice; and
- local tourism development.

These emergent properties would also influence one another through reinforcing and regulating feedback.

Based on the theoretical framework employed, this research was designed as a qualitative and embedded-multiple-case study (Yin, 2009) that made use of Grounded Theory methods (Charmaz, 2006; Corbin & Strauss, 2008) as the framework guiding data analysis. To help data organisation and visualisation through the lens of global–local social-ecological complex adaptive systems (SECASs), the research applied Strauss and Corbin's Coding Paradigm (Böhn, 2004, p. 272) to define its core variables as follows:

- *Context(ual Conditions)*: Developed and developing countries and globalisation.
- *Intervening Conditions*: Small oceanic island tourist destinations.
- *(Inter)Action Strategies*: Emergent system properties of localised conservation and tourism development.
- *Causal Conditions*: Events, incidents and actions leading to agency.
- *Phenomenon*: Self-organisation by means of competition and cooperation.
- *Consequences*: Sustainability challenges and degree of resilience.

Small oceanic islands were chosen as case studies for their capacity to portray intrinsic conditions and variables that could be more easily distinguished by being naturally amplified, for their clearly defined spatial boundaries and isolation. When the research was being designed (April 2009), Fernando de Noronha Archipelago and Lord Howe Island were the only small oceanic islands listed as natural World Heritage where tourism development included the provision of accommodation (at the time, Galápagos Islands were listed as an *in danger* property). Data collection was carried out through the triangulation of:

- direct observation;
- contextual and focused semi-structured interviews; and
- access to secondary sources.

A total of 71 qualitative interviews were conducted in Fernando de Noronha (26) and on Lord Howe Island (45) with representatives of:

- Federal (10), State (4) and Local Governments (19);
- the Local Community (7);
- NGOs (7); and
- Tourism (18) and complementary sectors (4) and universities (2), totalling 71 qualitative interviews.

In addition, 21 interviews were conducted with representatives of international organisations, including:

- those which are part of the United Nations System (11);
- NGOs (6); and
- Universities (4).

Context(ual conditions)

Contextual conditions are understood here as being 'slow' variables (Walker et al., 2012). Their study included the identification of:

- the evolution of global, national and regional tourism and related structures;
- the evolution of the global and national nature conservation movements and related structures; and

- the associated broader global and national historical backgrounds to which tourism development and the nature conservation movement were subject to.

Two complementary global structures influencing tourism development and environmental conservation were identified:

- one arising more prominently from national government relationships of competition and cooperation, namely the United Nations system of international governmental organisations; and
- the other arising from the efforts of non-governmental organisations worldwide.

Overall, this study identified the emergence and development of a global network of organisations influencing the management of tourism development and nature conservation worldwide, with organisations increasing specialisation and joint projects promoting integration as the network grows in complexity.

The investigation of national contexts led to the understanding that, despite both being federations of states, Brazil and Australia have experienced different pathways leading to the establishment of national, state and local levels that exercise different degrees of empowerment, resulting in the incorporation of global nature conservation and tourism development strategies in different ways. While Brazil first incorporated local tourism development (EMBRATUR, 2002) followed by regional tourism development (MTur, 2013) as strategies for decentralising tourism planning and management, in Australia:

> Despite a very well organised tourism sector at both federal and regional level, with a few exceptions, the organisation of tourism is weak at the local level. Nonetheless, local governments have considerable powers available for tourism, although these powers vary from State to State.
>
> (Cooper & Ruhanen, 2005, p. 47)

In regard to World Heritage, while this global conservation strategy has allowed the Australian Commonwealth Government to become more influential in the proclamation of national protected areas in state land (see Australia, 1997), in Brazil, considering the great empowerment of the Federal Government when it comes to the proclamation of protected areas, World Heritage is used more as a tool for raising awareness of the need to protect what is listed.

Analysis of global and national contexts reinforce the argument that tourism development and protected areas are both promoters and products of globalisation, thus affecting the level of dependence of local destinations to the global social-ecological complex adaptive system. These contexts shed light on the reinforcing feedbacks between global and national levels when it comes to tourism and environmental conservation, pushing for global, national and local (inter)action strategies of tourism development and localised conservation through the establishment of protected areas.

Intervening conditions

Environmental and demographic features affecting the two case studies were regarded as intervening conditions influencing self-organisation (by means of competition and cooperation) through tourism development and localised conservation. Fernando de Noronha Archipelago and Lord Howe Island are small oceanic islands laying respectively on the South Atlantic and South Pacific oceans. While Fernando de Noronha is located some 345 km off the northeast coast of Brazil, close to the Equator and subject to the warm nutrient-poor southern Equatorial current originating in Africa, Lord Howe Island is about 570 km off the east coast of Australia, below the Tropic of Capricorn and in the confluence of the warm eastern Australian current and cold streams coming from the south, resulting in greater biodiversity of marine life. With greater altitude differences (0 to 875 m above sea level), Lord Howe Island also portrays greater terrestrial biodiversity than Fernando de Noronha (0 to 323 m above sea level). Both islands are the result of eroded volcanic formations which were gradually colonised by different species that, in many cases, specialised and became endemic.

The Brazilian archipelago was discovered by the Portuguese in early 1500 with permanent residents arriving in 1737, while the Australian island was discovered by the British in 1788 with early settlers arriving in 1833. By 2011, Lord Howe Island was estimated to host some 360 inhabitants while Fernando de Noronha held about ten times that number. Despite Fernando de Noronha's main island being slightly larger (17.01 sq. km) than Lord Howe Island (14.54 sq. km) and both portraying similar ratios of land areas available for residence due to the protected areas implemented (30.0 per cent in the case of Fernando de Noronha and 29.6 per cent in the case of Lord Howe Island), they present very different demographic densities: 705 people per sq. km and 84 people per sq. km, respectively. Their growth trends are also very different, stability being reached by Lord Howe Island in the 1990s when it had 369 inhabitants and Fernando de Noronha experiencing significant growth since the early 1990s, more than doubling its already greater population during this period (1,700 to 3,600 inhabitants).

Overshooting of Fernando de Noronha's carrying capacity was already evident in 2000 when it was calculated as being able to host 1,433 people and its population had already reached 2,900 people (ADM&TEC, 2000). Development of infrastructure did not match population growth as eight years later the island was equipped to properly serve only 2,550 people, while residents and tourists accounted for a total daily average of 4,000 (ICMBio & Elabore Consultoria, 2008). Tourism growth followed similar patterns to those of the resident population, with numbers increasing more than ten times in Fernando de Noronha during the period 1991–2009 (5,911 to 62,823 people: ADEFN, 2005, 2010), while Lord Howe Island experienced little growth, going from 13,182 to 15,148 during the period 1999–2010 (LHIB, 2011).

In terms of localised conservation through the implementation of protected areas in the context of World Heritage listing, Lord Howe Island was proclaimed a World Heritage property in 1982, right after a State Permanent Park Preserve was declared to protect about 70 per cent of its terrestrial areas. Its surrounding marine areas were declared protected areas in 1989 with the proclamation of a State Marine Park, and expanded in 2000 and 2012 with the proclamation of Commonwealth Marine Parks.

In the case of Fernando de Noronha, World Heritage listing came last, when in 2001 it was proclaimed part of a serial natural World Heritage property also containing Rocas' Atoll. Prior to that, a Federal Environmental Protection Area was declared in 1986 containing both terrestrial and marine areas. In 1988, part of it was converted into a Marine National Park comprising both land and sea. In 1993 the existing protected areas were considered as core and buffer areas of part of the Atlantic Forest Biosphere Reserve, extending protection to a marine transition area. With no practical effect, in 1989 the whole archipelago was declared a State Environmental Protected Area and in 1995 a State Marine Park.

Overall, when environmental and demographic characteristics were analysed, geographical and social isolation were identified as both barriers and facilitators to globalisation, and pressured by the implementation of localised conservation and tourism development. Complementarily, local social cohesion and local empowerment have been impacted by regulations that recognise and grant specific rights to different social groups sharing isolated local areas. By exploring the tension between endemic and invasive species and different social groups in local social-ecological complex adaptive systems that experienced processes of increasing globalisation, this study highlights the interactions established between global and local systems and their results in terms of shifting regimes that lead different populations to stability, instability and extinction.

To contextualise the role of isolation in cultural differentiation and the emergence of different local social groups and the relationships of competition and cooperation established among them, the ecological concepts of traditional/endemic and invasive species provide helpful insights, helping to explain how *Noronhenses* and Lord Howe Islanders have specialised in terms of their cultural gene pool. In that respect, comparison of the two cases suggest that the resilience of the local social-ecological complex adaptive system is increased when local and global knowledge are shared among traditional islanders and migrants, allowing the establishment of processes that can make the best use of global resources to leverage the establishment of local processes that cater for a more autonomous local system.

(Inter)Action Strategies

Tourism development and the implementation of protected areas in Fernando de Noronha Archipelago and Lord Howe Island were the two (inter)action strategies researched, or the two emergent properties of the local and global social-ecological complex adaptive systems that were the focus of investigation. Results pointed to tourism development and the implementation of protected areas as being self-regulating and self-reinforcing properties that portray great influence on one another and on the resilience of local social-ecological complex adaptive systems by interfering with their level of empowerment and independence.

Differing models of tourism development seem to have arisen on Lord Howe Island and Fernando de Noronha Archipelago stemming from their respective histories and associated levels of local empowerment and self-sufficiency that contribute to their reinforcement. The two case studies suggest that the emergence of tourism on these isolated communities came as an alternative for these local economies to adapt to global economic shifts, by allowing greater connectivity, as well as the economic exploration of these islands' natural attractions. As for localised conservation through the implementation of protected areas, this has been the result of a combination of political, economic and conservation pressures that have influenced human and tourism development, and land control.

Causal conditions and phenomenon

The main historical events, incidents and actions influencing the way self-organisation has taken place in the two islands are the causal conditions for the researched phenomenon of competition and cooperation. Results pointed to three types of competition and cooperation that are determined by the relationships established:

- among local agents;
- among global agents; and
- between local and global agents.

Global and local empowerment and social cohesion were found to be highly influential to the way competition and cooperation takes place. The contrast and comparison of these findings coupled with more abstract analysis led to the development of a grounded theory for competition and cooperation in global–local social-ecological complex adaptive systems (See Figure 5.1).

Mediating the relationships between local and global agents are entry barriers that are affected by global and local relational and institutional structures. These structures arise from relationships of competition and cooperation that are linked with certain levels of empowerment and social cohesion experienced by global agents and local agents within their global and local systems. The resilience of global and local social-ecological complex adaptive systems and the sustainability of their emergent properties of tourism development and nature conservation through localised conservation (by means of the establishment of protected areas) are greatly dependent on the levels of empowerment and social cohesion experienced by both the global and local levels.

Consequences

Selected resulting sustainability challenges were investigated as the consequences arising from specific types of competition and cooperation taking place on Lord Howe Island and Fernando de Noronha Archipelago. These challenges were defined through a process of consultation with key stakeholders from both destinations, and based on UNWTO's (2004) proposed list of baseline issues grouping indicators of sustainable tourism for the types of tourist destinations being researched. The list was expanded with contributions from the interviewees, and later categorised to facilitate analysis (see Table 5.1).

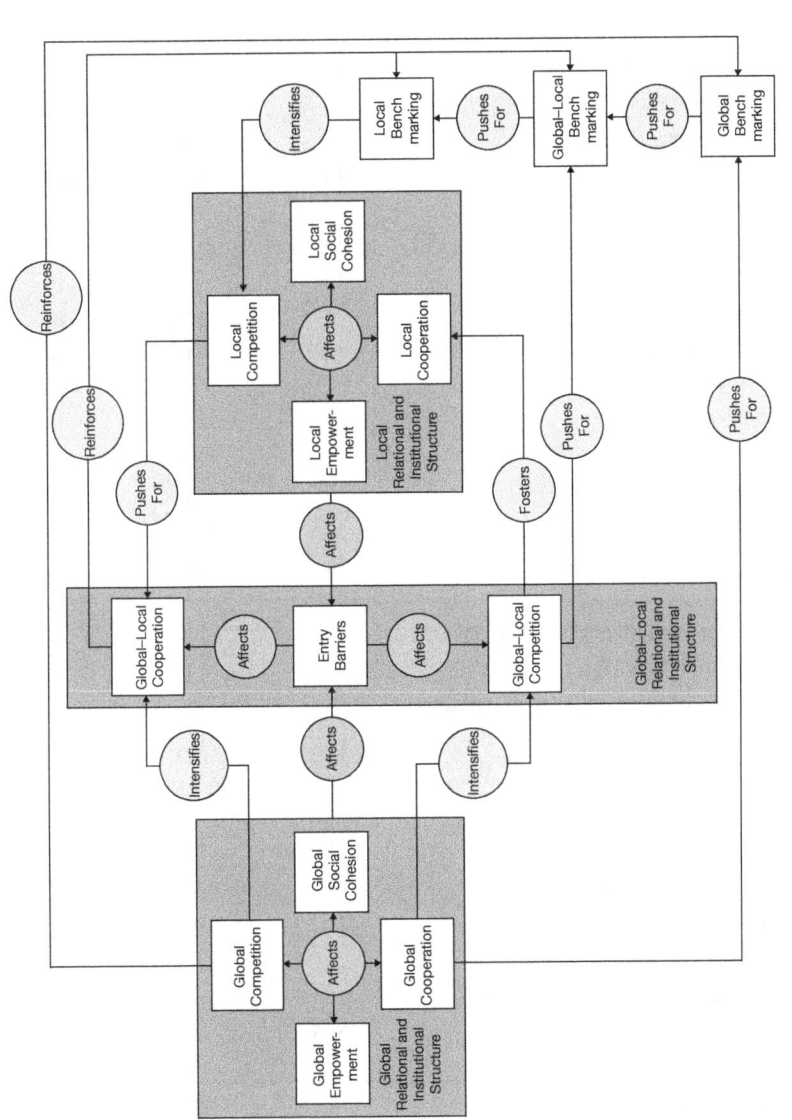

Figure 5.1 Resulting grounded theory for the researched phenomenon
Source: Nogueira de Moraes, 2014, p. 350

Table 5.1 Categorisation of sustainability challenges incidences on interviews

Sustainability Challenges	Fernando de Noronha	Lord Howe Island	Total
Waste Management (UNWTO)	17	12	29
Energy Consumption and Generation (UNWTO)	11	15	26
Sewage Treatment (UNWTO)	12	13	25
Water Management (UNWTO)	13	3	16
Island Supply Management (FDN)	1	10	11
Material and Energy Flows	**54**	**53**	**107**
Environmental Impact Management (UNWTO)	13	17	30
Access to Natural Resources (UNWTO)	13	3	16
Climate Change (UNWTO)	2	12	14
Environment and Conservation	**28**	**32**	**60**
Local Empowerment (FDN)	3	18	21
Social Cohesion (FDN)	1	19	20
Benefits Retention (UNWTO)	6	6	12
Sense of Ownership of Public Spaces (LHI)	N/A	3	3
Community	**10**	**46**	**56**
Tourism Development Management (UNWTO)	10	14	24
Tourism Intensity (UNWTO)	5	13	18
Access to Destination and Attractions (UNWTO)	2	9	11
Tourism	**17**	**36**	**53**
Demographic Density (FDN)	8	12	20
Housing (FDN)	4	15	19
Seasonality and Migration (UNWTO)	9	3	12
Population	**21**	**30**	**51**
Cultural Heritage Protection (UNWTO)	7	7	14
Skills Strategy (LHI)	N/A	6	6
Education (FDN)	1	4	5
Information and Knowledge	**8**	**17**	**25**
Total	**138**	**214**	**352**

Notes:
UNWTO – proposed by the World Tourism Organisation
FDN – emerged from interview in Fernando de Noronha Archipelago
LHI – emerged from interview on Lord Howe Island

Of all the sustainability challenges identified, social cohesion and local empowerment were identified as being of great importance by their intrinsic relationship with the researched phenomenon of self-organisation by means of competition and cooperation. Various tourism development and environmental impact management models seem to emerge from different levels of local social cohesion and local empowerment and at the same time they can reinforce these same levels. When it comes to large-scale commoditisation of the tourist destination, results point to a close relationship to lower levels of social cohesion and local empowerment, the same logic manifesting for small scale and differentiation linked to higher levels of local empowerment and social cohesion.

Conclusions

Competition, cooperation, and resilience

Competition and cooperation are understood here as driving forces of self-organisation leading to a particular organisational structure of social-ecological complex adaptive systems. Understanding systems as being hierarchical, and therefore portraying a contextuality property (Chu et al., 2003, p. 25), helped to clarify the role of different types of competition and cooperation in making social-ecological complex adaptive systems more resilient. The identification of two types of agents involved with tourist destinations arose – global and local – and three types of relationships of competition and cooperation established among them – global competition and global cooperation established among global agents, local competition and local cooperation established among local agents, and global–local competition and global–local cooperation established between global agents and local agents.

When Fernando de Noronha Archipelago and Lord Howe Island are contrasted and compared, results point to combinations of certain levels of the three types of relationship leading to resilience contexts that are influenced by local social cohesion and local empowerment in different ways. Stronger local competition and weaker local cooperation added to stronger global–local cooperation seem to drive decreased local social cohesion and local empowerment, while stronger global–local competition, stronger local cooperation and weaker local competition appear to increase local social cohesion and local empowerment.

While better access to global resources is improved through increased global–local cooperation, it also increases the capacity of global factors that are outside local control to impact the local social-ecological adaptive system. World wars and global economic and financial crises mark great shifts on local systems on both Lord Howe Island and Fernando de Noronha Archipelago. However, being able to tap into global resources can also make the local system more resilient by allowing it to survive when local conditions are less favourable. This way, it is inferred that the emergent process of globalisation needs to be counteracted by localisation for the global–local social-ecological complex adaptive system to become more resilient, localisation being dependent on how island stakeholders relate to these islands as places that help form their identity.

Influence of localised conservation and tourism development

Tourism development and localised conservation through the proclamation of protected areas in the context of World Heritage listing were the two (inter)action strategies (or system-wide properties) emerging from self-organisation by means of competition and cooperation that were the focus of the research. Considering the self-reinforcing and self-regulating roles of emergent properties on the same phenomenon of self-organisation, this study sought to understand how localised conservation and tourism development can influence the patterns of competition and cooperation among stakeholders of the two case studies.

Results point to localised conservation affecting the balance of tourism offer and demand in those destinations by limiting the access to resources and containing the establishment of tourism businesses directly related to the capacity of hosting tourists. Both islands experience some sort of bed-capping (or imposed limits on the offer of accommodation) aiming to keep tourist inflows within the destination's tourism carrying capacity. However, by regulating accommodation businesses and not doing likewise with local tour operators, these islands portray different levels of competition among the two types of tourism business. Limited land availability and unique land tenure systems in place also help contain the establishment of new tourism businesses. When it comes to cooperation, localised conservation tends to push for greater global–local cooperation as global investors are many times better able to deal with the great amount of regulations involved with establishing and running businesses in destinations that are part of protected areas. In this scenario, access to formal education by island residents seem to play a great role in making local businesses more independent and competitive when it comes to tapping into global resources and conforming to regulations without the need to resort to cooperation with global investors.

Localised conservation also influences competition between economic sectors in those islands. By affecting land use, protected areas tend to privilege tourism over more traditional activities such as agriculture and fishing. This is reinforced by increased connection to the mainland promoted by tourism, which brings down the costs of importing food produced, contributing to greater dependence of those islands on global food stocks, therefore making them less resilient. This situation is reinforced by mainland regulations that are implemented on these local areas that are easily adhered to by large businesses, but that may make small-scale local businesses become financially unfeasible.

As for the influence of tourism development in self-organisation, it has promoted a decrease of community values and an increase in individuality in both islands in different degrees. By allowing greater connectivity with the mainland, the tourism sector pushed for greater local–global cooperation and greater local competition in the appropriation of scarce local resources by local businesses. What helps to explain the lower degree of individualisation experienced by Lord Howe Island is its longer history of greater local empowerment, allied to the longer existence of a local traditional population that portrays greater attachment to the island and greater local identity. Overall, the greater control over tourist numbers exercised by Lord Howe Island has played a considerable role in containing the effects of globalisation on the island, bringing the conclusion that tourism carrying capacity must be analysed in direct connection with social-ecological carrying capacity, for excessive numbers of tourists seem also to impact on the local population's capacity to face the challenges brought by globalisation.

Role of local environment and local human population

Intervening conditions were considered in the case studies as comprised of intrinsic environmental characteristics and local human population demographics of both islands. Findings from this study highlights the role of scale and of geographical and social isolation as promoters of differentiation, serving also as barriers and facilitators to globalisation and linked with the implementation of localised conservation and tourism development. Moreover, this research outlined that local empowerment and social cohesion can be largely impacted by laws and regulations that define and acknowledge rights exercised by different social groups and affecting the way they relate to their local environment.

Intervening conditions also help determine the ratio of resource offer and demand, therefore influencing patterns of competition and cooperation and resulting in sustainability challenges that can attenuate or reinforce certain processes. As a result, understanding social-ecological carrying capacity is essential to tourism carrying capacity studies focused on the sustainability of tourism development in destinations. By helping shape the way agents compete and cooperate, environmental and demographic characteristics indirectly affect local empowerment, social cohesion, attachment to place and local identity and, consequently, the sustainability of human development in these islands.

Role of global and national contexts (or 'slow variables') and that of history

Study of the global and national contexts affecting Lord Howe Island and Fernando de Noronha Archipelago highlight self-reinforcing relationships between global and national levels and between national and local levels that result in tourism and nature conservation (relational and institutional) structures that continue to grow in complexity. In return, this growth seems to contribute to greater globalisation fuelled by the implementation of local tourism development and localised conservation through the proclamation of protected areas. Local tourism development is understood here as tourism development that takes place at the local level in any form and not necessarily following a community-based model. As for localised conservation, it is understood as conservation that is geographically situated in a specific (protected) area.

Additionally, context influences local mindset (or embodied structures) in self-reinforcing and self-regulating ways, historical levels of local empowerment and local social cohesion helping to determine the outcomes of the struggle between globalisation and localisation. Moreover, historical events (or causal conditions) help determine the patterns of competition and cooperation that are established in local areas, leading to certain levels of local empowerment and local social cohesion. It is understood here that sustainability should be sought through a process of transitioning capable of tapping into self-reinforcing and self-regulating processes that can change the way the relational structure is established locally and the way it connects to the global relational structure.

Overall, when the two case studies are contrasted, history seems to play an extremely relevant role in setting paths of dependence leading to different sustainability outcomes. Additionally, reaching greater degrees of resilience requires processual structural changes and the reinforcement of competition and cooperation among local and global agents.

Crosscutting conclusions

Results obtained for the two case studies portraying the relationship (1) between environmental conservation, localisation and collectivism/well-being, and (2) that between environmental degradation, excessive globalisation and materialism/individualism, seem to corroborate those by Hurst et al. (2013), who argue that processes of localisation, identity and attachment to place emerge alongside the development of a network of self-reinforcing values that highlight the central role of community, the importance of collective ownership of public spaces and the understanding that individuals are part of systems.

Finally, future research on the topic could focus on places portraying combinations of different levels of global, local and global–local competition and cooperation so as to critique, complement and adjust the conclusions in this chapter. The identification of possible negative effects of extreme levels of local empowerment to the resilience of local social-ecological complex adaptive systems appear to be a promising avenue, Norfolk Island being a potential case study in that regard.

References

ADEFN (2005). *Comparativo do Perfil do Visitante de Fernando de Noronha 2001–4*. Fernando de Noronha: Administração do Distrito Estadual de Fernando de Noronha.

ADEFN (2010). *Controle Diário de Visitantes 2009*. Fernando de Noronha: Administração do Distrito Estadual de Fernando de Noronha.

ADM&TEC (2000). *Plano de Gestão do Arquipélago de Fernando de Noronha, Ecoturismo e Desenvolvimento Sustentável – Fase 1 (Capacidade de Suporte): Relatório Final*. Recife: Instituto de Administração e Tecnologia.

Allison, H. E., Moore, S. A. & Strickland-Munro, J. K. (2010). Using resilience concepts to investigate the impacts of protected area tourism on communities. *Annals of Tourism Research*, 37(2), 499–519. http://doi.org/10.1016/j.annals.2009.11.001

Australia (1997). *Heads of agreement on Commonwealth and State roles and responsibilities for the Environment*. Canberra: Council of Australian Governments.

Böhn, A. (2004). Theoretical Coding: Text Analysis in Grounded Theory. In U. Flick, E. von Kardoff & I. Steinke (Eds.), *A Companion to Qualitative Research* (pp. 270–75). London: Sage.

Charmaz, K. (2006). *Constructing grounded theory: a practical guide through qualitative analysis*. London: SAGE.

Chu, D., Strand, R. & Fjelland, R. (2003). Theories of complexity: common denominators of complex systems. *Complexity*, 8(3), 19–30. http://doi.org/10.1002/cplx.10059

Cooper, C. & Ruhanen, L. (2005). The Organisation of Tourism in Australia. In C. Cooper & C. M. Hall (Eds.), *Oceania: A Tourism Handbook*. Clevedon, UK: Channel View Publications.

Corbin, J. M. & Strauss, A. L. (2008). *Basics of qualitative research: techniques and procedures for developing grounded theory* (3rd ed.). Los Angeles: SAGE Publications.

EMBRATUR (2002). *Retratos de uma caminhada: PNMT 8 anos*. (Gerência de Programas Nacionais – Supervisão de Projetos de Descentralização, Ed.). Brasília: Instituto Brasileiro de Turismo.

Hamzah, A. & Hampton, M. P. (2012). Resilience and Non-Linear Change in Island Tourism. *Tourism Geographies: an International Journal of Tourism Space, Place and Environment*, 15(1), 1–25. http://doi.org/10.1080/14616688.2012.675582

Hurst, M., Dittmar, H., Bond, R. & Kasser, T. (2013). The relationship between materialistic values and environmental attitudes and behaviors: A meta-analysis. *Journal of Environmental Psychology*, 36(C), 257–69. http://doi.org/10.1016/j.jenvp.2013.09.003

ICMBio & Elabore Consultoria (2008). *Estudo e Determinação da Capacidade de Suporte e Seus Indicadores de Sustentabilidade com Vistas à Implantação do Plano de Manejo da APA-FN: Produtos 3 e 4 – Relatório Final* (pp. 1–316). Brasília: Instituto Chico Mendes de Conservação da Biodiversidade.

Lew, A. A., Ng, P. T., Ni, C-C. & Wu, T-C. (2016). Community sustainability and resilience: similarities, differences and indicators. *Tourism Geographies: an International Journal of Tourism Space, Place and Environment*, 18(1), 18–27. http://doi.org/10.1080/14616688.2015.1122664

LHIB. (2011). *Tourist Arrivals to Lord Howe Island 1998–2011*. Lord Howe Island, NSW: Lord Howe Island Board.

López, J. & Scott, J. (2000). *Social Structure*. Buckingham: Open University Press.

Mitchell, M. (2009). *Complexity: A Guided Tour*. Oxford: Oxford University Press.

MTur. (2013). *Programa de Regionalização do Turismo: Diretrizes*. Ministério do Turismo. Brasília.

Newman, M. (2010). *Networks: An Introduction* (1st ed., Vol. 1). Oxford, England: Oxford University Press. http://doi.org/10.1093/acprof:oso/9780199206650.001.0001

Nogueira de Moraes, L. (2014). Inheriting sustainability: world heritage listing, the design of tourism development and the resilience of social-ecological complex adaptive systems in small oceanic islands: a comparative case study of Lord Howe Island (Australia) and Fernando de Noronha (Brazil). (Doctoral Thesis). The University of Melbourne Digital Repository, Melbourne. Retrieved from http://hdl.handle.net/11343/48400

Norberg, J. & Cumming, G. S. (2008). *Complexity theory for a sustainable future*. New York: Columbia University Press.

Plog, S. (2001). Why Destination Areas Rise and Fall in Popularity: An Update of a Cornell Quarterly Classic. *Cornell Hotel and Restaurant Administration Quarterly*, 42(3), 13–24. http://doi.org/10.1177/0010880401423001

Scheyvens, R. & Momsen, J. (2008). Tourism in Small Island States: From Vulnerability to Strengths. *Journal of Sustainable Tourism*, 16(5), 491–510.

SRC (2014). *Applying Resilience Thinking: Seven Principles for Building Resilience in Social-Ecological Systems*. Stockholm Resilience Centre. Stockholm: Stockholm University.

UNWTO (2004). *Indicators of sustainable development for tourism destinations: a guidebook.* Madrid: United Nations World Tourism Organization.

Walker, B. H., Carpenter, S. R., Rockstrom, J., Crépin, A.-S. & Peterson, G. D. (2012). Drivers, 'Slow' Variables, 'Fast' Variables, Shocks, and Resilience. *Ecology and Society*, 17(3): 30. http://doi.org/10.5751/ES-05063-170330

Yin, R. K. (2009). *Case study research: design and methods* (4th ed.). Thousand Oaks, California: Sage.

6 Eco-tourism, climate change and rural resilience in Trinidad and Tobago

Tisha Holmes

Introduction

The Caribbean is a prime tourist destination where 3S (sun, sea, sand) tourism is a major revenue source for these small island economies. Critiques of the social and environmental burdens that mass tourism activities place on Caribbean states are well documented (Dehoorne, Murat & Petit-Charles, 2010; UNWTO, 2002). Additionally, not all Caribbean islands have geographic features to market a traditional 3S tourism experience (Weaver & Schulter, 2001). Eco-tourism emerged as an economic development alternative to diffuse tourism activities away from the coastal resort sites, support natural resource conservation and empower local communities to pursue secure livelihoods (Honey, 2008). In contrast to mass tourism, eco-tourism includes activities which require less capital infrastructure and have limited impacts on the physical and cultural environments. In turn, revenues derived from eco-tourism are directed to the continued protection of the local resources and cultural heritage (Dehoorne & Augier, 2011).

Research on the impacts of and responses to extreme weather events and climate change in the tourism industry continue to grow (Becken & Hay, 2012; Becken & Hay, 2007; Jones & Phillips, 2010; Uyarra et al., 2005). Due to their small land areas and sensitive environments, small island states have constrained capacities to absorb environmental shocks and the effects on tourism markets are more pronounced (Ghina, 2003; Kaly, Pratt & Howorth, 2002; Munro, 2010). However, there are also narratives of social-ecological adaptability and resilience embedded in the sustainable tourism experiences of island nations that are underexplored (Scheyvens, 2008; Scheyvens & Momsen, 2008). In a similar vein, there is a need to assess the responses of the eco-tourism industry to weather disasters in SIDS. The IDB estimates that in 2012 the eco-tourism industry contributed US$15.7 billion (4.6 percent of total GDP) to the Caribbean region and 647,000 jobs (3.9 percent of total employment) (Wilson, Sagewan-Ali & Calatayud, 2014). This market share is projected to increase, as are the risks of weather disasters as a result of climate change. The chaos and uncertainty in the climate system present negative externalities to the viability of the tourism industry. Tourism agents with interests in maintaining the industry should work to proactively avoid impacts and internalize these externalities (Becken & Hay, 2007). Assessing the practice of developing adaptive projects to build resilience can generate lessons capable of informing the operations and decision-making processes in small island eco-tourism communities.

This study presents a qualitative analysis of the efforts of an eco-tourism organization to build resilience to weather hazards in the small island developing state of Trinidad and Tobago. The chapter begins with a brief review of resilience theory and how it can be applied to analyze climate change and disaster planning efforts. The case of Sans Souci Climate Adaptation Pilot project details the risk perspectives of eco-tourism organizations and evaluates the process of developing a local-scale climate change adaptation and disaster risk-reduction project. The chapter concludes with implications for resilience theory, policy recommendations and potential avenues for future research.

Assessing community resilience to weather disasters and climate change

Resilience characterizes a system's ability to 'bounce back' to a reference state after a disturbance, persist and maintain certain structures and functions (Adger, 2003; Klein, Nicholls & Thomalla, 2003; Pelling, 2011). The Stockholm Centre 2014 outlines seven principles of resilient systems:

1   diversity and redundancy;
2   connectivity;
3   managing slow variables and feedbacks;
4   complex adaptive systems thinking;
5   encouraging learning;
6   broadening participation; and
7   promoting polycentric governance systems.

Other conceptualizations of resilience identify the resilience as qualities or characteristics residing in individuals, communities and societies (Norris, Stevens, Pfefferbaum, Wyche & Pfefferbaum, 2008; Ride & Bretherton, 2011). Community resilience is characterized as the capacity to cope with an emergency associated with a disaster or crisis, to rebuild and to learn from the experience. Rather than returning to a prior state, this perspective advances the idea that people and institutions in communities undergo and transform as they bounce forward to a new state.

Community-based approaches to reducing vulnerability and building resilience to climate-related disasters have grown, with attention focused on valuing local knowledge and capacities, adaptable institutions and ecosystem flexibility (Allen, 2006; Bahadur, Ibrahim & Tanner, 2013; Gero, Meheux & Dominey-Howes, 2011). Focusing on disaster response and prompt rebuilding after a disaster has transitioned to new adaptive approaches which couples stronger imperatives for mitigation, reflexive action and proactive planning (Ahmad, 2007; Rose, 2011). Local institutions shape how rural residents respond to environmental change by influencing how households respond and mediate the flow of external interventions after an event. Formal institutions such as community-based organizations can direct information gathering and dissemination, resource mobilization and allocation, and provide leadership and networking with other decision makers and institutions (Agrawal, 2008).

The pursuit of these responses vary and are dependent on the institutions' endowments, location, access to power and resources, social and institutional capital, connectivity and networks that facilitate coordination and cooperation for mutual benefit. Lew et al.'s (2016) resilience indicators for tourism: (1) building community capacity for change; (2) creating new environmental knowledge; (3) improving living conditions and employment; and (4) supporting social collaboration can be used to assess whether a tourism system is resilient as well as how tourism organizations are working to become resilient. This chapter applies these tourism resilience indicators to assess the efforts of eco-tourism organizations in promoting resilience in the small island state of Trinidad and Tobago.

Case Study: The Turtle Region of Trinidad and Tobago

Geographical context

Trinidad is the larger of the two-island state of Trinidad and Tobago. It is located off the north coast of Venezuela in the Caribbean Sea, between 10° N and 11.5° N latitude and between 60° W and 62° W longitude. The island has an area of approximately 4,800 sq. km and a population of some 1.2 million (www.cso.gov. tt/data). Trinidad's economy is primarily supported by petroleum and petro-chemical extractive industries, which employs less than 5 percent of the population (Artana, Auguste, Moya, Sookram & Watson, 2007; Braveboy-Wagner, 2010). The tourism industry contributes approximately 3–4 percent to the country's US$24.43 billion annual Gross Domestic Product (GDP) and accounts for 5 percent of the labor force (Lewis & Jordan, 2008). As a result of the volatility in energy markets and the reality of declining reserves, there is growing emphasis on promoting diversification to sectors such as tourism, financial services and manufacturing (CBTT, 2014).

The Turtle Region (also known as the Matura to Matelot Region) is a remote coastal region in the county of St. David in northeast Trinidad (Figure 6.1). Fifteen villages are connected by a 55.3 km long (34.36 miles) winding coastal road. Once considered the food basket of the country, the area suffered a collapse of large-scale agricultural estate productions and subsequent lack of investment from successive national governments (McIntosh, 2002).

In the Tourism Master Plan Study for Trinidad and Tobago (ARA Consulting Group, 1995), the northeast coast was identified as a valued location to grow the tourism market. The region houses an ecological reserve area and is noted for outdoor recreation, boutique hotels, eco-resorts and get-away second homes (Figure 6.2). The interior montane forests are rich in natural resources and habitats including rivers, waterfalls, springs, pools and endemic flora and fauna. Visitors seeking to escape the hustle of urban life can engage in eco-tours, hiking, camping,

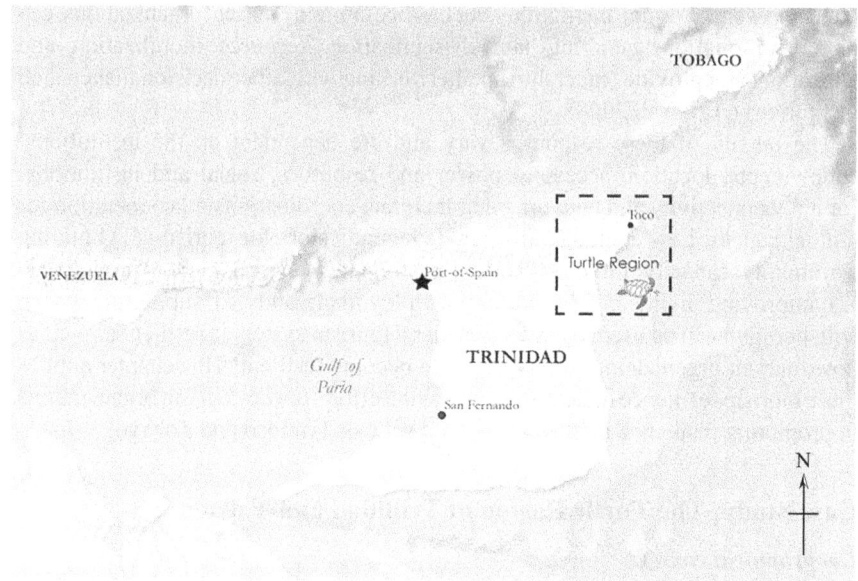

Figure 6.1 Map of Trinidad and Tobago highlighting the Turtle Region
Source: Generated by Melanie Marques, February 22, 2017. Using ArcGIS for Desktop [GIS].
Version 10.1. Redlands, CA: Esri 2012

river bathing, hunting and research. During the months of May to October, there are moderate-energy waves on the coast which allows for recreational swimming, surfing and rock fishing near the coral reef.

The Turtle Region is also known as one of the world's densest nesting sites for marine turtles (Eckert, 2013). Three of the world's seven species of sea turtles that nest on Trinidad and Tobago's shores are listed on the IUCN Red List of Threatened Species. A natural resource co-management arrangement with villagers, community-based organizations and government agencies was established in 1989. In 1990, three beaches – Matura, Grand Riviere and Salyibia – were designated as Prohibited Beaches. Permits are required to gain access to these beaches during nesting season and hunting or taking of turtles and/or their eggs are punishable by law (Lee Lum, 2002).

The popularity of turtle nesting eco-tourism has grown over time. In 2013, permit sales to enter the Prohibited Beaches increased from approximately US$57,000 for 540 permits in the prior year to US$94,000 for 4,504 permits (Poon, 2013). The growth of the eco-tourism market also provides a link between regional natural resource conservation programs while creating employment opportunities for the local communities (Lee Lum, 2002). Other economic activities include fishing, subsistence agriculture, temporary and permanent service work.

Figure 6.2 Hotel on Grande Riviere Beach and turtle nesting area
Source: Photos taken by author

Risk perspectives on community vulnerabilities

The eco-tourism industry in the Turtle Region is supported by the presence of eight tourism development organizations (TDOs) (Table 5.1). These organizations are community-based non-profit and for-profit organizations which source primary membership from the host communities. They provide eco-tourism and hospitality services and are closely linked to community economic development, training and education, arts and craft and environmental conservation initiatives.

Participatory workshops and supplementary interviews were conducted with four TDOs to determine the most pressing risks and impacts of disasters on their operations. The workshops revealed that the Turtle Region faces several hazards which simultaneously impact eco-tourism activities and community infrastructure. Much of the vulnerability from flooding, landslides and coastal erosion hazards are enhanced by the physical features of the region—variable mountainous and low lying topography, saturated soil conditions, a densely forested watershed and dynamic beach systems. However, factors related to land use and quality of infrastructure amplify the overall exposure.

Direct impacts to tourism activities are mainly related to loss of access to the region during heavy rain events which trigger flooding and landslips along the main road. The rainy season period extends from January to June and overlaps with most of the turtle migration and nesting season (March to August). This period is considered the 'peak time' for the industry and repeated closures along the single road paralyzes access to the region for extended periods of time, leading to declines in revenue and income. There is also concern regarding the tourists' perception of safety when visiting the Turtle Region during the rainy season due

Table 6.1 List of tourism development organizations in Turtle Region

Name	Community	Established	Traditional Activities
Grande Riviere Nature Tour Guide Association	Grande Riviere	1999	Tours Conservation Advocacy
Grande Riviere Tourism Development Organization	Grande Riviere	2000	Tourism market development Information center
Natureseekers Guides and Tours	Matura	1990	Tours Conservation Education and training Arts and craft
Pawi Sports Culture and Eco Club	Matelot	–	Tours Community events
Sans Souci Wildlife and Tourism Development Organization	Sans Souci	2011	Tours Conservation
Toco Foundation	Toco	–	Radio Agro-tourism hotel and conference center Conservation Education and training

Source: Interviews conducted by author

to the increased flooding and landslide risks. As more disaster events are publicized in the media, there is a potential for decreased visitors because of perceptions of heightened risk.

The proximity of hotels to the shoreline and impending property losses from flooding, storm surges and sea encroachment was raised as a significant concern. The value of ocean front property in the region is increasing as the market demand for eco-tourism activities grow. Although some adaptive design features are common across the region, such as stilt houses and terraced hills – the building codes, quality of construction and materials used vary widely. Consequently, development guest houses and resort properties are expanding along the coast despite the risk of incurring major financial losses. On the other end of the spectrum, households which rely on eco-tourism activities for income often occupy sub-standard housing on marginal lands without the protections of insurance.

These changes in coastal dynamics are coupled with noticeable erosion of sand from nesting beaches and loss of trees which serve as buffers for turtle nesting and reduces the wind intensities during storms. The ecological pressures on endangered turtle populations can also have significant economic ramifications on the viability of the 'bread and butter' eco-tourism activities revolving around the turtle season:

> A thing we have to do, to really protect the turtles here too is planting trees – because turtles really like trees. I could remember when they had all these trees here, it used to have more turtle than now and many different species too
>
> (TDO Member)

> It's very clear that we are experiencing serious coastal erosion over the last 10 years, I can identify areas that have been eroded to such an extent it is unbelievable. We have two roads that have disappeared. These are indicators of the sea moving into the coast.
>
> (NGO Member)

Some participants indicated that the annual frequency of these events, often in the same vicinity of previous events, can identify the problem areas that should receive priority remedial attention. However, there is general consensus that these are temporary fixes to a deteriorating rural infrastructure system. It was also determined that upgrades to the Turtle region's transportation, drainage and hazard mitigation infrastructure in conjunction with promoting more sustainable collective land use decisions would promote more resilience in the eco-tourism economy and communities in the Turtle Region.

San Souci Wildlife and Tourism Development Organization climate change adaptation project

The Sans Souci climate change adaptation project is aimed at increasing the capacity of community residents residing in the village of Sans Souci to respond to extreme weather events and climate change. Two specific objectives of the project are to:

- decrease vulnerability to sea level-rise, flooding and landslides; and
- increase the institutional response capacity, community knowledge and awareness of climate change and disaster risks.

The project manager is the Sans Souci Wildlife and Tourism Development Organization (SWATDO). The impetus for the project was the need for a response to recent weather events which were defined by some workshop participants as 'unprecedented' between 2011 and 2013. Additionally, there is a sense of urgency regarding the potential for the loss of a keystone species and local ecological knowledge which could dramatically affect the viability of the environmental and economic future of the Turtle Region:

> Growing up in this community, we have witnessed the onslaught of the Turtles … we said that it is necessary, our kids, our grandkids would not know what a turtle looked like. … however, out of this group, we saw there was a need for so many other things; When you look around this community from years gone by, you see that the community has deteriorated because of climate change and natural disasters … It is wreaking havoc in the rural areas.
>
> (SWATDO member)

> We have our young people who don't know what was happening in the past. And our older population who are not sharing information with younger members of population … however, there are some who are gradually coming to terms with these changes
>
> (SWATDO member)

The project was envisioned as a community-owned and -operated pilot effort which would eventually develop into a long-term regional program (Figure 6.3).

The pilot phase is funded by an international multilateral organization with field offices in Trinidad. The planning process was facilitated in two stages:

1   SWATDO received a planning grant to perform a vulnerability risk assessment (VRA) in order to develop a complete project proposal based on the community needs. The VRA was initiated with a community-wide meeting to explain the purpose of the assessment and supported by guidance from a stakeholder consultation group (see Table 6.2).
2   A community survey was subsequently conducted to assess awareness levels in the community regarding local hazards, climate change and adaptation options. The response rate for the VRA was low and did not generate much information due to reluctance of community members to answer questions.

Figure 6.3 SWATDO project site
Source: Photo taken by author

Table 6.2 Vulnerability risk assessment (VRA) stakeholder consultation groups

Stakeholder consultation groups
Ministry of Environment and Water Resources
Environmental Management Authority
Institute of Marine Affairs
Forestry Division
Fisheries Division
Local Health Authority
Community Reforestation Group
Regional Corporation (Infrastructure and Utilities)
Regional Corporation (Disaster Management)

Source: Interviews conducted by author

SWATDO developed and submitted the project proposal to the donor. The proposal identified six initiatives to pursue:

- providing disaster response and first aid training;
- developing a disaster/emergency supply repository;
- data collection and evaluation of environmental changes and risks;
- identifying and implementing soft mitigation strategies e.g. tree planting, enforcing set back limits;
- developing community outreach and education programmes; and
- institutional capacity and partnership development.

The next phase of the project will involve the implementation of these strategies over the next one or two years.

Building resilience in the Turtle Region: opportunities and challenges

Many themes emerge from SWATDO's efforts to capitalize on opportunities to build climate and disaster resilience which align with Lew et al.'s (2016) tourism resilience indicators. Barriers experienced during this planning process also provide important insights for resilience theory and practice.

Building community capacity for change

SWATDO was established to protect turtles and develop economic opportunities for the community through eco-tourism. This mandate evolved to respond to disasters impacting the community and much of the project planning process was beyond their realm of expertise. Their unfamiliarity with technical skills in grant applications and project development slowed down the project's progression:

> One of the problems we have had with the project is the capacity to implement a project. A lot of young people start there, and then leave. They start very excited about the project and then they go get a job outside. You didn't have a lot of people living there who had the skills to carry out the project.
>
> (NGO member)

However, mentoring partners helped group members navigate the process and build project management capacities and technical knowledge. This should enable SWATDO to pursue more complex planning activities beyond the pilot project phase:

> We realized we were not going forward. We were not getting around at all – so we reached out to similar partners who were more established to get advice on the routes to take.
>
> (SWATDO member)

They had a lot of capacity building to do, but they were committed and because they had a mentor – an established NGO that was working in the areas, so they understood the needs of the community.

(NGO member)

The group's sphere of influence grew with an increase in membership from 5 to 32 people. With the growth in membership, the range of planning activities increased and tasks were completed in a shorter time frame. A wider membership also attended to concerns about sustaining momentum and leadership succession beyond the pilot phase:

Community projects shouldn't have one or two people leading. There are always people who drive things. But they should have a number of persons – if one person left or got ill – at least there would be other beneficiaries who stay with the community group.

(NGO member)

SWATDO is developing a clear voice for the Sans Souci community with respect to environmental protection and disaster risk management issues. It remains to be seen whether this enhancement in organizational capacity can influence the provision of physical capital in the community.

Knowledge creation

Prior to the project's inception, the extent of the community's knowledge about weather hazards, disasters and climate change adaptation was unknown. The initial VRA process provided learning lessons about data collection strategies, as many people misunderstood the intent and purpose of the questionnaire. A broader awareness of the potential risks and adaptation opportunities evolved when community members engaged with professionals and government officials. The value of integrated sources of local, indigenous and expert knowledge is an important element for ensuring that adaptation strategies are technically sound and tailored to local needs. SWATDO continues to have community consultations where professionals are invited to answer questions from community members.

At first agencies were just observers and community members actually had to participate. The first aspect was them understanding what issues were affecting the community. They would provide suggestions of what was needed and what would opportunities they saw.

(NGO member)

The donor agency also promotes knowledge preservation and transfer by requiring grantees to reserve funding for knowledge management activities.

Supporting social collaboration

SWATDO's proactive outreach to and collaboration with non-governmental and government stakeholders enabled them to develop a strategic network of technical resources and funding support (Table 6.2). Because climate adaptation and disaster planning require technical expertise in different areas, the distribution of labour among collaborative partners allows SWATDO to serve as a facilitator, rather than a technical expert, in the planning process:

> with the infrastructure assessment, why do we have to take it up on ourselves when the Regional Corporation who is responsible for roads and drainage does that? We don't have to take it all upon ourselves.
>
> (SWATDO member)

Promoting internal strategic partnerships is equally important to ensure successful project approval and implementation. Some SWATDO members also serve on other community organizations, such as the Village Council – which is the main administration body for village affairs; and the Turtle Village Trust – the national turtle conservation organization: 'You can't do anything in the community without going through us' (Village council member).

Improvement in living conditions and employment

Livelihood security proved to be a significant incentive to join SWATDO. Community participation increased once grant funding became available. This speaks to a reality for communities with constrained livelihood options who view incentives to participate in monetary terms. Additionally, the employment in the eco-tourism industry is seasonal because activities are mainly tied to when turtles nest. During non-nesting months, the tourism activities slow down and available jobs decline. Consequently, villagers often seek more predictable year-round forms of employment:

> The reality of climate change to lots of rural people is not something that is of importance. Unless you can identify something that they can put a timeline or a dollar amount to and they can say but wait a minute, then you can get them thinking.
>
> (SWATDO member)

> We do a lot of volunteering for years before getting a dollar. If you didn't pay people to protect the turtles, nobody would do it.
>
> (SWATDO member)

> You have to try and work with the people, to get them to understand what you trying to do. But it's really for six months of the year. Everybody trying to eat a food and if the food eh sharing, they're gone.
>
> (SWATDO member)

Improvements in the community infrastructure and assets were identified as major priorities in the workshops. However, the scale of planning required to accomplish such upgrades is beyond SWATDO's capacities and sphere of influence. As a result, these communities will continue to rely on the government to implement high impact adaptation interventions.

Challenges

There are a number of barriers to the implementation of climate change planning measures by tourism businesses. These relate to lack of information, behavioural patterns, lack of resources and skilled personnel, the large number of small businesses and the international nature of tourism (Becken & Hay, 2007). The SWATDO project is the first of its kind in Trinidad and Tobago. The presence of the international donor as a source of funding and technical guidance was noted as a necessity for the success of the project. SWATDO is required to submit quarterly progress reports and held accountable in terms of adhering to activities, spending and timelines. This rigidity in reporting can streamline the planning process in a rational and comprehensive fashion. However, the reporting system may reduce flexibility and adaptability in the planning process, especially when deviations from original proposals must receive formal approval. Although the project implementation must be directed by the community group, there is potential for dilution of true project ownership if reporting requirements are perceived as burdensome. This challenge is particularly acute for small organizations such as SWATDO.

Commitment

A final theme, which can be characterized as an opportunity and a barrier, is the issue of commitment. The success of SWATDO's adaptation project rests on community buy-in and the commitment of members to the project goals over the long term. SWATDO members reside in the communities they serve, so there is a vested interest to improve the quality of community life, even in the absence of monetary benefits. There is also a culture of self-help – possibly propagated by being ignored by political and economic elites – which enables SWATDO to build grassroots support for the project:

> We decided as a group to start taking upon ourselves to see what we could do. As the community people, we have to help each other as people.
> (SWATDO member)

> This is not something that is done for selfish gains. There is so much that you can contribute to your community. It's all about trying to develop this community. Keep it here.
> (SWATDO member)

Finally, there is also the commitment to each other in times of crisis – a theme which has been well documented in the community disaster response literature:

> … everybody sprang into action. Even in the dark, they were trying to see what they could do. But what I admired most is that, everybody came together and everybody started searching for that man … When there is a disaster in this community, everybody comes together.
>
> (SWATDO member)

However, there are limits to commitment, even in communities with high levels of social bonds. Disillusionment with the process, bureaucratic roadblocks and/or community contention can erode the members' commitment to each other and to the project:

> Nothing is happening here and it is like an uphill struggle, an uphill battle for the community people who are trying. It tends to turn you off sometimes.
>
> (Anonymous)

The most significant threat to commitment is if the community's contributions are perceived as being taken for granted or ignored by those in power. The success of SWATDO's work depends on maintaining productive collaborative partnerships. When community morale and trust are broken, it is difficult to regain and move forward:

> By the time the fire service got there, the villagers had already found his body. But then on the news they never mentioned anything about the villagers. They said good work by the fire service. The villagers said, imagine that we did all the work and they get the credit. So sometimes I guess people turn off and don't bother with them. I guess sometimes we tend to live it in the negative.
>
> (Anonymous)

Conclusion

This chapter presented a case study of an eco-tourism organization working to build community resilience to climate change and weather-related disasters. Eco-tourism organizations are traditionally guided by environmental protection and sustainable development principles. SWATDO's climate change adaptation project has carved out a new space for eco-tourism organizations to engage in natural hazards resilience planning. This case provides insight into how tourism agents seeking to build community-level resilience can become institutionally resilient. The case also sought to demonstrate a natural link between protecting eco-tourism livelihoods and responding to environmental shocks and disasters, but it also shows that these adaptation strategies have impacts which reach beyond protecting a single sector.

There continues to be wide debate about the appropriate indicators to measure resilience. The Stockholm Resilience Centre's principles of social learning, broadening participation, adaptive systems thinking and polycentric governance systems are reflected in SWATDO's efforts. However, these principles and related indicators are applied universally and miss the unique interactions of different stages of the planning process: visioning, project development, implementation, monitoring, etc. These indicators also examine resilience as a normative state which has already been achieved – e.g 'high diversity', 'effective institutions' – rather than a pursuit of future actions or moving into a new state (resilience *of* what, rather than resilience *to* what). For rural eco-tourism contexts, the planning process itself can be considered a constant exercise of moving into new states. These indicators also miss the nuances of how community groups mobilize their reserves of social capital, commitment and knowledge bases. Additionally, resilience theory continues to struggle with explaining how power dynamics greatly influences the interactions of rural actors. Although not an exhaustive list, inclusion of development concepts such as 'equity', 'preparedness and planning' and 'diversity' (Bahadur et al., 2013) can enhance the analytical power of indicators used to evaluate the resilience of tourism systems, organizations and economies.

There are practice-based lessons which can be applied to other rural eco-tourism contexts facing similar threats. The lack of technical capacity was a limiting factor which influenced the progression of developing (and perhaps the level of sophistication) of the proposed adaptation strategies. SWATDO's proactive development of partnerships with governmental and non-governmental agencies helped to build their institutional capacities and distribute the technical work among experts. Seeking efforts to train members 'in-house and on the ground' can also develop the organization's capacities. Gaining community input and buy-in was a crucial step to inform the adaptation strategies selected. Attention should be paid to the most effective ways to engage and educate the public about the complexities of issues like climate change. More formalized opportunities for community group education and sharing of lessons learned can build the local knowledge bases. Finally, seeking sustained sources of funding such as long-term grants and consulting contracts, can help extend work beyond the pilot phase and scale up to incorporate other communities in the Turtle Region.

References

Adger, W. N. (2003). Social capital, collective action, and adaptation to climate change. *Economic Geography, 79*(4), 387–404. doi:10.1111/j.1944–8287.2003.tb00220.x

Agrawal, A. (2008). *The Role of Local Institutions In Adaptation To Climate Change.* Washington DC: The World Bank.

Ahmad, R. (2007). *Risk Management, Vulnerability and Natural Disasters in the Caribbean.* Retrieved from http://proventionconsortium.net/themes/default/pdfs/Forum08/Caribbean_Ahmad.pdf

Allen, K. M. (2006). Community-based disaster preparedness and climate adaptation: local capacity-building in the Philippines. *Disasters, 30*(1), 81–101. doi:10.1111/j.1467–9523.2006.00308.x

ARA Consulting Group (1995). *Trinidad and Tobago Tourism Master Plan*. Port of Spain: Trinidad and Tobago Ministry of Tourism.

Artana, D., Auguste, S., Moya, R., Sookram, S. & Watson, P. (2007). *Trinidad and Tobago: Economic Growth in a Dual Economy*. Retrieved from

Bahadur, A. V., Ibrahim, M. & Tanner, T. (2013). Characterising resilience: unpacking the concept for tackling climate change and development. *Climate and Development*, 5(1), 55–65.

Becken, S. & Hay, J. E. (2007). *Tourism and Climate Change: Risks and Opportunities*. Cleveland, Buffalo, Toronto: Channel View Publications.

Becken, S. & Hay, J. (2012). *Climate Change and Tourism: From Policy to Practice* London: Taylor & Francis.

Braveboy-Wagner, J. (2010). Opportunities and limitations of the exercise of foreign policy power by a very small state: the case of Trinidad and Tobago. *Cambridge Review of International Affairs*, 23(3).

CBTT (2014). *2013 Annual Economic Survey*. Port of Spain: Central Bank of Trinidad and Tobago. Retrieved from www.central-bank.org.tt/sites/default/files/AES%20Report%20 2013%20-%20Online%20Version%20Revised%202.pdf.

Dehoorne, O. & Augier, D. (2011). Toward a new tourism policy in the French West Indies: the end of mass tourism resorts and a new policy for sustainable tourism and eco-tourism. *Études caribbéennes*, 19. Retrieved from https://etudescaribeennes.revenues/org/5262.

Dehoorne, O., Murat, C. & Petit-Charles, N. (2010). International tourism in the Caribbean area: current and future prospects. *Études caribbéennes*, 16. Retrieved from http://etudescaribeennes.revenues.org/4713.

Eckert, S.A. (2013). An assessment of population size and status of Trinidad's Leatherback Sea Turtle Nesting Colonies. WIDECAST Information Document No. 2013-01. Bailwin, Missouri: Wider Caribbean Sea Turtle Conservation Network (WIDECAST).

Gero, A., Meheux, K. & Dominey-Howes, D. (2011). Integrating community based disater risk reduction and climate change adaptation: examples from the Pacific. *Natural Hazards and Earth System Sciences*, 11, 101–13. Retrieved from www.nat-hazards-earth-syst-sci.net/11/101/2011/nhess-11-101-2011.pdf.

Ghina, F. (2003). Sustainable development in small island developing states. *Environment, Development and Sustainability*, 5, 139–65.

Honey, M. (2008). *Ecotourism and Sustainable Development* (2nd ed.). Washington, DC: Island Press.

Jones, A. L. & Phillips, M. (2010). *Ecotourism Book: Disappearing Destinations: Climate Change and Future Challenges For Coastal Tourism*. Chippenham: CAB International.

Kaly, U., Pratt, C. & Howorth, R. (2002). A framework for managing environmental vulnerability in small island developing states. *Development Bulletin*, 58, 33–8.

Klein, R. J. T., Nicholls, R. J. & Thomalla, F. (2003). Resilience to natural hazards: how useful is this concept. *Environmental Hazards*, 5, 35–45.

Lee Lum, L. (2002). *Report on the Project to Monitor Beach Dynamics and the Risk Posed to Leatherback Turtle Egg Clutches at Grande Riviere Beach, Trinidad, West Indies*. Chaguaramas: Trinidad and Tobago Institute of Marine Affairs.

Lew, A.A., Ng, P.T., Ni, C-C. & Wu, T-C. (2016). Community Sustainability and Resilience: Similarities, Differences and Indicators. *Tourism Geographies*, 18(1), 18–27.

Lewis, A. & Jordan, L. (2008). Tourism in Trinidad and Tobago: carving a niche in a petroleum-based economy. *International Journal of Tourism Research*, 10, 247–57.

McIntosh, S. (2002). *Toco Charts Its Own Development: A Case Study from Trinidad and Tobago of Effective Local Advocacy and Participation*. Paper presented at the Islands of the World VII Conference, Prince Edward Island, Canada.

Munro, A. (2010). Climate change and SIDS. In J. J. R. a. Y. N. M. Shyam Nath (Ed.), *Saving Small Island Developing States: Environmental and Natural Resource Challenge*. London: Commonwealth Secretariat.

Norris, F. H., Stevens, S. P., Pfefferbaum, B., Wyche, K. F. & Pfefferbaum, R. L. (2008). Community resilience as a metaphor, theory, set of capacities and strategy for disaster readiness. *American Journal of Community Psychology*, 41, 127–50.

Pelling, M. (2011). *Adaptation to Climate Change: From Resilience to Transformation*. London: Routledge.

Poon, S. (2013). Wildlife, In *The 2013 Marine Turtle Project Report*. Port of Prince: Trinidad and Tobago Ministry of Land, Agriculture and Fisheries, Forestry Division Information Unit.

Ride, A. & Bretherton, D. (Eds.). (2011). *Community Resilience in Natural Disasters*. New York: Palgrave Macmillan.

Rose, A. (2011). Resilience and sustainability in the face of disasters. *Environmental Innovation and Societal Transitions*, 1, 96–100.

Scheyvens, R. (2008). Tourism in small island states: From vulnerability to strentghs. *Journal of Sustainable Tourism*, 16(5).

Scheyvens, R. & Momsen, J. (2008). Tourism and poverty reduction: Issues for small island states. *Tourism Geographies: An International Journal of Tourism Space, Place and Environment*, 10(1), 22–41.

Uyarra, M. C., Cote, I. M., Gill, J. A., Tinch, R. R. T., Viner, D. & Watkinson, A. R. (2005). Island-specific preferences of tourists for environmental features: implications of climate change for tourism dependent states. *Environmental Conservation*, 32(1), 11–19.

Weaver, D. B. & Schulter, R. (2001). Latin America and the Caribbean. In D. B. Weaver (Ed.), *The Encyclopedia of Ecotourism*. Wallingford: CAB International.

Wilson, S., Sagewan-Alli, I. & Calatayud, A. (2014). *The Ecotourism Industry in the Caribbean: A Value Chain Analysis*. Washington, DC: Inter-American Development Bank. Retrieved from https://publications.iadb.org/.../CMF_TN_Ecotourism_Industry_in_the_Caribbean.pdf.

UNWTO (2002). *The World Ecotourism Summit, Quebec City, Canada, 19 to 22 May 2002: Final Report*. Madrid: United Nations World Tourism Organization.

7 Cultural ecosystem services, tourism and community resilience in coastal wetland conservation in Taiwan

Alan A. Lew and Tsung-chiung Wu

Cultural ecosystem services

Ecosystems services (ES) comprise all the benefits that people obtain from nature-based ecosystems (MEA, 2005). Such services are typically categorized into four types, as outlined in Table 7.1. In general,

- *Supporting services* and *regulating services* have received the most research attention, due to their basis in the biological sciences, for which ecosystem services provides an applied perspective (Schaich, Bieling & Pleininger, 2010).
- These have been followed by *provisioning services*, which are mostly associated with human economic exploitations of natural processes (farming and fishing, for example). They are also well suited to economic valuation methodologies, which are widely used to assess ecosystem services (Boyd & Banzhaf, 2007).
- *Cultural services* have been the least studied because they require different methodologies, and for that reason some have recommended removing them entirely from the conceptualization of ecosystem services (Fisher, Turner, & Morling, 2009).

Cultural services can have economic value (when they are clearly paid for), but their broader psychological and social values are far more difficult to quantify (Daniel et al., 2012). Regulating and supporting services are also challenging to value economically, although they are less likely to vary by location and scale to the same degree that provisioning and cultural services are due to their deeper ties to human behaviour, perceptions and politics. For example, how a desert-based society interacts with and depends upon its environment will be significantly different from how a tropical island society would. From a geographic scale perspective, a large urban metropolitan area will relate to its dependence on natural processes in far more complex ways than small rural hamlet. Despite the many and nuanced variations that exist, a healthy supply of ecosystem services is essential for the healthy survival of the human society that relies on them. The management and resilience of essential ecosystem services is, therefore, a key requirement in creating sustainable systems and for building the resilience capacity of communities to respond to rapidly changing global and local conditions (Walker & Salt, 2006).

Table 7.1 Four types of ecosystem service

Type of Ecosystem Service	Description
Provisioning services	Products produced by natural environments that are collected and used by humans. Examples include food, water, various types of fuel, and medicinal and chemical products.
Regulating services	Natural processes and functions that humans use and benefit from. Examples include the planting of trees to moderate local weather extremes (climate regulation), insect pollination (especially by bees) of crops that enable the creation of foods for human consumption, and water purification processes that make polluted water potable for human uses.
Cultural services	The diverse ways that humans benefit socially and psychologically (in a non-material way) through engagement with natural environments. Examples include recreation and tourism, spiritual and aesthetic inspirations, cultural heritage, personal identity, and sense of place values.
Supporting services	The basic underlying processes that are essential for all other ecosystem services to exist. These include soil formation (soil cycle), nutrient cycle, and primary biological production, among others.

Source: based on MEA, 2005

Studies of ecosystems services have historically tended to emphasize the impacts of a single driver of change (such as changing water supply or fertilizer use) on a single service, which is usually a provisioning or regulating service (Bennett, Peterson & Gordon, 2009). This occurs despite calls for a more integrated approach that recognizes how an overemphasis of one resource (or set of resources) can cause a decline in others (MEA, 2005; Gordon, Peterson, & Bennett, 2008). The challenge has been in effectively modelling (simplifying) the complex ecological relationships of synergies and trade-offs that exist among different services within an ecosystem (Carpenter et al., 2006). Even when such relationships are the focus of a study, most research only addresses how two services affect each other (Tallis et al., 2008).

Bennett et al. (2009) proposed a model of how different ecosystem services (excluding supporting services) relate to one another, based on the degree of interaction between services and the breadth of impact across services caused by different drivers of change. Based on that model, they suggest that regulating services are the most important to the health of an ecosystem, and that cultural services are the least important. They equate regulating services to slow change variables in resilience theory (Walker et al., 2012), and based on that they recommend that regulating services should be the primary focus for building ecosystem resilience. As with many studies of ecosystem services, Bennett et al. (2009) give short shrift to cultural services, treating them in a peripheral manner without significant discussion.

Of the four types of ecosystem services, cultural services have been the least studied and, therefore, the least conceptualized and understood (Carpenter et al., 2006; Schiach et al., 2010; Daniel et al., 2012; Tengberg et al., 2012). This is due to the greater complexity caused by the social, behavioural, psychological and political underpinnings of cultural ecosystem services (MEA, 2005; Chan, Satterfield, & Goldstein, 2012; Reyers et al., 2013), and possibly because many of those who write about ecosystem services emanate from the fields of environmental economics, ecology and resource management (Schaich et al., 2010). Measuring and monitoring change in natural systems is complicated, with land use and land cover being the most common proxy for services themselves (Bennett et al., 2009). Applying the same techniques to human values rarely results in comparable measures of validity (Bennett, Peterson & Gordon, 2009). Several authors have noted the special difficulty in identifying and assessing the value of intangible and subjective cultural services (Chan et al., 2012; Daniel et al., 2012; Plieninger et al., 2013).

Daniel et al. (2012) suggest that cultural ecosystem services be evaluated on a scale ranging from those that are easily monetized through actual expenditure studies (e.g., tourism and recreation), to those that can be assessed by willingness to pay studies (e.g., natural landscape aesthetics), to cultural services that can be assessed through non-monetary perception studies (e.g., cultural heritage) and finally to those services that can only be assessed through multi-criteria studies, including simple observation (e.g., spiritual and religious). In a somewhat similar vein, Chan et al. (2012) suggest that the broader concept of human values and environmental ethics is at the core of understanding cultural ecosystem services, recognizing that some of these values are more readily suited to monetary valuation schemes than others. They review various ways in which such values can be assessed or interpreted, as well as the possibility of abandoning such attempts in favour of a direct democratic advocacy process.

Except possibly for that last suggestion, the political ecology of ecosystem services, in terms of their relationship to different political ideologies, power relationships and structures, and neoliberalism, is not discussed in either of these frameworks (Duffy, 2015). Neither are historical and ethnographic methodologies discussed (Schaich et al., 2010; Tengberg et al., 2012), and nor is resilience considered. One interesting point that has been raised in the discussion of cultural ecosystem services in general, and tourism and recreation in particular, is the likelihood that activities related to the use of these services can increase public sentiment and support for protecting all ecosystem services (Daniels et al., 2012).

Ecosystem services have primarily been viewed from an applied anthropocentric sustainable development perspective. The goal of an ecosystem services approach is to efficiently extract all the services from nature-based ecosystems (including those that are heavily managed by humans) that benefit humans, while also protecting the ability of the ecosystem to continue to maintain its functions and capacity to provide such services. In this way, it is almost identical to inter-generational equity espoused in the *Brundtland Report* (WCED, 1987) definition of sustainable development as using resources to meet 'the needs of the present without compromising the ability of future generations to meet their own needs'. The concept of resilience has recently gained considerable academic and policy interest (Walker & Salt, 2006; Ruiz-Ballesteros, 2011; Folke, 2016) as either a supplement or an alternative to the sustainable development paradigm (Lew, Wu, Ni & Ng, 2017). Resilience differs from sustainability in its focus on how systems respond to external pressure to change. Thus, an ecosystem that is highly sustainable in terms of the balance of relationships among its organisms and functions may have either high or low resilience to external change events, which is reflected in its resistance, flexibility, and adaptability (however those might be measured).

Research objectives and background

An examination of two wetland sites in Taiwan that have developed tourism ecosystem services can provide some insight into the relationship among the concepts of cultural ecosystem services, tourism development, sustainability, and resilience. Such an examination addresses the following research questions:

- How can tourism development contribute to the sustainability and resilience of cultural and other services offered by an ecosystem?
- How can tourism development of an ecosystem service contribute to the sustainability and resilience of a community?
- What are the differences in the sustainability and resilience considerations in the management of an ecosystem service?

The first of these questions is relatively standard in assessments of ecosystem services in that it evaluates human activities and their effects on either enhancing or degrading a desired benefit from an ecosystem. The second question is rare in ecosystems service studies in that the focus is on the impacts of ecosystem management on a community that is associated with the ecosystem service. This is a social-ecological question that is more common in cultural landscape studies (Schaich et al., 2010) and resilience research (Folke, 2016) than ecosystem services research. The final question is based on potentially confusing definitions of sustainability and resilience. For this current study, sustainability is narrowly defined as the conservation (or restoration) of resources at normative ideal levels. We recognize that broader definitions of sustainability exist, but they are also more vague (Peirce, Budd, & Lovrich, 2011), prompting our adoption of this narrower conceptualization. Resilience, on the other hand, is the capacity of a system to effectively respond to change through acceptable levels of resistance, adaptation or transformation (Lew, Ng, Ni & Wu, 2016).

Table 7.2 summarizes the range of potential cultural services that may be derived from wetland ecosystems. While some of the elements in this table reflect a pure nature-based benefit (e.g. enjoyment of the wetland environment), most are a deep social-ecological intermixing of human culture, livelihoods, and experiences within this type of natural ecosystem.

Table 7.2 Cultural ecosystem services provided by wetlands

Type of Cultural Ecosystem Service	Examples
Enjoyment of the Wetland Environment	Aesthetic scenery Wildlife viewing (especially birds) Walking and hiking
Inspiration	Art and music influences Literature and folklore
Health & Relaxation	Physical exercise and training – land or water-based Spa and stress treatments
Education	Environmental (science) education Cultural (heritage) education
Spirituality & Religion	Rituals and ceremonies Religious sites Meditation and contemplation
Culture & Heritage (Landscapes, Livelihoods, Identities, and Sense of Place)	Fishing, fisheries, and other forms of aquaculture Fishing and other wetland-base human settlements Built structures (dams, waterways, buildings, boats, boat harbors and yards)
Recreation & Tourism	Boating, canoeing, and kayaking Wildlife and plant viewing and photography Recreational fishing and hunting Ecotourism and educational tourism

Source: Based on Barbier, 2013; modified and expanded by the authors

From a tourism perspective, natural attractions are among the most visited and comprise some of the most iconic landscapes in the tourism offerings of most countries. Examples include the Grand Canyon in the US, Mt Fuji in Japan, Iguazu Falls on the border of Argentina and Brazil, and the Alps of Switzerland and Austria. Visitors are drawn to natural environments for the aesthetic, spiritual, inspirational, recreational and educational services that they provide. Possibly even more important are the cultural meanings that are attached to many natural landscapes and places, which in turn become part of the personal and social identities of those who visit them (Tengberg et al., 2012; Williams & Lew, 2014).

Although tourism is listed as one category among those in Table 7.2, it has a significant role to play across all the different types of cultural ecosystem services because they all have the potential to be tourist attractions. Tourism development, therefore, has the potential to either protect an ecosystem service or exploit and transform it.

Ecosystem services can be protected through management, regulation and incentive programs (MEA, 2005). The direct management of ecosystem services requires either full or partial ownership of those services. This is most easily done on publicly owned lands that are under direct management of governments. However, there are also examples of private lands trusts (such as the Nature Conservancy in the US) that purchase sensitive lands to protect them from human encroachment (Farley & Costanza, 2010). Governments are also able to use their legal power to create and enforce laws to regulate how ecosystem services are utilized on lands that are not directly owned by the government. Such direct regulation may come about through administrative decisions and actions, or through legislative processes, both of which can be shaped by political influences as much as scientific considerations. Maintaining stable atmospheric conditions (an ecosystem services that impacts seasonal climate patterns around the globe), for example, tends to be highly politicized, thus resulting in mixed successes, at best (Meadowcroft, 2011).

Political considerations may result in government incentive programmes as an alternative to the more controversial direct government regulation of ecosystem services. Incentives may be in the form of financial subsidies and reduced taxes. Government-run auctions of public spaces for private use is another form of government incentive, with the example of the creation of a carbon trading market to improve air quality (an ecosystem service) and reduce greenhouse gasses (again, for atmospheric stability). These are all forms of public–private partnerships in which the government offers its substantial resources to co-manage ecosystem services with private interests.

Another major approach that uses incentives to managing the supply and quality of ecosystem services is direct financial payments. This may be in the form of grants, usually from governments, to encourage and pay for (at least in part) effective management of ecosystem services. Another approach is known as 'payments for ecosystem services' (PES) which involves mechanisms in which a downstream user of the service pays a fee to the upstream source to ensure its proper maintenance (Bulte et al., 2008). City residents, for example, would pay rural residents to maintain a watershed to ensure their access to a stable and high-quality water supply. Entrance fees to national parks and to private nature areas and heritage sites are also a form of PES where visitors help pay to maintain the resource (Farley & Constanza, 2010).

Coastal wetland ecosystems generally provide regulating services that include protection from flooding and storm surges, erosion control, food and resting areas for migratory birds, food and habitat for other water and land wildlife, water purification and carbon sequestration (MEA, 2005; Barbier, 2013). Provisioning services from wetland include foods for fisheries and human consumption, and building and craft materials used by humans. Driftwood and wave smoothened stones, for example, may be used by artists and craftsmen in their work. Provisioning and regulating services are normally the focus of studies of PES and other approaches to managing ecosystem services, which reflects the field's foundation in environmental management, economics and environmental sciences.

Two wetlands in Taiwan

The Wuweigang Wetland and the Aogu Wetland and Forest Park are two wetland areas in Taiwan that provide insight into how ecosystem services can be managed in different ways in a tourism context (Figure 7.1). Tourism development in wetland areas can vary considerably in terms of commercial commodification and

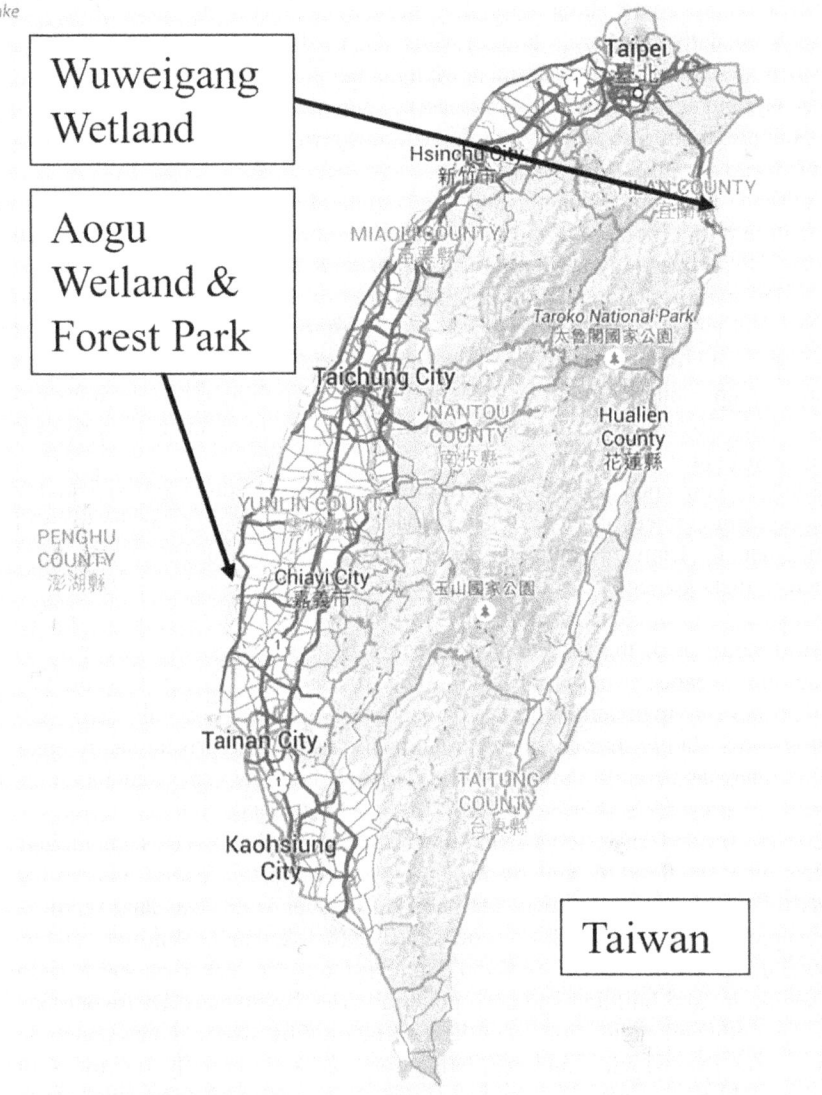

Figure 7.1 Map of Taiwan showing the Wuweigang Wetland and Aogu Wetland and Forest Park locations

Source: Authors (Map Data © 2017 by Google, ZENRIN, used with permission)

educational programming, and these two examples are not meant to address the full range of such possibilities. Despite that limitation sufficient differences exist between the two cases to offer generalizations on the research questions addressed in this paper.

Wuweigang Wetland

Wuweigang (tailless river) Wetland is a 102 hectare protected bird sanctuary and wetland on the northeast coast of Taiwan, facing the Pacific Ocean, in Yilan County (Figure 7.1). Its resources consist of a river that no longer reaches the sea due to sedimentation (the tailless river), a lake, coastal forests, swamps, sand dunes and a beach, some upland areas and migratory birds. The villages surrounding the Wuweigang Wetland formerly relied on net fishing and hunting for their livelihoods. These forms of livelihood began a gradual decline in the 1970s, and today the villages mostly serve as commuter communities, where people live while working in larger towns nearby.

The Wuweigang communities point to 1986 as a turning point in their self-identity and sense of place. That was the year that Taiwan's central government proposed the construction of a coal fire power plant in the Wuweigang Wetland. The communities rallied in protest and in 1991 the government withdrew the power plant proposal. In its place, the Wuweigang Wetland Protected Area was established in 1993 (Figure 7.2). Village leaders continue to proudly point to their success in both stopping the power plant and establishing the protected area.

In 1997, local environmental leaders founded the Wuweigang Culture and Educational Association. This conservation-oriented NGO successfully obtained funding for a community-based wetland monitoring and management system. They also obtained funding to train locals in guiding and birding for visitors to the wetland. Because of its relatively small size, the Wuweigang Wetland was more conducive to direct local involvement in its management and development, although outside experts were often brought in to assist in reshaping the wetland into the resource that exists today.

In 2013, the central government designated one of the Wuweigang villages as a government-approved Environmental Education Centre (Figure 7.3). A facility was subsequently built to provide environmental education services and facilities for visitors and groups from around Taiwan. The primary tourism-related activities in Wuweigang consist of environmental education and general recreation programmes provided to visiting groups. Nature trails and bird-watching platforms supplement the Environmental Education Centre's offerings, and a large brick pizza oven was built in an open area of the community centre where cooking and arts and crafts activities take place. Being better connected to nearby urban centres, the Wuweigang villages themselves do not have any restaurants and there is only one bed and breakfast establishment that does not operate on a regular basis.

Grant activities over the years, by both the local NGO and by the elected leaders of the villages' Community Associations, have often been supported by

Figure 7.2 View of the Wuweigang Wetland, with the Pacific Ocean in the distance
Source: Alan A. Lew

external advisers (other NGOs and universities). These have enabled Wuweigang to build its reputation as an environmental education leader in Taiwan. However, as other rural communities have increased their grant writing skills, competition has increased, funding has been reduced, and some political divisions have subsequently arisen among the different community organizations in Wuweigang.

Wuweigang is not a typical tourism destination. Before 1986, it was more like a forgotten village in the south of Yilan County, with only an occasional bird watcher visiting. At that time, few people were aware of the existence of Wuweigang Wetland. However, their public protests over the proposed coal-fired power plant, followed by their wetland conservation efforts, gained them national attention and fame as a bird-watching destination. Some private organizations, student groups and communities organized education tours to learn about Wuweigang's environmental conservation experiences and to share their appreciation of the wetland habitat. These tours increased especially after the Environmental Education Centre was approved in 2013. To properly host the increasing numbers of visitors, villagers planned and trained themselves to provide various sets of environmental interpretation programmes and enhanced their knowledge and skills for birds watching.

Figure 7.3 Wuweigang Wetland map display in its Environmental Education Centre
Source: Alan A. Lew

Wuweigang also benefited from bigger changes in Taiwan that promoted tourism. The construction in 2006 of the Taipei–Yilan Highway (with the longest mountain tunnel in Taiwan) greatly enhanced the accessibility to this part of the island from Taipei at the same time that domestic tourism (increasing 166 per cent from 2006 to 2016) and inbound tourism (increasing 152 per cent from 2006 to 2016) was taking off. Some tourists visit Wuweigang as an alternative destination in order to experience local authenticity. Their stay usually includes visiting the wetland landscape and participating in the recreation activities offered by the community.

Aogu Wetland and Forest Park

Aogu (sea turtle drum) Wetland and Forest Park is a protected area located on the southwest coast of Taiwan in Chiayi county. It is 1,470 hectares in size and has marshes, lagoons, mud beaches, forests and migratory birds. It is surrounded by active fish ponds, fisheries and oyster beds, which are attributes of the longstanding traditional livelihoods of the area's residents. Because of the significant size of the wetland area, local involvement in its planning and management has been marginal. Instead, the Taiwan Forest Bureau brought in university consultants to develop a master plan for the Aogu Wetland and Forest Park (Figure 7.4). Residents were only included in this process through educational outreach programs.

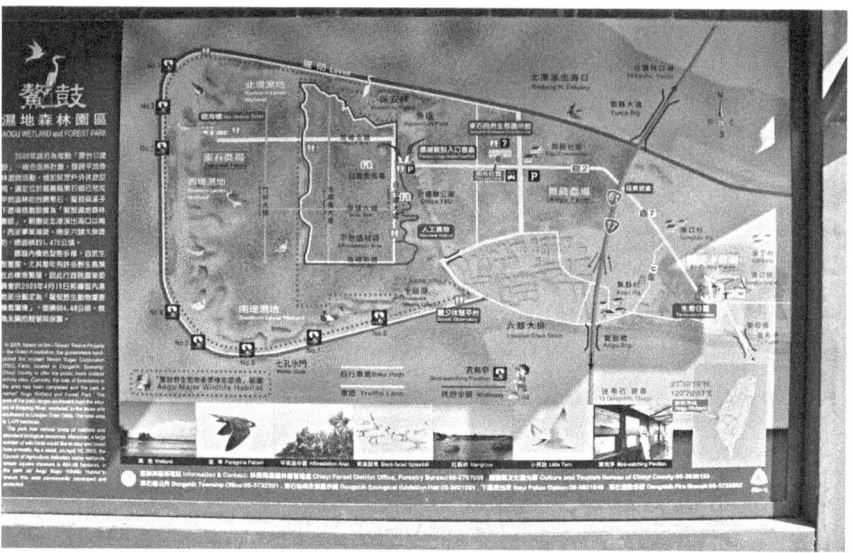

Figure 7.4 Aogu public information map sign
Note: The wetlands lie immediately adjacent to the interior portions of the levee (seawall) that
surrounds Aogu. The forest park is in the centre. Bird-viewing platforms are shown by the boxes along
the levee. Villages are on the east side of the park lands.
Source: Alan A. Lew

The Aogu Wetland that exists today was created in 1964 when a sea wall was built
around a large mud flat that was drained to create a sugar cane plantation and a pig
farm (Figure 7.5). In the 1990s, the Taiwan Sugar Company (a government-owned
monopoly) gradually abandoned this site due to land subsidence, the encroachment
of the sea and the high cost of sugar production in Taiwan compared to other
Asian countries. The closing of one of the major employers in this rural area
encouraged younger people to move away from the villages to the larger cities.
Fish and oyster farming, however, are still widely practised in the lands and waters
adjacent to the Aogu Wetland and Forest Park.

Today, the Taiwan Sugar Company produces no sugar, but is instead a major
land-owning branch of the Taiwan central government. In 2001, the Taiwan
Forest Bureau started a major afforestation programme (creating a forest where
none existed before) on the former plantation site. The purpose was to help with
carbon sequestration (to assist the prevention of climate change) and to provide
recreation space for Taiwan residents. In 2010 the Forest Bureau started a new
planning process that resulted in the Aogu Wetland and Forest Park Master Plan
(ASLA 2011). The plan includes visitor centres, bird-watching stations, hiking
and biking trails, and bird guide training. Several themed recreation areas are also
in the plan, but have not yet been undertaken due to funding limitations.

The Aogu Wetland and villages are relatively remote from the major urban
centres on Taiwan's west coast. Current tourist facilities consist of two visitor

Figure 7.5 The Aogu levee (seawall) that protects the wetland from the Taiwan Strait
Note: Oyster beds can be seen in the shallow sea waters facing the Taiwan Strait on the left, and a bird-viewing platform is shown in the wetland area in the lower right.
Source: Alan A. Lew

centres (one in the park, the other in a gateway village), improved roadways and bird-watching stands. There are a couple of restaurants in the villages, and at least one private fishery pond has been retrofitted to serve meals and provide other services for visitors. No accommodations are available within easy access of the wetland park (as of 2016). The Forest Bureau's outreach programme has been training residents to be bird-watching guides, which has proven popular and provides supplementary incomes to those who participate in a bird-guiding cooperative. Some residents have also developed a strong identity with the wetland and voluntarily monitor and document endangered black-faced spoonbill populations, as well as looking after quality of wetland as a bird habitat.

The Forest Bureau has also initiated community education efforts to fishery and oyster farmers who saw the migratory birds more as pests than as an ecosystem resource. Most of the traditional economy farmers have mixed feelings about the potential development of the wetland park, although most agree that the pig farm that is still (in 2016) located within the park must be removed.

Before the establishment of the Aogu Wetland and Forest Park in 2010, Augo was already well established as a bird-watching destination and birding associations would organize bird-watching trips and events there, especially during the peak migration seasons (Figure 7.6). However, Augo was not on the map for more general tourists. Since 2010, development efforts by the Taiwan Forest Bureau have made Augo into a well-known Taiwanese ecotourism destination. Visitors come mostly in small groups, with an occasional bus of up to 30 people. Unlike experienced birders, general tourists tour around the wetland area with locally trained guides, learning about wetland habitats of animals and plants (such as birds, bats and the afforestation), as well as local religious traditions and the history of the wetland and surrounding settlements. The guides will also introduce them to the hobby of bird-watching, which some of them may adopt.

Figure 7.6 Bird enthusiasts with their viewing equipment at the Aogu Wetland and Forest Park
Source: Alan A. Lew

Similarities and differences

The two wetland communities have a number of similarities:

- The wetlands are largely human-made, or have otherwise been altered significantly through human intervention.
- Migratory birds are their single most significant attraction, although natural scenery is also important.
- They are both highly dependent on funding from the central government to maintain the wetland resource.
- Each community consists of several villages (three in Wuweigang, four in Aogu), with one being more significant than the others.
- For both communities, tourism contributes more to building local identity and pride, rather than simply being a source of income for residents.

In terms of the four types of ecosystem services (Table 7.1), the wetlands of Wuweigang and Aogu also share a lot in common. Their ecosystem services may be characteristic of most coastal wetland areas around the globe. They include:

- *Provisioning Services:* food for humans (fishing, foraging, hunting); raw materials for humans (driftwood, plants).

- *Regulating Services:* climate regulation (vegetation, trees, water); carbon sequestration; flood prevention (from typhoons); soil stabilization.
- *Cultural Services:* recreation and ecotourism; aesthetics and inspiration (photography, bird viewing); sense of place; education; destination image.
- *Support Services:* fish habitat; soil formation; nutrient recycling.

The two communities, however, also have some significant differences. These are due to differences in local resources, historical development and relationships to the central government in Taipei. They include:

- Wetland protection in Wuweigang was initiated through successful community protests against alternative central government plans for the area, whereas Aogu's wetland came into being when the site was abandoned as a sugar cane plantation due to encroachments of the sea, among other reasons.
- Because of their different origins, Wuweigang has been a mostly bottom-up development initiative led by a local environmental NGO and other local leaders, whereas Aogu has been more clearly a top-down initiative, with a master planning process initiated by the Taiwan Forest Bureau.
- The Wuweigang villages have been engaged with their wetlands' development issues since the mid-1980s, whereas the Aogu villages have been only peripherally involved in their wetland's tourism development since about 2010, resulting in different levels of maturity in community expectations and points of view.

Wuweigang and Aogu lessons

The lessons from the tourism development experiences of these two wetland sites are significant, although they may not apply to all coastal wetlands. In terms of ecosystem services management, both sites show the crucial importance of government funding in achieving successful development. Even the bottom-up initiative of Wuweigang was very dependent on government grants, and the recent shortage of such assistance was the source of a major complaint by community leaders. From a broader perspective, this is because the central government of Taiwan is the owner of these wetland ecosystems. As such, the central government has ultimate responsibility to manage the ecosystem resources to ensure the well-being of both the wetlands and the social systems that engage with the wetlands. The different histories of Wuweigang and Aogu show how government policies respond within the context of different contextual circumstances and local residents' expectations and desires.

In both study sites, tourism was closely related to community pride in their wetland resource. However, tourism was not a major income source for either community, nor was it at a level that would overwhelm a local sense of resource ownership. Local participation in the resource was significant in Wuweigang, and gradually increasing in Aogu. Residents who were more closely involved in conservation and tourism were stronger proponents of the resource, although there seemed to be mostly positive feelings toward their wetlands across both communities. This may be due, in part, to the relatively low level of tourism development that has occurred. The situation might be different if the Aogu Master Plan were to be fully implemented with its multiple themed recreation areas.

Community resilience analysis

Rate of change

Both Wuweigang and Aogu face various challenges resulting from external drivers of change, to which they have been forced to respond and adapt, and to manage and plan for. Nature-based tourism is one of the adaptive tactics they have adopted in responding to their changing contexts, and tourism seems to provide a good (and increasingly common) reflection of community resilience in the face of external change. Note that the emphasis here is on how the conservation and development of the wetlands has enhanced the resilience of the community. Conservation efforts also enhances the resilience of the wetlands, but that aspect is only touched upon peripherally in this discussion.

In reviewing the literature on resilience thinking as it relates to tourism, Lew (2014) identified spatial scale and the rate of change as the two key factors in understanding community resilience. The rate of change was conceptualized as a continuum that extended from slow to fast events, with the latter often being referred to as a 'disaster'. Table 7.3 summarizes the drivers of change in Wuweigang and Aogu using Lew's Scale, Change and Resilience (SCR) model.

Neither of these communities has experienced a major disaster event. However, each has experienced a signature event that has acted as a catalyst in defining the development path that it is experiencing today. In evolutionary economic geography (EEG) terms, these would be considered 'tipping points' between a previous path dependence and an emerging path creation (Sanz-Ibáñez & Clavé, 2014). In resilience thinking, these would also be considered tipping points, but

Table 7.3 Fast-, medium- and slow-change drivers in Wuweigang and Aogu

	Wuweigang	*Aogu*
Fast Change Drivers – Economic Restructuring	Proposed power plant (1986–1991)	Withdrawal of sugar plantation (1990s)
Medium Change Drivers – Socio-Cultural, Socio-Economic & Political Changes	Decline in net fishing (1970s–2000s)	Afforestation (starting in 2001) Park Master Plan (2013)
	Changing demographics (fewer youths, since 1970s and 1980s) Increase in wetland recreation and education interests (2000s) Changing attitudes of residents toward wetlands (ongoing)	
Slow Change Drivers – Environmental & Climate Change	Landscape and vegetation changes in the wetland ecosystems (ongoing)	

Source: Authors

between two different regime states. In that sense, although the closing of the sugar plantation took place over almost a decade, it is still considered a 'fast' event because of this tipping point characteristic. It would be inconceivable for the wetlands to exist today, in either community, both visually and in their relation with the residents and outside world, were it not for the fast-change driver events.

In the context of these two wetlands, there have not been any other natural or human disasters to create a turning point in community and resource development. Instead, the fast-change driver events emanated from institutional or organizational responses to changing circumstances. The government proposed the power plant to meet growing energy needs, while the closure of the sugar plantation was caused by changing global sugar economics. Both events were in response to global economic restructuring, which probably reflects a common driver of change in non-disaster scenarios throughout the world.

Medium-change drivers were a mix of local economic restructuring, combined with socio-cultural changes. Together these have resulted in new expectations for livelihoods and quality of life among local youths and urban residents who come to visit the wetlands. Also in this medium-change category are central government-driven environmental and social policies and programs that seek to address broader climate change and quality of life goals.

The principle slow driver of change in both wetlands is primarily related to the natural environment itself. While nature can be the source of major disaster events (earthquakes, tsunamis and flooding), the environment generally provides a relatively stable background against which more rapid changes occur in the human realm. In resilience thinking, this is referred to as the system's *controlling variable* and is considered one of the most important aspects to monitor to ensure system resilience and survivability (Walker, et al., 2012). Walker and Salt (2006, p. 148) suggest that 'A resilient world would include all of the unpriced ecosystem services in development proposals and assessments' because these controlling variables are too important to be overlooked.

However, the natural environment is also constantly changing. Coastal systems are impacted by ocean-related processes that expand, contract and reshape sand dunes and beaches, with similar changes occurring in river estuary systems, such as in Wuweigang. In addition, the slow process of climate change and global warming is gradually shifting plant and animal ecosystems, although the rate is often imperceptible to most residents.

Spatial scale

In addition to the rate of change, the second issue considered in the SCR model is the spatial or geographic scale of the system whose resilience is under examination. In the case of these wetlands, there are two types of spatial system that deserve consideration:

- The ecosystem service, which in this case is the wetland itself. The administratively defined protected area boundary was the primary criteria for its definition. However, in practice, it is not possible to completely isolate the ecosystem services within such boundaries because they affect, and are affected by, land use and other activities that surround it. This was especially true of the Aogu Wetland and Forest Park, where the boundary takes a circuitous route around settlements and fish ponds. From a cultural ecosystem services (and especially tourism) perspective, the villages and traditional livelihoods that are associated with the wetlands are as much a part of its ecosystem services as are the provisioning and regulating services that are associated more directly with the resources themselves.

- The issue of identifying the concept of 'community' as it relates to community resilience. Because the central government essentially owns both wetlands, there are both local interests and nation-wide community interest that are at play in the management and development of the wetlands. National interests are expressed through central government programmes, funding and either direct (in Aogu) or indirect (in Wuweigang) management of the wetlands. Sometimes these can have political overtones, for example if a local legislative representative is from a minority political party.

Geographic space also comes into play in how the multiple communities (subgroups of a village) have different perspectives based on their location and historical relationships to the designated protected areas. On some issues, they have a shared voice, while on others they have divergent perspectives. For example, in the Augo wetland, both the Sigu and Augo communities belong to Village of Augo. The Sigu community is situated within the boundaries of Aogu Wetland and Forest Park and serves as a gateway to it, housing the main administrative and interpretive visitors centre for the park. Because of this, the Sigu community is much more enthusiastic about ecotourism development plans. In comparison, the Augo community is well outside the park's boundaries and has a reduced sense of historical attachment to the wetlands (fewer families were employed in the original sugar cane plantation) and today fewer Aogu guides have been trained. A similar situation exists in Wuweigang, where the Gangkou and Lingjiao (also known as Gangbian) communities are better situated as gateways to the Wuweigang Wetland's trails and bird-watching facilities, whereas the other two communities are more peripherally situated and involved in the wetland. This has also influenced different degrees of engagement in the early power plant protest, participation in environment conservation and monitoring programmes, and relations with external expert groups (NGOs and universities). Views on future directions for the wetlands and community development, therefore, vary among the four communities of the Wuweigang Wetland (as they do on political directions for Taiwan in general).

Ecosystem services assessment

Provisioning and regulating services

Through the examples of these two wetlands, the four types of ecosystems service (Table 4.1) relate to resilience in distinct ways. To better conceptualize a more innovative resilience approach, Lew et al. (2016) suggest that it is juxtaposed against a more stable and conservation-oriented sustainability approach to resource planning and management. Neither approach is considered better than the other, as each offers benefits that the other lacks. Provisioning ecosystem services from these wetlands were found to include food directly consumed by humans (fishing, foraging, hunting) and the collection of raw material for human use (mostly driftwood and plants). The management of these provisioning ecosystem services primarily involves a conservation (or sustainability) approach to ensure their available for both current and future generations. Local villagers utilize these primarily, rather than outsiders, and they tend to change very slowly over time. Regulating ecosystem services from these wetlands includes climate moderation that is both direct (through vegetation and water) and indirect (through carbon sequestration by planting forests), as well as for soil and beach stabilization. Local climate moderation and soil protection contribute to sustainability goals and can be enhanced through the scientific management of the wetland resource. In this way, a local wetland can also contribute to global climate change sustainability goals.

Cultural and supporting services

Cultural ecosystem services from these wetlands are considerable, including recreation, ecotourism, aesthetic inspirations (bird watching and photography) and environmental education, as well as a sense of place and local image and identity. Management of these resources requires a perspective that is more based in the social sciences than environmental science as key concerns are primarily about planning how humans engage with the wetland resources, both today and in the future. As such, it requires an open approach that is more creative and innovative, and more associated with resilience thinking than conservation and sustainability (Lew, 2014). While some cultural resources will be managed from a sustainability perspective, such as the 'museumization' (Relph, 1976) of heritage sites, most are more forward-looking in meeting the changing interests and needs of residents and visitors.

Supporting ecosystem services from these wetlands includes fish and wildlife habitats, soil formation and nutrient cycling. In both wetlands, there has been a period of resource recovery in which previous neglect or other uses had downgraded the habitat and vitality of the wetlands. This is also a conservation and sustainability approach that is best management by scientists who are specifically trained in wetland conservation. Table 7.4 summarizes how the four types of ecosystem service relate to resilience (innovations in resource management and use) and sustainability (recovery and conservation of resources at a desired level).

Many of the cultural ecosystem services in Wuweigang and Aogu were inseparable from the communities themselves, because they were either created by them or formed an integral part of their community identities. Cultural ecosystem services appear to be more closely tied to the broader community than are other ecosystem services which are only related to parts of a community. In this way, more so than for other types of ecosystem service, cultural services

Table 7.4 Wetland ecosystem services, sustainability and resilience

Service Type	Wetland Services	Sustainability (conservation)	Resilience (adaptation)*
Provisioning services	Food; raw materials	Conserving provisioning services for local consumption	Adapting provisioning services for new markets
Regulating services	Climate regulation (local and global); soil stabilization	Conserving regulating services for local environmental stability	Adapting regulating services for regional and global stability
Cultural services	Recreation and tourism; education; aesthetics; place identity	Conserving cultural services (heritage landscapes, sites, and practices; natural landscape views and experiences)	Adapting cultural services to meet changing human needs
Supporting services	Fish and wildlife habitats; soil formation; nutrient cycle	Conservation supporting services to meet traditional ecosystem and human needs	Adapting supporting services to increase ecosystem effectiveness and resource productivity

* In this context, sustainability is defined as resource recovery or conservation and protection; resilience is defined as resource adaptation or innovation, as based on Lew et al., 2016.

Source: Authors

could serve as an indicator of general community resilience (Brand, 2008), although this study did not specifically examine that hypothesis. If true, then the strengthening of a wetland's cultural ecosystem services would become a means of strengthening the general resilience of communities associated with that wetland. This is seen through tourism and recreation development, which is based on the educational and aesthetic experience that the wetland landscapes and wildlife provide. Through tourism those benefits are made available to a larger population who come to appreciate, and hopefully then support, the conservation of the wetland (Daniels et al., 2012). The communities involved also build greater pride in their wetland resources in response to having more visitors. This pride benefits the wetlands, which they want to protect, but also strengthens their community spirit, sense of purpose and cohesion, which are essential elements of resilience thinking (SRC, 2014). More directly, those community members who are involved in the tourism development of their wetlands have access to broader opportunities to diversify their economic well-being through the new avenues of employment and the new skills that they have learned.

The provisioning, regulating and supporting ecosystem services are, of course, prerequisites for (and closely interrelated with) cultural services. As a social-ecological ecosystem, all four of the service types must be considered equally for an effective management regime. As noted on p. 104, because they are based on the social sciences, cultural services tend to receive insufficient attention in the application of ecosystem services to real world scenarios, where the emphasis is more from the environmental sciences (Schaich et al., 2010; Wu & Petriello, 2011). Thus, assessments of cultural ecosystem services require different epistemologies and methodologies (Tengberg et al., 2012). However, it is likely that the relationship of the four types of ecosystems service to sustainability and resilience, as defined in Table 7.4, would be consistent when applied to other geographic contexts. This could be a model in which, for example, considerations for the enhancement of the cultural services offered by an ecosystem would focus on heritage conservation as the primary sustainability goal, while adaptations to changing human needs (e.g. economic diversification) would be the principal resilience goal.

Conclusions

Coastal wetlands are often targeted as key ecosystems for conservation, including both protection and rehabilitation, because of the wealth of ecosystem services that they provide. Much of the research on ecosystem services has focused on the sustainability of supporting services, provisioning services and regulatory services, with only passing acknowledgement of the cultural services crucial to the economic and social well-being and resilience of communities. The two coastal wetland examples from Taiwan show that, as an attraction, wetlands hold a broad iconic image for tourists because of their prominence in the landscape. However, in practice they function more as a niche speciality product for bird watchers and recreational fishing enthusiasts. As such, the direct economic benefits of wetland tourism are often modest. On the other hand, tourism development as a form of enhancing cultural ecosystem services in protected wetlands has a broad utility for both sustainability and resilience goals. This is reflected in the three research questions posed at the start of this chapter (p. 105).

1. Tourism development to enhance cultural and other ecosystem services.

In both wetlands studied, the primary process of tourism development involved the enhancement of cultural ecosystem services. These include:

- Trails and viewing platforms to enable visitors to enjoy aesthetic landscapes, view and photograph wildlife (mostly birds), for walking and hiking in nature: both wetland sites.
- Educational infrastructure with a focus on environmental education about the wetlands, but also cultural education about wetland communities: both wetland sites.
- Viewing and learning about traditional aquaculture landscapes, built structures and practices in areas adjacent to the wetland: Aogu wetland.
- Guiding and interpretive services for ecotourism and nature-based tour groups: both wetlands.

Developing these wetlands for recreational and educational purposes provides greater financial support from Taiwan's national government. Government support comes through a variety of policy and financial mechanisms and is fundamental in building local community capacity to develop entrepreneurial and interpretive activities, as well as management skills. This is most readily evident in infrastructure improvements, including trails and educational facilities, but also in the management of the resources to achieve environmental sustainability objectives. If visitor facilities and services are going to be created, it is, of course, imperative that the resource is worth visiting. In this way, tourism development also enhances supporting and regulating ecosystem services. As protected natural areas, provisioning services are only utilized at a low level by some residents, mostly in the collection of wild plants for food and medicinal purposes. These too, however, are enhanced by the overall quality of the wetland ecosystem.

2. Tourism development of cultural ecosystem services to enhance the sustainability and resilience of a community

Tourism development in association with wetland conservation is seldom significant from a purely economic standpoint. However, from ecosystem services and community resilience perspectives, wetland conservation can contribute significantly to local community well-being (Reyers et al., 2013) through the development of local education and skill capacities, local pride and sense of place, and environmentally sensitive changes in attitude and behaviour.

Interpretive activities (environmental education) for both local communities and tourists are a crucial part of creating buy-in and support for the successful conservation of wetland resources. The development and enhancement of cultural and other ecosystem services through tourism was found to generate a high level of pride and sense of ownership in their resource among residents. This is another intangible cultural benefit that tourism has brought to the community, and it was directly attributed to the interests that visitors had expressed in 'their wetland'. While this was strongest among residents who were more closely associated with the tourism infrastructure and activities, it was still evident throughout the villages studied. It was also seen despite differences in the bottom-up history of the older Wuweigang Wetland and the top-down history of the newer Aogu Wetland and Forest Park, where educational outreach programmes have specifically targeted aquaculture operators who have traditionally considered wild birds a pest to their operations.

This change in attitude among residents, which was especially true in the Aogu communities, is significant for the environmental sustainability of the resource. People are more likely to care about, conserve and protect environments with which they have a deep sense of attachment. For Wuweigang, this was the motivation behind their wetland's formal establishment. For Aogu, this is a more recent development as the site was repurposed from industrial agriculture to its current use. Either way, once established, that sense of ownership can be a basis for political support for sustainability policies and practices.

Community resilience is also enhanced in the communities that are involved in the tourism development of their wetlands. The bottom-up history of the Wuweigang Wetland created a context in which locally based initiatives (with NGO and other assistance) created a tradition of independence, cooperation, activism and self-organization that are important traits of resilient communities (Ruiz-Ballesteros, 2011). The involvement of NGOs, including university researchers, as outside consultants and advisers, and usually funded by the government, is another key element in training and capacity building in traditional rural communities.

Tourism also contributes to this by diversifying the economy, offering employment opportunities for young people to reduce out-migration rates, and enhancing entrepreneurial ventures among residents. For some individuals, the wetland provides a foundation for tourism business activities (often food-related) and other tourism-related career options (such as interpretive guiding). While tourism was not a major activity in either of these communities, it was present at a sufficient level to offer these benefits, and thereby increase the resilience of the local economies.

3. Sustainability and resilience considerations in the management of an ecosystem service

One of the major implications of this study was the differences between sustainability (conservation) and resilience (adaptation) approaches to ecosystem management. In general, sustainability approaches are most appropriate for managing supporting and regulating ecosystems services in a protected natural environment (Table 7.4). Wetlands, however, are somewhat different in that natural wetlands are subject to frequent ecological changes (after each major flood), and what exists today is often the result of considerable human intervention. This was true for both wetlands in Taiwan, where restoration and conservation were based on ideal models of what these wetlands could be. In that sense, efforts to maintain an ideal state of supporting and regulating services are still very much aligned with the philosophy of sustainability. Sustainability was also the preferred approach for managing provisioning services in the two Taiwan wetlands, mostly because they were so meagre.

Provisioning services, however, are often commodified for sale, in which case they may be the primary focus around which other ecosystem services are managed (Power, 2010; Posthumus et al., 2011). Agriculture and timber production, for example, are commodifications of ecosystem services, which when intensified can push other services beyond their natural levels. All four ecosystem services can be exploited and commodified by humans in this way, which makes them opportunities for resilience (diversification, innovation, adaptation) from a community perspective, as shown in Table 7.4. (It should be noted that such exploitation may reduce natural resilience from the standpoint of the resources, itself.) However, these Taiwan cases show that some ecosystem services are mostly, but not exclusively, addressed from a sustainability perspective and are focused on the resources themselves, while others are addressed more in a resilience manner, with a primary focus on the community. The two key points in this may be summarized as:

- Supporting and regulating services are more likely to be managed with the sustainability of the resource as the primary objective, whereas provisioning and cultural services are more likely to be managed with community resilience as the main goal.
- Sustainability approaches to ecosystem services management are most likely to focus on the needs of ecosystem services, whereas resilience approaches to ecosystem services management are more likely to focus on the needs of the communities that utilizes the ecosystem services.

How well these generalizations apply to other wetlands and other ecosystems may be addressed as intersections between the concepts of resilience, sustainability and ecosystem services are further explored. From this study, the management and enhancement of cultural ecosystem services were found to have a greater direct impact on the adaptive and innovative resilience of communities to meet their need in a changing world, which should be a central component to any successful ecosystem management plan.

References

ASLA (2011). An emerging natural paradise: Aogu Wetland Forest Park master plan. American Society of Landscape Architects. *2011 ASLA Professional Awards*. Online at: https://www.asla.org/2011awards/217.html

Barbier, E.B. (2013). Valuing ecosystem services for coastal wetland protection and restoration: progress and challenges. *Resources*, 2(3), 213–30, doi:10.3390/resources2030213.

Bennett, E.M., Peterson, G.D. and Gordon, L.J. (2009). Understanding relationships among multiple ecosystem services. *Ecology Letters*, 12: 1–11.

Boyd, J. and Banzhaf, S. (2007). What are ecosystem services? The need for standardized environmental accounting units. *Ecological Economics*, 63: 616–26.

Brand, F. (2008). Critical natural capital revisited: ecological resilience and sustainable development, *Ecological Economics*, 68(3), 605–12, doi:10.1016/j.ecolecon.2008.09.013

Bulte, E., Lipper, L., Stringer, R., and Zilberman, D. (2008). Payments for ecosystem services and poverty reduction: concepts, issues, and empirical perspectives. *Environment and Development Economics*, 13(3), 245–54.

Carpenter, S.R., DeFries, R., Dietz, T., Mooney, H.A., Polansky, S., Reid, W.V., and Scholes, R.J. (2006). Millennium Ecosystem Assessment: Research Needs. *Science* 314, 257–8.

Chan, K.M.A., Satterfield, T. Goldstein, J. (2012). Rethinking ecosystem services to better address and navigate cultural values. *Ecological Economics*, 74, 8–18.

Daniel, T.C., Muhar, A., Arnberger, A., Aznarc, O., Boyd, J.W., Chan, K.M.A. … von der Dunk, A. (2012). Cultural ecosystem services: potential contributions to the ecosystems services science and policy agenda. *Proceedings of the National Academy of Sciences of the USA*, 109, 8812–19.

Duffy, R. (2015) Nature-based tourism and neoliberalism: concealing contradictions. *Tourism Geographies* 17(4), 529–43.

Farley, J. & Costanza, R. (2010) Payments for ecosystem services: From local to global. *Ecological Economics*, 69, 2060–68.

Fisher, B., Turner, R.K., Morling, P. (2009). Defining and classifying ecosystem services for decision making. *Ecological Economics* 68, 643–53.

Folke, C. (2016). Resilience. In *Oxford Research Encyclopedia of Environmental Science* (pp. 1–68). New York: Oxford University Press. DOI:10.1093/acrefore/9780199389414. 013.8

Gordon, L.J., Peterson, G.D. & Bennett, E.M. (2008). Agricultural modifications of hydrological flows create ecological surprises. *Trends in Ecological Evolution*, 23, 211–19.

Lew, A.A. (2014). Scale, change and resilience in community tourism planning. *Tourism Geographies*, 16(1), 14–22. doi:10.1080/14616688.2013.864325.

Lew, A.A., Ng, P.T., Ni, C-C. & Wu, T-C. (2016). Community sustainability and resilience: similarities, differences and indicators. *Tourism Geographies*, 18(1), 18–27, doi:10.10 80/14616688.2015.1122664

Lew, A.A., Wu, T-C., Ni, C-C. & Ng, P.T. (2017). Community tourism resilience: some applications of the scale, change and resilience (SCR) model. In R. Butler, ed., *Tourism and Resilience* (pp. 23–31). Wallingford: CABI.

MEA (2005). *Ecosystems and Human Well-Being: Wetland and Water, Synthesis.* Millennium Ecosystem Assessment. World Resources Institute, Washington, DC. Online at: http://www.millenniumassessment.org/documents/document.358.aspx.pdf

Meadowcroft, J. (2011). Engaging with the politics of sustainability transitions. *Transitions*, 1(1): 70–75. doi:10.1016/j.eist.2011.02.003

Peirce, J. C., Budd, W. W. & Lovrich, N.P. (2011). Resilience and sustainability in US urban areas. *Environmental Politics*, 20(4), 566–84

Plieninger, T., Dijks, S., Oteros-Rozas, E. & Bieling, C. (2013). Assessing, mapping, and quantifying cultural ecosystem services at community level. *Land Use Policy*, 33, 118–29.

Posthumus, H., Rouquette, J.R., Morris, J., Gowing, D.J.G. & Hess, T.M. (2011). A framework for the assessment of ecosystem goods and services; a case study on lowland floodplains in England. *Ecological Economics*, 69, 1510–23.

Power, A.G. (2010). Ecosystem services and agriculture: tradeoffs and synergies. *Philosophical Transactions of the Royal Society B*, 365, 2959–71.

Relph, E. (1976). *Place and Placelessness*. London: Pion.

Reyers, B., Biggs, R., Cumming, G.S., Elmqvist, T., Hejnowicz, A.P. & Polasky, S. (2013) Getting the measure of ecosystem services: a social-ecological approach. *Frontiers in Ecological and the Environment*, 11(5), 268–73, doi:10.1890/120144.

Ruiz-Ballesteros, E. (2011). Social-ecological resilience and community-based tourism: an approach from Agua Blanca, Ecuador. *Tourism Management* 32, 655–66.

Sanz-Ibáñez, C. and Clavé, S.A. (2014). The evolution of destinations: towards an evolutionary and relational economic geography approach. *Tourism Geographies*, 16(4), 563–79, doi:10.1080/14616688.2014.925965

Schaich, H., Bieling, C., Plieninger, T. (2010). Linking ecosystem services with cultural landscape research. *GAIA* 19 (4), 269–77.

Stockholm Resilience Centre (SRC) (2014) *Applying Resilience Thinking: Seven Principles for Building Resilience in Social-Ecological Systems*. Stockholm: Stockholm University.

Tallis, H., Kareiva, P., Marvier, M. & Chang, A. (2008). An ecosystem services framework to support both practical conservation and economic development. *Proceedings of the National Academy of Sciences USA*, 105, 9457–564.

Tengberg, A., Fredholm, S., Eliasson, I., Saltzman, K. & Wetterberg, O. (2012). Cultural ecosystem services provided by landscapes: assessment of heritage values and identity. *Ecosystem Services* 2, 14–26.

Williams, S. & Lew, A.A. (2014). *Tourism Geography: Critical Understandings of Place, Space and Experience*, 3rd edn. London: Routledge.

Walker, B. & Salt, D. (2006). *Resilience Thinking: Sustaining ecosystems and people in a changing world.* Washington, DC: Island Press.

Walker, B. H., Carpenter, S. R., Rockstrom, J., Crépin, A.-S. & Peterson, G. D. (2012). Drivers, 'slow' variables, 'fast' variables, shocks, and resilience. *Ecology and Society*, 17(3), 30, doi: 10.5751/ES-05063-170330.

World Commission for Environment and Development (WCED) (1987). *Our Common Future: The Brundtland Report.* Oxford: Oxford University Press.

Wu, T., and Petriello, M.A. (2011). Culture and biodiversity losses linked. *Letters to Science*, 331, 30.

8 Managing for resilience in the face of climate change

The adaptive capacity of U.S. ski areas

Natalie Ooi

Introduction

Since the 1980s, a growing body of literature has examined current and future impacts of climate change for ski areas in the U.S., with much of the industry having already experienced significant change. This includes increases in average winter temperatures, less natural snow cover and a decline in snowpack and average snow depth (Burakowski & Magnusson, 2012; Scott, McBoyle, Minogue & Mills, 2006). While the ski industry has always been subject to the vagaries of natural snowfall and fluctuating temperatures, warmer, low-precipitation winters have become an increasingly common phenomenon. This places growing pressure on valuable water and energy resources due to increased snowmaking demands, with decreases in skier visits associated with low snowfall having also cost U.S. ski areas over US$1 billion in aggregated revenue from November 1999 to April 2010 (Burakowski & Magnusson, 2012). While some ski areas have struggled to adjust, resulting in early season closures and a significant loss in economic revenue, others are seeking innovative ways to adapt to their newfound reality.

This chapter examines the current situation of U.S. ski areas through the use of resilience as a theoretical framework. Defined as "the long-term capacity of a system to deal with change and continue to develop" (Stockholm Resilience Centre, 2015, p. 3), many ski areas of all sizes appear to be demonstrating resilience through innovative measures that improve their adaptive capacity to deal with the slow-change variable of climate change (Becken, 2013). Such a focus on resilience is important for the ski industry, given the dynamic interaction between climate change and the fast-change variables of fluctuating visitor numbers, revenue, and thus profitability, all of which have significant economic impacts for ski areas and their surrounding communities. By examining how resilience is managed and operationalized within the U.S. ski industry, preliminary insight into factors that influence the adaptive capacity of ski areas towards climate change can be gained. This provides a basis for understanding how ski areas can best build their adaptive capacity towards climate change to enable their continued function and success in what is an increasingly stressed alpine environment.

Understanding and defining resilience

According to Holling (1973, p. 17), resilience is a measure of the ability of systems to "absorb changes of state variables, driving variables, and parameters, and still persist." This differs from stability, which is "the ability of a system to return to an equilibrium state after a temporary disturbance" (Holling, 1973, p. 17). The more rapid the return to equilibrium and the least amount of fluctuation, the greater the stability of the system. This differentiation of terms represents a shift in thinking from the importance of ecological system stability to system resilience, with a focus on persistence rather than fluctuation from equilibrium providing a more realistic understanding of a system's behavior (Holling, 1973).

Much of the understanding of resilience throughout the academic literature has since followed along the lines of Holling's (1973) differentiation between stability and resilience. For example, Gunderson (2000) differentiates between *engineering resilience* and *ecological resilience*, with engineering resilience referring to the time required for a system to return to a single, global equilibrium following a perturbation, while ecological resilience is determined by the amount of disturbance that can be absorbed before a system moves to an alternative stability domain. Whereas the former can be associated with Holling's (1973) definition of stability, the later equates to his understanding of resilience.

More recent studies have built upon the dynamic and variable nature of stability domains. This has resulted in the emergence of *evolutionary resilience* or *social-ecological resilience,* where the notion of equilibrium and a return to normality is challenged. Rather, systems are seen to be constantly in a state of flux as they respond, adapt, and change in response to various shocks and disturbances. As acknowledged by Walker, Holling, Carpenter and Kinzig (2004), resilience is only one of three related attributes (resilience, adaptability, and transformability) that have the ability to affect the future trajectory of a social-ecological system (SES). Within this framework of understanding, resilience is defined as "the capacity of a system to absorb disturbance and reorganize while undergoing change so as to still retain essentially the same function, structure, identity, and feedbacks" (Walker et al., 2004, p. 5). Here, resilience is understood as consisting of three critical components: latitude, resistance, and precariousness:

- latitude refers to the maximum amount that a system can be altered before it loses its ability to recover;
- resistance is the ease or difficulty of changing the system; and
- precariousness is how close the state of the system is to a threshold change.

(Walker et al., 2004)

Adaptability then refers to the capacity of actors in a system to influence resilience (Walker et al., 2004). Given the dominant role held by humans within SESs, adaptability can more specifically be understood as the collective capacity of humans to manage resilience, whether intentionally or unintentionally, to determine whether a system stays within certain thresholds or crosses over, bringing significant changes to system function and structure (Walker, Gunderson, Kinzig, Folke, Carpenter & Schultz, 2006). Such human capacity to bring about efficient change to address dynamic circumstances, thereby encouraging flexibility, creativity, and innovation (Lew, Ng, Ni & Wu, 2016), may result in transformation—the creation of a fundamentally new system where existing ecological, economic, and social conditions make the existing system no longer tenable (Walker et al., 2004).

This understanding of resilience, and the related notions of adaptability and transformation, shifts the perspective from controlling change within assumed stable systems, to sustaining and enhancing "the capacity of social-ecological systems to cope with, adapt to, and shape change" over time (Folke, 2003, p. 227). Resilience is therefore about the continued ability of a system to use shocks and disturbances as an opportunity to spur innovation and renewal (Stockholm Resilience Centre, 2015). Thus, it can be argued that systems do not become resilient in spite of shocks and disturbances, but, rather, because of them (Davoudi, 2012). Obviously, the more resilient a system, the larger the disturbance it can absorb before it undergoes a transformation (Walker et al., 2006). This ability of a system to build and increase its capacity for learning and adaptation in a way that does not constrain or erode future opportunities is a central aspect of resilience (Folke, 2003).

Climate change and resilience

While shocks and disturbances to SESs are often acute, they can also present themselves slowly over time in the form of slow-change variables. These react to drivers that shape the response of faster-change variables within a system (Walker, Carpenter, Rockstrom, Crepin & Peterson, 2012). Climate change is one such example. Although climate undergoes natural cycles of variability, with its daily manifestation referred to as weather, long-term systemic changes to the earth's climate have brought about warmer temperatures, changes to precipitation patterns, and more frequent and intense weather events (Becken, 2013). These changes have profoundly affected the economic, social, and environmental dynamics affecting ski areas and other tourism-related businesses (Becken, 2013). Understanding, managing, and investing in resilience, therefore, becomes particularly crucial in environments where change is occurring and is inevitable. By understanding how and why SESs change, ski area managers and other relevant stakeholders can better avoid, manage, and/or engineer such changes. The goal here is to increase the robustness of a given SES's capacity to reduce its vulnerability to changing slow- and fast-change variables.

However, it is important to realize that there is no "one size fits all" approach, given the future uncertainty surrounding the effects of human-induced climate change. Rather, "learning through experimentation and innovation is necessary to develop and test knowledge and understanding for coping with change and uncertainty" (Walker, et al., 2006, p. 20). This can be achieved through adopting new and novel ways to increase the latitude and resistance of a system, while also limiting its precariousness. Such efforts to manage resilience can help provide a form of "insurance" against future shocks and disturbances, as ecological diversity is maintained and critical resources safeguarded, thereby increase the adaptive capacity of the SES (Stockholm Resilience Centre, 2015).

Managing and operationalizing resilience in the face of climate change

One of the difficulties associated with the management and operationalization of resilience in a real-world context is the application of concepts, such as latitude, resistance, and precariousness, to help examine complex SESs (Becken, 2013) (see Table 8.1). This has limited the development of a more precise understanding of how resilience can be applied to specific contexts. In response, Becken (2013) identified a number of surrogates that describe the stability landscape (latitude, resistance, and precariousness) within the context of tourism destinations facing climate change. Given that these surrogates encapsulate relevant features of a tourism system's dynamics in the context of climate change, many were deemed relevant and promising for understanding the adaptive capacity of ski areas, as they seek to improve and manage their resilience within fragile alpine environments.

Beginning with a description of the current context in which North American ski areas operate, the efforts of various North American ski areas to improve their resistance, increase their latitude, and limit the precariousness of their situation, through the application of these surrogates, are examined. This application of the stability landscape to the North American ski industry can provide greater awareness of how the ski industry as a whole can enhance their adaptive capacity towards climate change.

Table 8.1 Resilience surrogates for resistance, latitude, and precariousness

Stability landscape aspect	Surrogate
Resistance	Weather sensitivity Coping range
Latitude	Diversity of tourism offerings Diversity of markets and segments Degree of operational flexibility Dependence on a particular location or resource Information for business decision making Connectedness of ski area
Precariousness	Frequency of climatic disturbance Extent of climate change disturbance

Source: Adapted from Becken (2013)

Climate change and the North American ski context

The impacts of climate change on the North America ski industry have already been felt across the U.S., from California to the North-East. Over the past 20 years, annual snowfall in Vail and Tahoe City has trended downward (Bebb, 2015), while the North-East region has experienced warmer than average temperatures (Scott, et al., 2006). In the period 2012–14, California resorts experienced a period of drought that resulted in annual snowfall totals of less than 60 percent of average (Bebb, 2015). This caused many California resorts located within the Sierra Nevada, such as Sugar Bowl, China Peak, and Sierra-At-Tahoe, to close in February and March over the 2014–15 winter season due to a lack of snow and warmer temperatures. While this three-year timeframe does not necessarily represent a trend, such temporary weather phenomena is likely to continue, with human emission of greenhouse gases (GHGs) increasing the probability of such events by at least a factor of three (Swain, Diffenbaugh & Rajaratnam, 2014).

In light of these challenges presented by low snowfall years and higher-than-average temperatures, both large and small ski areas across the U.S. have invested heavily in snowmaking technology. According to the National Ski Areas Association (NSAA, 2013), 100 percent of ski areas in the eastern U.S., 94 percent in the mid-west, and 91 percent in the Rocky Mountain region utilize snowmaking systems. This has reduced the reliance of ski areas on natural snowfall and the effects of warmer than average temperatures, allowing them to provide a consistent skiing surface throughout the season, as well as extend the operating season to increase visitor numbers (NSAA, 2013).

Scott et al. (2006) examined the vulnerability of eastern North American ski areas to climate change by assessing how it would affect season length, the probability of operating during key tourism periods, the cost of snowmaking, and water requirements in both low- and high-impact climate change scenarios. It was determined that when a resort's ability to make snow is taken into consideration, even in the high impact scenario for 2020, the majority of ski areas will still be able to remain open. However, under all scenarios, the amount of natural snowfall did reduce significantly. In particular, projections for the 2050 high impact scenario raise questions regarding the long-term ability of ski areas to open for important holiday tourism periods during which a significant amount of revenue is earned. Additionally, while snowmaking is likely to help offset declining snowpack, the required costs, energy demands, and access to limited water resources, particularly at the start of the season, will also limit the ability of many resorts to remain competitive (Scott et al., 2006).

Even with snowmaking, climate change has taken an economic toll on the ski industry. In the U.S., ski area operations currently contribute US$2.9 billion a year to the national economy, providing 75,900 jobs in many rural resort communities where ski areas are commonly the largest employer (Burakowski & Magnusson, 2012). However, due to average increases in temperature and decreases in snowfall, the U.S. ski industry has lost an estimated US$1.07 billion in aggregated revenue between low and high snowfall years from November 1999 to April 2010. This is the result of over 15 million fewer visits, with annual snowfall highly correlated to the number of skier visits (Bebb, 2015; Burakowski & Magnusson, 2012).

Snowmaking also requires access to large bodies of water. In much of the American West where water is a scarce and precious commodity, this can be both expensive and have significant environmental effects. For those ski areas that do not have their own reservoirs, the withdrawal of large amounts of water from local streams can inhibit the upstream movement of trout, increasing the chance of fish eggs being exposed and freezing (Hudson, 1999). For those that do have their own reservoirs, these expand the ecological footprint of a resort, with the construction of pump-houses, underground pipes, and the reservoir itself, causing additional environmental damage (Wolfsegger, Gossling & Scott, 2008). Thus, while snowmaking is undoubtedly the most important climate adaptation tool used by ski areas now and likely into the future, the energy and water demands it exacts on the natural environment raise questions regarding its long-term environmental impacts. Therefore, without further adaptation, continued economic loss and strain on energy and water resources can be expected to accompany projected decreases in snow cover, snowfall, and length of seasons.

Increasing ski area resistance

Resistance refers to the ease or difficulty of changing a system (Walker, et al., 2004). For ski areas, this is largely linked to climatic disturbances, with snow-based activities and businesses highly sensitive to a range of weather conditions (Becken, 2013). A lack of snow or too much snow can be detrimental, as can high winds, high temperatures, poor visibility, and rain. This results in critical thresholds that if exceeded (e.g., high winds and heavy snowfall), can limit the ability of a ski area to operate. Conversely, if thresholds are not met (e.g., insufficient snowfall), operations may also need to close. While most ski areas have strict operational and workplace safety policies regarding maximum wind speeds or minimum temperatures before lifts shut down (Becken, 2013), the sensitivity of ski areas to variable weather conditions has significantly decreased over the years. This is because the large majority of ski areas have invested in snowmaking technology, as discussed above.

This investment in snowmaking by most U.S. ski areas has dramatically increased their resistance, and, in turn, their latitude for operations. This is because they are less susceptible to rising temperatures and lower-than-average snowfall, allowing them to operate for longer, particularly at the beginning and end of the winter season. However, given the resource intensive nature of snowmaking, many ski areas, in partnership with local and state entities, have sought to invest in high-efficiency snowmaking technology and other innovative endeavors to maximize snowmaking capacity and minimize environmental damage caused by scarce water levels. This can also produce significant economic savings due to reduced water and electricity usage.

In recent years, U.S. ski areas have made significant investments in:

- the installation of high-efficiency snowguns that use a 1:1 air/water ratio;
- the creation of gravity-fed snowmaking systems;
- upgrades to computer automation to provide real-time controls, sensors;
- monitoring to optimize snowmaking; and
- the replacement of air compressors with newer, more efficient models.

(NSAA, 2017).

One such example comes from Lake Placid, New York, where a US$5.7 million energy efficiency initiative in 2006 helped replace most of the air compressors used for snowmaking at Whiteface and Gore Mountains with higher efficiency models. This resulted in a reduction of annual electricity use by approximately 600,000 kWh, 4,500 barrels of oil, and a decrease in annual gas emissions of over 2,000 tons; a US$430,000 savings (AlpineZone News, 2006).

While most investment into high-efficiency snowmaking technology has occurred sporadically throughout the industry, a novel initiative put forward in 2014 by Efficiency Vermont, a privately operated non-profit organization funded through a statewide electric efficiency utility fee, led to the largest investment in high-efficiency snowguns. Five million dollars in incentives were provided by Efficiency Vermont to ski areas that agreed to scrap 80 percent of their old technology snowguns (Hawks, 2015). This encouraged Vermont ski areas to spend more than US$15 million of their own capital, resulting in the installation of 2,787 high efficiency snowguns in 2014. This is expected to save enough electricity and diesel fuel to power 1,500 homes and heat another 340 every year; a US$2 million dollar annual energy saving (Hawks, 2015). Many other states have since implemented similar programs, with Efficiency Maine having allocated US$540,000 worth of incentives to four ski areas to upgrade their fleets, and New Hampshire, Colorado, New York, Utah, and California offering similar initiatives on a smaller scale (Hawks, 2015).

At the same time, many ski areas have invested in innovative ways to conserve water for snowmaking. This includes:

- installing water coolers to lower the ambient temperature of water used for snowmaking;
- using dirt to construct freestyle features to reduce the amount of snow required;
- installing flow meters on snowmaking systems for more accurate measures of water usage; and
- the use of high quality, reclaimed water for snowmaking.

(NSAA, 2017).

For example, Arizona Snowbowl uses 100 percent reclaimed and treated sewage effluent for its snowmaking activities (Dicker, 2012). Another unique example of water conservation innovation comes from Mt. Baldy Ski Resort in California, being the first ski area in the world to install black, hollow plastic conservation balls on their 10 million gallon snowmaking reservoir to prevent water evaporation. These 4 inch balls act as a floating cover and reduce water evaporation by up to 90 percent, saving the ski area approximately US$10,000 a month in pumping costs, and enhancing snowmaking capabilities throughout the season (Emerson, 2013). This has increased the resort's economic capacity while also conserving valuable water resources required for snowmaking.

Each of these efforts demonstrate concerted attempts by ski areas to reduce their sensitivity to changes in both individual weather events, but also longer weather patterns. In relation to the stability landscape, this can be seen as increasing the coping range of ski areas, particularly as it relates to snowfall and temperature. By increasing their resistance, higher levels of perturbation are required to change the state of these alpine systems and the ability of ski areas to operate. Thus, while ski areas are highly dependent on specific weather conditions for their operations, efforts to make more snow with less resources, and conserve limited resources, has saved ski areas large amounts of money and allowed many to remain in operation, in spite of climate change.

Increasing ski area latitude

Latitude, in the context of the stability landscape, refers to the maximum amount a system can be altered before it is unable to recover (Walker, et al., 2004). According to Becken (2013), there are a number of factors that can affect the latitude of tourism businesses operating in alpine environments in the face of climate change. These include:

- degree of operational flexibility (as discussed above in regards to snowmaking);
- diversity of product offerings;
- diversity of markets;
- available information;
- connectedness of the ski area; and
- dependence on location and resource.

Ski areas across the U.S. have made significant efforts to increase their latitude by addressing each of these factors. In particular, many ski areas have sought to diversify product offerings to extend the tourism season beyond the winter. What used to be considered as the "off-season", summer operations are now an integral part of a ski area's success, given the changing circumstances of winter. While many ski areas have offered summer activities for many years, such as downhill mountain biking, hiking, and scenic gondola rides, it has only been in the past decade that there have been concerted efforts to develop a critical mass of summer tourism offerings to attract resort visitation. The largest summer tourism investment to date can be attributed to Vail Resorts, Inc., in its development of "Epic Discovery." Opening summer 2016, this will be an engaging, educational, and interpretive summer mountain adventure offered at Vail, Breckenridge, and Heavenly resorts, and includes mountain coasters, zip lines, challenge courses, canopy tours, hiking, mountain biking, segway tours, and interpretive centers for environmental education (Vail Resorts, 2013).

Through such summer tourism development and investment, visitation has grown by approximately 37 percent over the past five summers (Esralew & Belin, 2014). Although the average daily summer spend across U.S. resorts per person is less than that compared to winter (US$76 as opposed to US$102), summer tourism revenue has grown in terms of average ski area revenue from 9 percent in 2009 to 11 percent in 2013 (Esralew & Belin, 2014). This diversification of product offerings and the resulting revenue and cash flow can therefore be seen as providing ski areas with greater operational flexibility and relief during low snowfall years where visitation drops, allowing many to stay financially afloat.

Another example of ski areas increasing their latitude is their attempts to reach out to a more diverse skiing/snowboarding market. While skiing/snowboarding has traditionally been a Caucasian-centric, middle–upper class, male endeavor, changing demographic trends, including an ageing Baby Boomer market, and the growing US$2 trillion Hispanic, African American, and Asian markets, has forced the ski industry to invest time and resources into building a more diverse customer base (Rufo, 2014). For many ski areas, particularly those located near major urban areas, this has meant reaching out to minority groups that have been historically unrepresented on the slopes.

For Mountain High in California, where 52 percent of its customer base is non-Caucasian, multicultural marketing is central to its success (Rufo, 2014). In particular, the existence of large Korean and Chinese markets located in nearby Los Angeles have encouraged the ski area to provide weather forecasts in Korean and Chinese. At Camelback Mountain Resort in Pennsylvania, a significant number of Spanish-speaking guests has resulted in Spanish radio advertisements in nearby urban centers, as well as the hiring of Spanish-speaking staff in the call center, at the ticket office, guest services, and also the ski school (Rufo, 2014). These efforts are slowly being replicated across North American ski areas; however, there is still significant room for improvement, with minorities still only representing 13 percent of skiing/snowboarding visits over the 2014–15 season, relative to their 37.9 percent share of the U.S. population (Belin & Becher, 2015).

The importance of having up-to-date information has also become central to determining a ski area's latitude, as weather variability, road and traffic conditions, and any number of factors can affect ski area visitation. Most ski areas now rely on a wide range of data to determine staffing and resource needs, with one such example coming from Wachusett Mountain Ski Area, Massachusetts. Their development of an internal A-F scheduling system takes into account a number of variables, such as snowfall, temperature, holidays, week day/night vs. weekend, events being held at the ski area, etc., and grades each night and day accordingly in regard to expected visitation. This is communicated daily to all department managers and supervisors to allow for informed scheduling of staff and resources to ensure supply matches demand, thereby improving operational efficiency and flexibility.

A final factor affecting the latitude of ski areas is the acquisition of multiple ski areas within geographically diverse locations by resort corporations. This can be seen as a way to connect ski areas and "insure" against seasonal variations across different regions. Using Vail Resorts, Inc. as an example, while its Tahoe-owned resorts of Heavenly, Northstar, and Kirkwood have faced significant drought conditions over the past three years, more favorable conditions at its Rocky Mountain resorts have allowed for the corporation to minimize its losses overall. Furthermore, cross-marketing efforts that encouraged California skiers to travel to Colorado to ski have further reduced the loss that would have been experienced if the corporation only owned ski areas within northern California. This same strategy has also been utilized by Powdr Corporation, which owns Killington Resort in the northeastern U.S., Copper Mountain in Colorado, and Mt. Bachelor in Oregon, among others. By spreading out their risk geographically, these ski resort corporations are better positioned to absorb "shocks" in visitor numbers that result from decreased snowfall and increased temperatures, thereby increasing their latitude.

Limiting ski area precariousness

Precariousness denotes how close a system is to a threshold, after which it is likely to change state (Walker, et al., 2004). Many ski areas operate quite close to their limits due to fluctuating weather conditions, which are predicted to further increase with climate change (Becken, 2013). This has led to increased efforts across the ski industry to reduce GHG emissions and increase resort and guest awareness regarding how climate change can be stopped or managed, to best protect the sensitive alpine environment in which they operate.

Over the past decade, there has been growing acknowledgement and effort at an industry level to combat climate change. Beginning in 2000, the NSAA introduced an environmental charter commonly referred to as "Sustainable Slopes", which identified climate change as a potential threat to the alpine system in which ski areas operate. This was followed by the launch of the "Keep Winter Cool" campaign in the 2002–3 season to educate ski areas and their guests about the impacts of climate change on skiing and alpine environments. In 2011, the NSAA developed its most comprehensive approach to addressing climate change with the Climate Challenge program. This is a voluntary program designed to help participating ski areas reduce their GHG emissions and reduce energy use and costs (NSAA, 2016). What started with 8 participating resorts in 2011 has since grown to 30, with a reduction of 4,009 metric tons of CO_2 emissions and the purchase of 41,192 renewable energy credits across participating resorts over the four years (NSAA, 2015).

Additionally, an increasing number of ski areas have undertaken their own measures to minimize their carbon footprint and educate guests on the importance of addressing climate change. An innovative idea that has grabbed public attention has been the conversion of methane to power lift and other ski area operations. One such effort has been adopted by Killington Resort, which signed onto Green Mountain Power's Cow Power Program in 2012. This program converts cow manure into energy on nearby dairy farms through an anaerobic digestor, displacing the use of fossil fuels and minimizing the amount of GHGs entering into the atmosphere (Menzel, 2015). Purchasing enough cow power to operate the resort's K-1 Gondola in its first year (300,000 kWh), Killington Resort has since increased its commitment, purchasing 1 million kWh of cow power, accounting for approximately 4 percent of the ski area's yearly energy use (Menzel, 2015). Although arguably a symbolic gesture, given the small percentage of overall energy use that can be attributed to biofuel, it has gained significant media attention and provided Killington Resort with a platform from which they can educate guests on climate change and the effects on the ski industry.

In 2012, Aspen Skiing Company (ASC) announced a much larger biofuel project that captures waste methane from the Elk Creek Mine into 24 million kWh of power: enough to power all of its 4 ski areas, 13 restaurants, and 3 hotels (Blevins, 2012). This was the first project of its kind within the U.S. at this scale, with ASC having spent US$5.4 million to develop the processing plant where the methane is cleaned and compressed, before it passes through generators and substations to eventually reach the Holy Cross Energy network grid as electricity (Blevins, 2012). Similar to the "Cow Power Program," a long-term benefit of this project is not only the clean power generated, but its novelty and visibility and, thus, the publicity it lends to the broader climate change debate.

Taking this education role one step further, ASC has also become a leading voice within the U.S. ski industry on climate change advocacy and policy formation. Encouraging action on a national level, ASC has invested significant time and money in lobbying Congress on the importance of reducing GHG emissions and making climate change a priority issue, in partnership with the non-profit organization Protect Our Winters (POW). Founded by professional snowboarder Jeremy Jones, POW has attracted the attention and support of other professional winter athletes, resorts, and winter sports companies dedicated to creating "the political will for meaningful action by state and federal policymakers" (POW, 2016). This includes representation at the recent UN Climate Change Conference/COP21 in Paris, and the creation of several platforms over which the 21 million sports enthusiasts within the U.S. are able to unite and voice their concerns regarding climate change. While such individual efforts by ski areas and non-profit organizations are not sufficient on their own to reverse or even slow the effects of climate change, it is hoped that the visible nature of the ski industry will help influence public policy, increase guest awareness, and ultimately change guest behavior, as well as garner attention from other larger industries to mitigate the effects of climate change more broadly. This would help limit the precariousness of ski areas and the SESs in which they operate, over time.

Factors affecting ski area resilience

Having examined the various ways in which ski areas and the U.S. ski industry have sought to increase their resistance and latitude, while also limit their precariousness, it is clear that there are significant factors that have helped certain ski areas to bolster their resilience. Those resorts that appear best poised to adapt to the increasing uncertainty resulting from climate change appear to have recognized the importance of partnerships to seek, fund, and supply innovative initiatives; the link between innovation and return on investment; and the value and power of communicating their efforts to stakeholders.

As noted above, innovation in the ski industry has been demonstrated through a wide range of initiatives, all of which have involved partnerships with suppliers and energy companies that have not traditionally been sought within the ski industry. These partnerships are indicative of an adaptive management approach (Stockholm Resilience Centre, 2015), with many ski areas recognizing that their capacity to successfully operate in fragile alpine environments affected by climate change is dependent upon their ability to collaborate with different organizations. While some of these partnerships are short term (e.g., the partnership between Vermont ski areas with Efficiency Vermont for the purchase of snowguns), others are more long term (e.g., the partnership between ASC and Elk Creek Mine). Regardless, each of these strategic partnerships have helped coordinate relevant actors at different scales to bring about meaningful action before critical thresholds are met (Stockholm Resilience Centre, 2015).

Furthermore, ski areas' efforts to increase resistance have reduced the amount of water and energy resources used for snowmaking and general resort operations, while also producing significant economic savings. This is extremely important for a capital-intensive industry that is commonly highly leveraged, with the growing inconsistency surrounding average winter temperatures and natural snowfall levels brought about by climate change, further exacerbating the difficulty of financing resort operations (NSAA, 2013). As climatic changes push some ski areas precariously close to critical thresholds, the return on investment in new innovative ventures is just as important as the need to adapt to changing social-ecological conditions. What is demonstrated from the examples discussed in this chapter, is that innovative measures to increase resistance and latitude and reduce precariousness can bring about a positive return on investment. This is in addition to any positive publicity and branding that may accompany efforts to adapt to, and manage, climate change.

By making the value of ecosystem services visible in society (Stockholm Resilience Centre, 2015), ski areas can also educate their guests and others on the fragile nature of alpine environments and the importance of innovation to best mitigate and adapt to climate change. This can be achieved at a resort level, with ski areas increasingly communicating their environmental efforts to guests, as well as through broader advocacy efforts in association with the NSAA and organizations such as POW at an industry level and beyond. This is vital for ski areas if they are to limit the precariousness of their situation, as, despite the best efforts of the ski industry, broader societal impetus to change is required in order to successfully mitigate the effects of climate change.

The importance of communication for ski areas also extends beyond environmental initiatives to the attraction of diverse markets. As discussed above, the ability of a ski area to increase its latitude is dependent on being able to attract diverse markets in terms of their seasonal interests, ethnicity, and geographic dispersion. Thus, by experimenting with new product and service offerings, whether related to summer tourism development or the hiring of bilingual staff, and effectively communicating their existence to relevant markets, ski areas can best position themselves to adapt to changing climatic conditions and broader societal and demographic trends.

Conclusion

This chapter has highlighted how the current situation of the U.S. ski industry can be examined through the application of a resilience framework, with many ski areas demonstrating their adaptive capacity to the increasingly unpredictable conditions brought about by climate change. In particular, those ski areas that have developed strategic partnerships at a resort and industry level, demonstrated the ability of innovation to reduce both the consumption of scarce ecological and economic resources, and have sought to educate and inform guests on the necessity of action for the future survival of skiing in alpine environments, appear to be best poised to adapt to, and manage, the slow-change variable of climate change.

As the ski industry is one of the most visibly affected industries from climate change, empirical research beyond Becken (2013) that examines the different factors that affect the degree to which ski areas are able to adapt to changing climatic conditions can provide instructive lessons for the industry. Given that many ski areas are not maximizing their adaptive capacity, the identification of strategic ways to help increase ski area resistance and latitude and limit the precariousness of their situation, as relevant to their geographic location, visitor-base, economic capacity, and other associated considerations, would be extremely beneficial. Such research could help contribute to the future success and viability of the U.S. ski industry, as the effects of climate change will only continue to become more pronounced over time without significant global intervention.

References

AlpineZone News (2006). $5.7 million initiative will reduce energy usage for Whiteface and Gore snowmaking. *AlpineZone News.*

Bebb, D. (2015). *Climate Exposure Impact on Equity Valuation: Case Study of Vail Resorts, Inc.* Stanford: Steyer-Taylor Center for Energy Policy and Finance

Becken, S. (2013). Developing a framework for assessing resilience of tourism sub-systems to climatic factors. *Annals of Tourism Research*, 43, 506–28.

Belin, D. & Becher, D. (2015). 2014–15 NSAA National Demographic Study results. *NSAA Journal, Early Winter*, 12–28.

Blevins, J. (2012). Aspen Skiing Co. Partners with coal mine for methane power. *The Denver Post.* Retrieved from http://www.denverpost.com/ci_21966674/aspen-skiing-co-partners-coal-mine-methane-power

Burakowski, E. & Magnusson, M. (2012). *Climate Impacts on the Winter Tourism Economy in the United States.* New York: Natural Resources Defense Council.

Davoudi, S. (2012). Resilience: a bridging concept or a dead end? *Planning Theory & Practice*, 13(2), 299–333.

Dicker, R. (2012). Artificial snow from sewage water at Arizona Snowbowl Resort: all systems go. *Huffington Post.*

Emerson, S. (2013). Mt. Baldy ski area to use technology to combat water evaporation. *Daily Bulletin News.*

Esralew, S. & Belin, D. (2014). Summer business: what's the return? *NSAA Journal, Summer*, 31–7.

Folke, C. (2003). Socio-ecological resilience and behavioral responses. In B. Hansson, A. Biel & M. Martensson (Eds.), *Individual and structural determinants of environmental practice*. Burlington, VT: Burlington, VT: Ashgate.

Gunderson, L. H. (2000). Ecological resilience: in theory and application. *Annual Review of Ecology and Systematics*, 31, 425–39. doi:10.2307/221739

Hawks, T. (2015). The great snow gun roundup. *NSAA Journal, Convention*, 34–41.

Holling, C. S. (1973). Resilience and stability of ecological systems. *Annual Review of Ecology and Systematics*, 4, 1–23. doi:10.2307/2096802

Hudson, S. (1999). *Snow Business: A Study of the International Ski Industry*. London: Cassell.

Lew, A., Ng, P.T., Ni, C-C. & Wu, T-C. (2016). Community sustainability and resilience: similarities, differences and indicators. *Tourism Geographies*, 18(1), 18–27.

Menzel, C. (2015). Killington's big dumps. *Powder Magazine*. Retrieved from www.powder.com/stories/news/killingtons-big-dumps/#h45tHEEmEy22476Z.97

NSAA (2013). *Economic Analysis of United States Ski Areas*. NSAA in conjunction with RRC Associates, LLC.

NSAA (2015). *2015 Climate Challenge Report*. Lakewood, CO: NSAA

NSAA (2016). "Climate Change". National Ski Areas Association. Retrieved from www.nsaa.org/environment/climate-change.

NSAA (2017). *Facts on snowmaking*. Lakewood, CO: National Ski Areas Association. Retrieved from www.nsaa.org/media/248986/snowmaking.pdf.

POW (2016). "About Us". Retrieved from http://protectourwinters.org/about-us/

Rufo, S. (2014). Multicultural marketing: how to ensure cultural relevance. *NSAA Journal*, 22(2), 8–10.

Scott, D., McBoyle, G., Minogue, A. & Mills, B. (2006). Climate change and the sustainability of ski-based tourism in eastern North America: a reassessment. *Journal of Sustainable Tourism*, 14(4), 376–98. Retrieved from http://www.informaworld.com/10.2167/jost550.0

Stockholm Resilience Centre (2015). *What is resilience? An introduction to social-ecological research*. Stockholm: Stockholm University.

Swain, D. L., Diffenbaugh, N. & Rajaratnam, B. (2014). Atmospheric conditions associated with the 2013–14 California drought are "very likely" linked to human-caused climate change. *Stanford Woods Institute for the Environment, Fall*, 1–2.

Vail Resorts (2013). "Epic Discovery". Retrieved May 2, 2016, from http://www.epicdiscovery.com/

Walker, B. H., Carpenter, S. R., Rockstrom, J., Crepin, A.-S. & Peterson, G. D. (2012). Drivers, "slow" variables, "fast" variables, shocks, and resilience. *Ecology and Society*, 17(3), 30.

Walker, B., Gunderson, L., Kinzig, A., Folke, C., Carpenter, S. & Schultz, L. (2006). A handful of heuristics and some propositions for understanding resilience in SESs. *Ecology and Society*, 11(1). Retrieved from http://www.ecologyandsociety.org/vol11/iss1/art13/

Walker, B., Holling, C. S., Carpenter, S. R. & Kinzig, A. P. (2004). Resilience, adaptability and transformability in social-ecological systems. *Ecology and Society*, 9(2). Retrieved from www.ecologyandsociety.org/vol9/iss2/art5/

Wolfsegger, C., Gossling, S. & Scott, D. (2008). Climate change risk appraisal in the Austrian ski industry. *Tourism Review International*, 12(1), 13–23.

9 (Re)production of resilient tourism space in the context of climate change in coastal Quebec, Canada

Dominic Lapointe and Bruno Sarrasin

Introduction

Recent decades have seen climates transformed by human actions resulting in ocean and atmospheric warming, rising sea levels and a decline in ice cover at the poles (IPCC, 2014b). According to the IPCC (2007; 2014a) we are heading towards an increase in average global temperature of 2.6°–4.8° C by the end of the century. This increase would result in a rise in sea level of 45–82 cm. These changes are associated with an increase in extreme weather events, ocean acidification, coastal erosion, flooding, and invasive species. In Quebec, Ouranos (2014) has predicted an increase of 2°–4° C by 2050 and 4°–7° C by the end of the century. The level of the Gulf of St. Lawrence and its estuary will increase by 30–75 cm. This increase will create erosion, flooding and an overall transformation of those coastal ecosystems (Bernatchez et al., 2008). These changes will affect the entire scope of society and industry, including the tourism industry. The aim of this chapter is to use a production of space framework with a socio-ecological resilience perspective to analyze the transformation and (re)production of coastal tourism space within a capitalist accumulation process. We will consequently identify how the discourses of tourism development and climate change adaptation work together to transform space and place, focussing on how local communities can interact with those discourses, to adapt or not, in a way that is resilient or not. The adaptation and the resilience of coastal tourism spaces are somewhat contingent on the capitalist accumulation process. First, we will present how climate change impacts tourism. Then, we will present the knowledge gaps in our understanding of climate change and tourism. This will bring us to focus on the importance of addressing the issue of climate change and tourism in peripheral coastal areas. This focus raises the following questions:

- How will climate change and discourses on climate change adaptation alter tourism industry development in peripheral coastal destinations?
- How can a socio-ecological resilience (SER) perspective contribute to improving our understanding of the adaptation dynamic?

We will answer these questions by applying a space analysis framework to a case study of Notre-Dame-du-Portage, a community in the lower St. Lawrence region of Quebec, Canada.

Climate change and tourism

The tourism industry is vulnerable to climate change because many of its resources, including coastal areas and natural attractions, will be strongly affected by it (Jones & Phillips, 2011; Becken & Hay, 2012; Nicholls, 2014). The two major aspects of the industry that will likely be most affected are the destinations themselves and tourism operations (Nicholls, 2014; IPCC, 2014b.). Coastal destinations are likely to be most affected by rising sea levels and extreme weather events that will threaten coastal infrastructures, submerge beaches and erode coasts. Nature tourism, whether by sea or not, will be affected by the transformation of ecosystems and loss of biodiversity, and winter destinations will have to deal with a reduction in snow cover (Becken & Hay, 2007; Simpson, Gössling, Scott, Hall & Gladin, 2008; Hall & Higham, 2005; Germain & Bleau, 2014; Kaján & Saarinen, 2013). All of these changes and the impacts they generate force tourism areas to adapt. However, compared to other sectors, research on climate adaptation for the tourism industry lags behind, both internationally (Becken & Hay, 2012; Scott, de Freitas & Matzarakis, 2009; Saarinen, 2014) and in Quebec (Germain & Bleau, 2014).

Adaptation is both a challenge and a priority for the tourism industry, which is of vital importance for many communities. However, the adaptation of local communities is influenced by spatial processes where larger geographic scales (e.g., national and international agreements and policies) (Lépy et al., 2014) intersect with the response capacity of local actors (Plante, Chouinard & Martin, 2011). Current research on the adaptation to climate change of the tourism industry shows limited interest in the spatial dimension of the adaptation process and their non-climatic factors. The role played by local stakeholders in the transformation of locally valued tourism resources is also poorly understood (Kaján & Saarinen, 2013). Existing research has focussed more on the technical dimensions of adaptation. It is therefore important that the social and spatial components of climate change adaptation in tourism communities, particularly coastal communities, be addressed.

The impacts of climate change on tourism have yet to be fully recognized and the understanding of practical adaptation to climate change is still limited (Hall, 2012). In the short term, the social dimension of adaptation in tourism destinations needs to become a research priority (Buckley, 2012). In North America, where the issue of climate change has been studied the most extensively, the adaptation of coastal destinations has yet to be addressed (Buckley, 2012). The main focus of existing research has been climate change adaptation for ski resorts and other winter destinations. (Germain & Bleau, 2014). Understanding the social dimension of climate change adaptation in peripheral coastal destinations is therefore necessary. Climate change adaptation at the local level needs to integrate the social representations of local actors (developers, planners, businesses and residents) (Lépy et al., 2014) in relation to the dominant discourses on climate change and tourism development (Jopp, DeLacy, Mair & Fluker, 2013). The management and development of tourism areas are already very complex, and climate change adaptation adds yet another dimension to this complexity.

Why is adaptation problematic for communities?

Tourism is a major international industry that accounts for 9 percent of global GDP and generates over US\$6 trillion in revenue (Nicolls, 2014). For territories outside of major cities, tourism is a sensitive industry and its main resources are land and natural attractions. Indeed, tourism is a very active industry in the process of restructuring the economies of peripheral territories, especially those with desirable and attractive environmental amenities (Hall, 2006). For coastal, mountain and forest areas, tourism is an economic alternative that can offset the loss of traditional industrial activities. However, this strategy is not without its difficulties. As mentioned by Müller and Jansson (2006), despite the dominant discourse of tourism as an engine of development for peripheral communities, this strategy is accompanied by a number of challenges, including obstacles to accessing capital, a lack of manpower training and tourism development policies that are ill-suited to their context. Despite these difficulties, some peripheral communities do strive in the tourism industry, and it is likely that climate change will profoundly affect the primary resource of these communities. Tourism industry adaptation in those communities is essential to support and maintain their development; this is especially true for the peripheral coastal areas we will be focussing on.

Coastal areas are a major tourism resource: access to water, coastlines and their associated tourism practices are one of the main incentives for travel (Duhamel & Violier, 2009). Climate change will have a significant impact on these areas, especially with regards to water regimes, rising and falling sea levels, and flooding and erosion (Jones & Phillips, 2011). The transformation of coastlines is a major challenge for coastal tourism areas, as it will affect not only the physical space and access infrastructures, but also the nature of possible tourism activities in a given coastal space. For example, the loss and transformation of coastal ecosystems on the US eastern seaboard alone will endanger a multi-billion dollar industry and radically transform the environment and the economy of thousands of coastal communities (Hughes, 2011).

Space is the main resource for tourism (Lozatot-Giotard, 2008; Hall & Page, 2006), and destinations do not have the ability to relocate, hence the crucial role of the adaptation process (Jopp et al., 2013). This is especially true in the case of coastal areas where space is a scarce and coveted resource (Plante, 2011) shared by a multitude of stakeholders: residents, environmental groups, user associations, government agencies, private companies, etc. These stakeholders all have their own representations and discourses on the use and development of their resource and, by extension, their own representations and discourses on climate change adaptation. Although tourism's potential to transform space is a known phenomenon (Ashworth & Dietvorst, 1995), its ability to adapt to the transformation of physical space due to climate change is little discussed. When the subject has been broached, it has mainly been in terms of natural and technical sciences (Kaján & Saarinen, 2013; Jones & Phillips, 2011), leaving aside social and spatial dimensions. Climate change adaptation strategies are chosen based on various stakeholders' perceptions of climate change, their own impact on the territory and what they consider to be viable options. These representations are expressed in formal and informal discourses on space, tourism, climate change, and adaptation

Space, place and resilience as an analytical framework for climate change adaptation

There is a current and widespread acknowledgement of the complex interdependence of humans and their environment and the uncertainty of natural resource management in a constantly changing world (Holling, 2001). These elements converge in the concept of a socio-ecological system (SES). In its simplest form, an SES is a system in which human beings and nature interact (Liu et al., 2007; Berkes & Folke 1998). This approach suggests that maintaining the quality of the natural environment cannot be achieved by eliminating the change or disturbances, but rather that variability and diversity contribute to a system's vitality (Berkes, Colding & Folke, 2003; Holling 2001). This is what Davoudi (2012) defines as evolutionary resilience. For many researchers, the key to sustainability lies in strengthening SES resilience, and not in optimizing the performance individual and isolated system components (Oström, 2009; Walker & Salt, 2006; Scheffer, Carpenter, Foley, Folke & Walker., 2001). Under these conditions, a resilient SES able to cope with change is synonymous with social, ecological and economic sustainability (Lew, Ng, Ni & Wu, 2016).

On the other hand, resilience is not always about the dynamic adaptation of complex systems to change: it can also refer to a return to a pre-disturbance equilibrium, in what is called engineering resilience (Davoudi, 2012). Engineering resilience designs to bounce back to an original state and protect existing assets and values (Fünfgeld & McEvoy, 2012). While systemic perspectives provided by SES embrace complexity, engineering resilience tends to simplify the dynamics at work in a complex system in order to focus on the time required to return to the original state, without questioning the underlying dynamics of the original state.

The exploratory perspective of our approach does not allow us to go beyond the socio-political dimensions and the socio-geographical system. In this context, we will pay particular attention to adaptation, defined as the ability of a system to respond to new situations without limiting its options for the future (Armitage, 2005; Folke et al., 2002). The socio-political dimension of adaptation is largely based on the presence of institutions and networks of actors that create open and flexible spaces for problem solving and power-sharing among a variety of interest groups (SRC, 2014; Armitage, 2005). Local-scale adaptive capacity is influenced by factors such as the existing infrastructure, political influence and access to financial, technological and information resources (Smit & Wandel, 2006).

Addressing the issue of climate change adaptation and resilience from a socio-geographical perspective requires an analysis applying social constructivist approaches to space, specifically to the production of space. This intersection between geographical and sociological gaze leads to an epistemological posture where space is a social output (Harvey, 1996; Lefebvre, 1974; Di Méo & Bulléon,

2005). Thus, space is not just a passive receptacle of human action; it is also a result of this action as interpreted and appropriated by local stakeholders. These interpretations, appropriations and social constructions by stakeholders territorialize action (Klein, 2008; Di Méo & Bulléon, 2005). In the case of tourism, they transform public, non-market values and amenities associated with space into commodities, through product development and land market speculation (Britton, 1991; Overton, 2010).

Our analysis will use a tripartite conceptualization of space:

- absolute space (material);
- relative space (that which flows through space); and
- relational space.

We will focus on relational space, that is, all social and political relations that give meaning to a given space for territorial stakeholders (Harvey, 2010) and all relationships expressed through representations of space and social practices (Lefebvre, 1986). These relationships allow us to focus on the development of the forces that produce the historical result of socio-economic activity in space (Smith, 1984; Soja, 1989), including tourism, without neglecting the role of physical constraints, which are being transformed by climate change. This is in line with the complex relational view of space and place included in SER (Davoudi, 2012; Fünfgeld & McEvoy, 2012). We will apply this theoretical approach in concert with the concept of resilience to address the issue of adaptation to climate change in coastal tourism areas.

This chapter reports the first exploratory phase of a larger project on the production and reproduction of tourism space in the context of climate change. In this phase, we restricted the analysis to the coastal community and to four discourse types:

- local municipal discourses;
- regional environmental technical discourses;
- local and national tourism development discourses; and
- climate change discourses.

The stakeholders producing these discourses include governmental and municipal bodies and local and regional groups and organizations. These discourses are conceptualized as one moment of the social processes internalizing all others social moments (power, values, institutions, material practices and social relations) (Harvey, 1996). In the context of climate change adaptation, we could say that these discourses are expressions of the complex interaction of environmental processes, social behaviour and contested values (Fünfgeld & McEvoy, 2012).

The different sources of discourses were selected on the basis that they came from recognized stakeholders in tourism, land planning and climate change. These discourses are found in the form of official documents, and they aim to define the use of the shore as well as the types of action that stakeholders will take in regards to tourism, land planning and adaptation to climate change. They were analyzed using NVivo thematic textual query software. The emergent themes guiding the queries were tourism, development, climate change, resilience, erosion and shores. We used this method to analyze what those discourses say about the local community of Notre-Dame-du-Portage in the lower St. Lawrence region in Quebec, Canada, and to study how those discourses act on the community. The discourse analysis was twofold, where the six emergent themes were aggregated in two main discourses:

- tourism development discourses; and
- climate change discourses.

Notre-Dame-du-Portage: a summer home resort at the forefront of climate change

The community of Notre-Dame-du-Portage is a small community by the St. Lawrence River. It is characterized by its history as a location for second homes and coastal tourism activities. Since the nineteenth century, Notre-Dame-du-Portage has been a summer home resort. In the beginning, it was mainly members of the upper class from Quebec City and Montréal who spent their summers in the area. The development of steam cruisers on the St. Lawrence followed by railroad development transformed the entire region into a popular tourism destination (Gagnon, 2003). At the height of this era, around 5,000 people were spending the summer in the resort towns of Notre-Dame-du-Portage, St-Patrice and Cacouna (Choko, Léger & Lefebvre, 2013). The main attractions were cruises on the St. Lawrence and the beaches. Today, Notre-Dame-du-Portage is the only town that has maintained some of its tourism functions.

In 2011, the village of Notre-Dame-du-Portage had 1,193 residents. Its average income and schooling level are the highest in the county. Nearly a third (31 percent) of all residents hold a university degree, compared to the 13 percent county average, and the average household income in 2011 was CA$66,574 compared to CA$43,066 for the county in general (Statistics Canada in MRC de Rivière-du-Loup, 2012). Notre-Dame-du-Portage is mainly a white-collar community, the place of residence of a local elite attracted by the views of the St. Lawrence and the access to its shore. There are still a good number of summer houses, even if the phenomenon is not as widespread as it was in the nineteenth century. The village has two hotels by the shore that offer spa and health services. Even if the community is no longer solely dependent on tourism, its identity revolves around its summer home resort characteristics and ambiance.

The core of tourism activities of Notre-Dame-du-Portage is concentrated on a small linear road located at the bottom of a cliff and running parallel to the shore of the St. Lawrence estuaries. This linear occupation of the shore, which is physically limited by the cliff, makes space a rare and valuable resource, and puts it at risk of extreme weather events. The effects of climate change on the estuaries and its shore have been documented since 1998 (Shaw, Taylor, Forbes & Solomon, 1998), but the questions took on a new importance for the residents of Notre-Dame-du-Portage in December 2010 when a storm surge hit the community. During the storm, a part

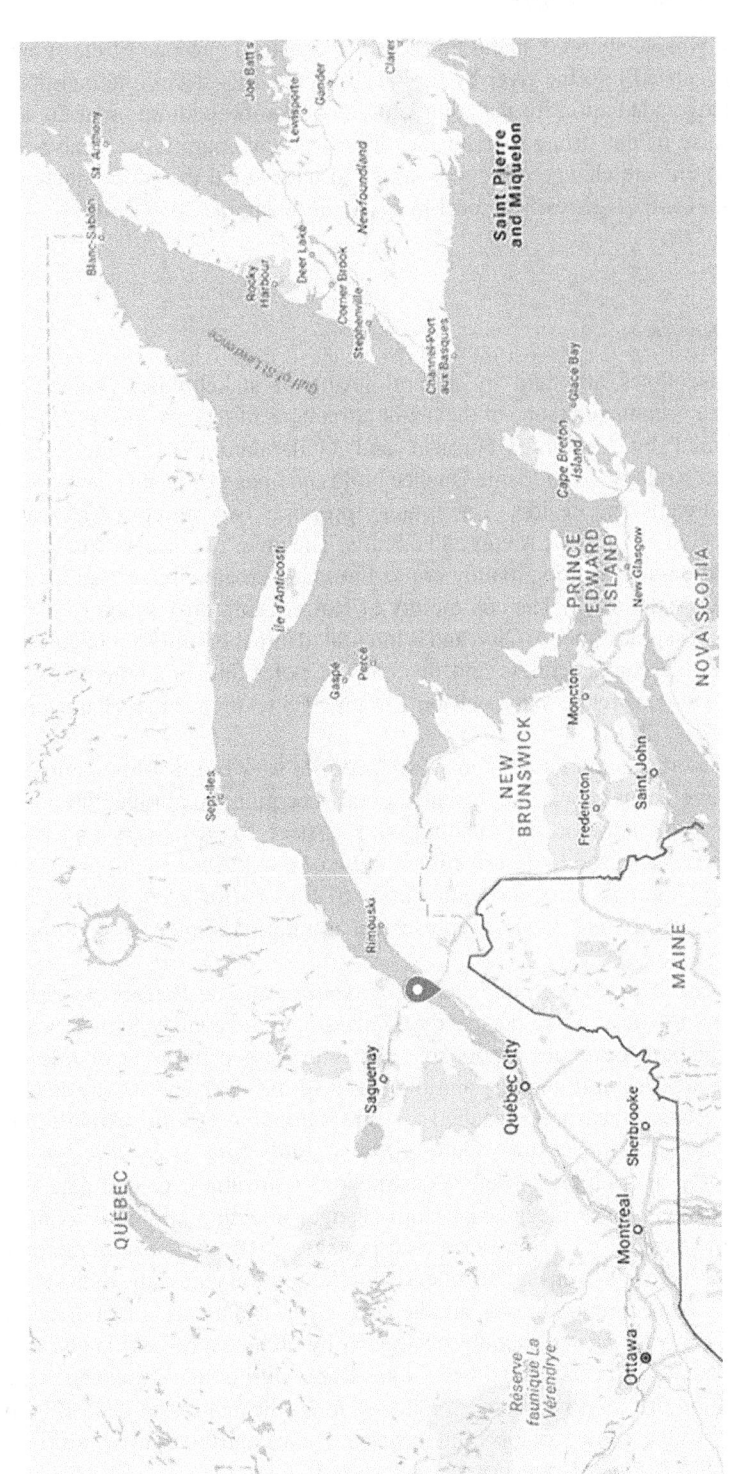

Figure 9.1 Map position of Notre-Dame-du-Portage
Source: Google Maps ©2016

of the road was submerged, many houses were damaged and one of the summer homes was carried into the river by the waves. Since this storm, the question of climate change adaptation for the community has been brought up on at the local level. Because of the strong tourism identity of the community, we analyzed its tourism and climate change discourses to try to understand the (re)production of Notre-Dame-du-Portage resilient tourism space following the 2010 storm.

Findings

Tourism discourses

Tourism discourses are held by several levels of stakeholders. Indeed, the St. Lawrence is considered one of the major attractions of the province of Quebec. As mentioned by Lapointe, Sarrasin and Guillemard (2015) the tourism development strategy of Tourisme Québec builds an image of the St. Lawrence as a place of dreams and desires. The strategy proposes two structured recreation poles connected by touristic routes and circuits. The poles are designed to capture international passenger flow, mainly cruise lines, and create circuits that divert it towards the coastal areas. The community of Notre-Dame-du-Portage is 200 km from the nearest pole, Quebec City, and is integrated into the Route des navigateurs circuit. The interconnectedness and diversity of poles and functions along the banks of the St. Lawrence may contribute to the SES resilience that characterizes this space (SRC, 2014).

The St. Lawrence and its coast are also favourite sites of tourism activities on a regional level. This space is represented as an exceptional natural environment, a place for contemplation and healing, and a space for discovery and active leisure. The images used to describe it are those of a wild ecosystem, one almost untouched by humans. It is represented as an SES in and of itself, as defined by Carl Folke (2006). This discourse represents it as natural heritage being proposed as a tourism resource.

At the local level, in the municipality of Notre-Dame-du-Portage, tourism is presented from a vacationing perspective. This discourse revolves mainly around the historical heritage of the resort area and is expressed by a set of rules for maintaining architectural heritage and the views of the river and its surrounding landscape. These measures resulted in the adoption of the Architectural Development and Integration Plan (*Plan d'implantation et d'intégration architecturale*) that supervises all "construction, renovation, colour palette or landscaping (shed, fence, hedges, parking) changes affecting all buildings in the Route du Fleuve sector," (Notre-Dame-du-portage, 2010, p. 3, our translation). The plan's objective is to ensure the integrity of the cultural landscape of the Route du Fleuve by respecting traditional architectural styles and assuring that the views on the St. Lawrence are not unduly obstructed by architectural and landscaping changes. This policy is associated with Notre-Dame-du-Portage's membership of the Association of the Most Beautiful Villages in Quebec and is embedded in the management of the "slow variables and feedbacks" that contributes to the resilience potential of the SES that Notre-Dame-du-Portage is a part of (SRC, 2014).

Another element of the tourism discourse lies in the identification of areas with public beach access. This last point reflects the extent to which land ownership and private appropriation of the coastal area is, as identified by Gagnon (2003), a process in the development of tourism positions in Quebec. Indeed, the coastal area is now highly artificial and densely built-up. Land holdings are privately owned and act as barriers to the use of the public part of the coast. This observation explains the local-level importance accorded to identifying, and in some cases to negotiating, passages for access to the coast. In some ways, this privatization and artificialization of the coastal ecosystem is an attempt to minimize the disturbances that major weather events could wreck upon on the economic system, which is based on land value. It aims to strengthen the resilience of the socio-economic system, in a bounce back engineering resilience strategy, but is detrimental to the capacity of the ecosystem to play its role in the dynamic of coastal erosion. This process is weakening the overall SES resilience in favour of a reactive engineering resilience strategy based on a simplification of the system through the market, and by aiming to protect and enhance land values with artificialized coastal protection (Figures 9.2 and 9.3).

In short, this tourism discourse represents Notre-Dame-du-Portage's space as an access point to the St. Lawrence River, which is a unique natural area harbouring ecosystems to be protected and discovered. Heritage conservation values are invested in space along the Route du Fleuve to limit impacts that private residential use, one of the main uses, could have on the resort-like atmosphere. In addition, the local desire to preserve public access to the coast in order to maintain the social and material practices associated with it (i.e. walking, swimming, boating and landscape contemplation) emphasizes the importance of coastal access in local tourism practices. These observations also reveal the extent to which this highly valued space is privately owned. It is also important to note that the analyzed tourism discourses only mention the issue of climate change in terms of how it could impact the seasonality of tourism, and never as a force capable of transforming the actual tourism space. In this sense, the tourism discourses analyzed create a dialectic situation. On the one hand, they contribute to SES resilience, especially by giving value to the protection and reproduction of the ecosystems and the social practices valued in those ecosystems. On the other hand, the protection of the economic values created by tourism reduce the resilience of the whole SES in favour of a reactive engineering resilience focus on the protection of private land values.

Climate change adaptation discourses

The discourse on climate change is held, in part, by national stakeholders, mainly in the form of the Quebec strategy to fight climate change (Quebec, 2012. As we pointed out in a previous article (Lapointe et al., 2015), this discourse presents the St. Lawrence as a focal area through multiple economic and social flows, but also as an area vulnerable to climate change. Adaptation is presented as dependent on the economic value of these areas, thereby implicitly suggesting regulation by the market. Here we are facing a typical top-down, risk-management focussed governmental strategy that simplifies the complexities needs of the socio-ecological system and the different stakeholders involved in order to ultimately have them managed by the market economy.

Regionally, the discourse on climate change adaptation is integrated into land planning and environmental planning discourses. The land planning discourse presents adaptation to climate change as a regulatory issue, one that manages the protection of shores and their surrounding infrastructure in those areas by restricting the planning regulations for landowners located in hazardous zones. Adaptation becomes the responsibility of private owners within a regulatory framework which identifies structures that can be built to protect the land from erosion and flooding. This discourse simplifies the complexity of the system into a technical issue that can be understood according to static planning rules. At this scale, the issue of climate change adaptation is also framed by engineering resilience ideologies aimed at protecting existing assets and values from climate risk.

The other regional discourse is the environmental discourse, held by integrated coastal zone management organizations. It presents the estuary as a natural space being transformed by the impacts of climate change and argues that this transformation will be amplified by the degradation of coastal ecosystems. This discourse presents adaptation in an ambiguous way: as something that is simultaneously private and public, individual and collective. This ambiguity is best demonstrated by the naturalization of coastal ecosystems, where public and collective actions play their respective roles in the control of coastal erosion. The environmental discourse emphasizes the restoration of coastal ecosystems through consultation and intervention, both of which foster participation, one of the key requirements for building resilience. However, it does not disqualify the importance of individual land protection interventions in coastal areas. Although this discourse prescribes artificial interventions with questionable long-term effectiveness, such as walls, it also offers a set of good practices to accompany these interventions. In fact, this discourse does not take sides, and it supports SES resilience by focussing on the ecosystem's role in controlling erosion, all the while suggesting ways to minimize the negative impacts of engineering resilience-based protective projects carried out by private owners, such as sea walls and rip rap.

This double discourse reflects a desire for efficiency. Private owners reacting to extreme weather events support interventions that aim to block the water and minimize its effects on their property. The main objective of private owners is to bounce back to the previous state as quickly as possible. In the case of Notre-Dame-du-Portage, where most of the coast is built-up and artificialized, this discourse is spatially expressed through an intervention to replant native species in the marshes of the northeast bay. This action is taking place in an area where there has been almost no private appropriation of the coast and where the coastal land is not on the market, therefore bestowing it with a higher tolerance for an adaptive strategy that is based on the dynamics of the ecosystem with a higher level of uncertainty.

Figure 9.2 Individual sewalls 1
Source: Lapointe et al., 2015

Figure 9.3 Individual sewalls 2
Source: Lapointe et al., 2015

Our analysis shows that the management approach advocated for a given system may increase or decrease resilience according to the resilience capacity of the system, which is organized in response to the actions taken. In this interpretation of resilience, managing an SES to control its sustainability does not push or force the system beyond its limits (Lew et al., 2016). Instead, management maintains diversity, variability and adaptability while seeking to understand when, where and how it is possible to intervene in the system (SRC, 2014, Folke, 2006; Berkes, Colding & Folke 2003; Folke et al., 2002). Considering the nature of socio-ecological systems, one of the major challenges is to develop institutional arrangements that take social and ecological processes into account at different spatial and temporal scales, without neglecting the links that connect these scales (Oström, 2009; Folke et al., 2002). In this approach, collaborative management associated with the concept of adaptive governance, commonly called co-management, contributes to the resilience of socio-political dimensions of SES. It can be defined as "A collaborative arrangement in which the community of local resource users, local and senior governments, other stakeholders, and external actors share responsibility and authority for management of the natural resource in question" (Tyler, 2006, p. 81). Through this partnership, several social actors can work together to negotiate, define and implement a number of functions, benefits and responsibilities for a particular space (Borrini-Feyerabend et al., 2004). Our case study shows that, although the analyzed discourses echo certain shared resilience principles, they do not show an operational convergence that could lead to coherent socio-geographical results in the (re)production of an evolving, resilient tourism space in the context of climate change adaptation. Instead, the discourses take an engineering resilience stance where the coastal community's main adaptation strategy is to return to its prior state through individual (private) landowners' initiatives. This is especially true with regard to the contradictions inherent in discourses that champion ecological adaptation but fail to challenge the artificialization of the shore. These discourses set out to protect land value despite the extent to which this compromises the ecological integrity of the whole SES.

Conclusion: (re)production of resilient tourism space in the context of climate change

Our objective was to explore how tourism development and climate change adaptation discourses were involved in (re)producing tourism space in relation to resilience. We analyzed national, regional and local discourses and their effects on the production of the tourism space of the community of Notre-Dame-du-Portage. Because of the exploratory dimension of our research, we limited the scope to those three discourses and used a basic threefold space matrix: absolute, relative and relational spaces as internalized relations of all social space and every social moment (Harvey, 2010). Where possible, we have identified the principles of resilience that underpinned the discourse being analyzed.

Going back to the idea of the production of the Notre-Dame-du-Portage's tourism space, it is worth noting that the different discourses on the development of tourism in this area, as well as those on adaptation to climate change, present relational space through a strong dialectic between non-market tourism resources (e.g. views, heritage, the St. Lawrence, coastline under the rising water levels, etc.) and private action (mainly land ownership and its ability to transform non-market tourism resources into exchange values in the relative space produced by economic flows). This dialectic illustrates how the development of tourism through the preservation and promotion of distinctive cultural and symbolic elements creates an increase in land market values and private ownership (Overton, 2010). However, in our case, private land ownership of the coastline limits collective action and public adaptation to climate change, consequently affecting SES resilience. Our case study revealed a preference for engineering resilience strategies focused on protecting and keeping actual assets and values safe through private initiatives and a disregard for systemic collective actions rooted in SER.

Our analysis revealed that public actions were focussed on protecting place ambiance (architectural integration plan, protection of views of the river, identification of "public" coastal areas—down to the high water marks). These measures can potentially reinforce SER by giving value to elements of the ecosystem and cultural systems that are embedded in relational space. On the other hand, they create a dialectical situation where the (re)production of those non-market tourism resources leads to an increase in privatization and land monopolies The values expressed in relational space are internalized in the land market as economic values, thereby simplifying the SES through exchange values in the market. In this situation, the more the non-market resources that make this space attractive are protected, the more the land itself becomes valued as rare and desirable. This reinforces the material and individual practices of land protection and promotes reactive engineering resilience strategies and outcomes, mainly the erection of walls, to protect against erosion and rising sea levels and an overall weakening of the whole system's SER.

The measures in place to mitigate the private appropriation of the coast and to maintain the quality of tourism and the attractiveness of the St. Lawrence and its coastline justify an investment in economic and symbolic values to protect coastal land. Protecting the quality of the tourism ambiance (re)produces the space of the coastline in Notre-Dame-du-Portage as a unique place capable of transforming this ambiance into value through the land market. This situation creates a feedback loop where protection and ownership exist co-dependently in a dialectical situation. In this context, the current discourse on climate change adaptation transforms the space into an at-risk area with an ever-present threat to the land justifying the actions of individual and private interventions that protect the land in a short-term reactive manner (engineering resilience) rather than by acknowledging complexities in a concerted public manner (social-ecological resilience). Tenure of space and its tourism development through private land ownership by second home markets are never challenged in the analyzed discourses, suggesting that these loops of production and (re)production of space by tourism will be dominant until the next extreme weather event and will continue to undermine the resilience of social-ecological system in place despite the presence of certain elements in tourism discourses that value intact ecosystems.

References

Armitage, D. (2005). Adaptive capacity and community-based natural resource management. *Environmental Management*, 35 (6): 703–15.

Ashworth, G.J. & A.G. Dietvorst (1995). *Tourism and spatial transformations.* Walingford: CAB international.

Becken, S. & J. Hay. (2007). *Tourism and Climate Change: Risks and Opportunities.* Clevedon: Channel View.

Becken, S. & J. Hay. (2012). *Climate Change and Tourism: From Policy to Practice.* New York: Routledge.

Berkes, F. & C. Folke (eds) (1998). *Linking social and ecological systems: management practices and social mechanisms for building resilience.* Cambridge: Cambridge University Press.

Berkes, F., J. Colding & C. Folke. (2003). *Navigating social-ecological systems: building resilience for complexity and change.* New York: Cambridge University Press.

Bernatchez, P., C., Fraser, S. Friesinger, Y. Jolivet, S. Dugas, S. Drejza & A. Morissette (2008). *Sensibilité des côtes et vulnérabilité des communautés du golfe du Saint-Laurent aux impacts des changements climatiques.* Rimouski: Université du Québec.

Borrini-Feyerabend, G., M. Pimbert, M. T. Farvar, A. Kothari & Y. Renard. (2004). *Sharing power. Learning-by-doing in co-management of natural resources throughout the world.* IIED and IUCN/CEESP/CMWG, Tehran.

Britton, S. (1991). Tourism, capital, and place: towards a critical geography of tourism. In *Environment and Planning D: Society and Space,* 9, 451–78.

Buckley, R. (2012). Climate Change: Tourism Destination Dynamics. In T.V. Singh (ed.) *Critical Debates in Tourism* (pp. 342–4). Clevedon: Channel View.

Choko, M.H., D. Léger & M. Lefebvre (2013). *Destination Québec: une histoire illustrée du tourisme.* Montréal: Les éditions de l'Homme.

Davoudi, S. (2012). Resilience: a bridging concept or a dead end? *Planning Theory and Practice* 13(2): 299–307

Di Méo, G. & P. Buléon. (2005). *L'espace social: Lecture géographique des sociétés.* Paris: Armand Collin.

Duhamel, P. & P. Violier (2009). *Tourisme et littoral: un enjeu du monde.* Paris: Bélin.

Folke, Carl. (2006). Resilience: the emergence of a perspective for social-ecological systems analyses. *Global Environmental Change,* 16: 253–67.

Folke, C., S. Carpenter, T. Elmqvist, L. Gunderson, C. S. Holling & B. Walker. (2002). Resilience and sustainable development: building adaptive capacity in a world of transformations. *Ambio,* 31(5): 437–40.

Fünfgeld, H. and D. McEvoy (2012) Resilience as a useful concept for climate change adaptation? *Planning Theory and Practice* 13(2), 324–8.

Gagnon, S. (2003). *L'échiquier touristique québécois.* Ste-Foy, Presses de l'Université Laval.

Germain, K. & S. Bleau.(2014). Tourisme et loisirs. In Ouranos, *Vers l'adaptation: Synthèse des connaissances sur les changements climatiques 2014.* Montréal: Ouranos.

Jones, A, & M. Phillips, (2011). Introduction—Disappearing Destinations: Current Issues, Challenges and Polemics. In A. Jones & M. Phillips (Eds.), *Disappearing Destinations: Climate Change and Future Challenges for Coastal Tourism* (pp. 1–9). Wallingsford: CABI international.

Jopp, R., T. DeLacy, J. Mair & M. Fluker (2013). Using a regional tourism adaptation framework to determine climate change adaptation options for Victoria's Surf Coast. *Asia Pacific Journal of Tourism Research,* 18(1–2), 144–64.

Hall, C. M. (2006). North–South perspective on tourism, regional development and peripheral areas. In D. K. Müller & B. Jansson (Dir.). *Tourism in Peripheries: Perspectives from the Far North and South* (pp. 19–37). Cambridge: CABI.

Hall, C.M. (2012). Tourism and climate change: knowledge gaps and issues. In T.V. Singh (ed.) *Critical Debates in Tourism* (pp. 318–37). Clevedon: Channel View.

Hall, C.M. & J. Higham(2005). Tourism, recreation and climate change. In C.M. Hall and J. Higham (eds.). *Tourism, Recreation and Climate Change* (pp. 3–28). Clevedon: Channel View.

Hall, C.M. & S.J. Page (2006). *The Geography of Tourism and Recreation: Environment, Place and Space.* New York: Routledge.

Harvey, D. (1996). *Justice, Nature and the Geography of Difference.* Oxford: Blackwell.

Harvey, D. (2010). L'espace comme mot clé. In D. Harvey, *Géographie et capital: Vers un matérialisme historico-géographique* (pp. 53–82). Paris: Editions Syllepse.

Holling, C. S. (2001). Understanding the complexity of economic, ecological, and social systems. *Ecosystems,* 4(5), 390–405.

Hughes, Z. (2011). Tourism and climate impact on the North American eastern seaboard. In A. Jones & M. Phillips (Eds.), *Disappearing Destinations: Climate Change and Future Challenges for Coastal Tourism* (pp. 161–76). Wallingford: CABI international.

IPCC (2007). *Climate Change 2007: Synthesis Report.* Geneva: Intergovernmental Panel on Climate Change.

IPCC (2014a). *Climate Change 2014: Synthesis Report.* Geneva: Intergovernmental Panel on Climate Change.

IPCC (2014b). *Climate Change 2014: Synthesis Report: Summary for Policymakers.* Geneva: Intergovernmental Panel on Climate Change.

Kaján, E. & Saarinen, J. (2013). Current Issues in Tourism Tourism, climate change and adaptation: a review. *Current Issues in Tourism*, 16:2, 167–95.

Klein, J. L. (2008). Territoire et développement: du local à la solidarité territoriale. In G. Massicotte (Dir.). *Sciences du territoire: perspectives québécoises* (pp. 315–34). Quebec: Presses de l'Université du Québec.

Lapointe, D., B. Sarrasin & A. Guillemard (2015) Changements climatiques et mise en tourisme du fleuve St-Laurent au Québec, *VertigO: la revue électronique en sciences de l'environnement*, Hors-série 23, November. Retrieved October 12, 2016, from http://vertigo.revues.org/16575.

Lefebvre, H. (1974). *La production de l'espace.* Paris: Anthropos.

Lépy, E., H.I. Heikkinen, T.P. Karakalainen, K. Tervo-Kankare, P. Kauppila, T. Suopajarvi, … A. Rautio (2014). Multidisciplinary and participatory approach for assessing local vulnerability of tourism industry to climate change. *Scandinavian Journal of Hospitality and Tourism*, 14:1, 41–59.

Lew, A.A., P.T. Ng, C-C. Ni & T-C. Wu. (2016). Community sustainability and resilience: similarities, differences and indicators. *Tourism Geographies*, 18(1): 18–27.

Liu, J. G., T. Dietz, S. R. Carpenter, C. Folke, M. Alberti, C. L. Redman, S. H. Schneider, E. Ostrom, A. N. Pell, J. Lubchenco, W. W. Taylor, Z. Y. Ouyang, P. Deadman, T. Kratz & W. Provencher (2007). Coupled human and natural systems. *Ambio*, 36(8): 639–49.

Lozato-Giotart, J.-P. (2003). *Géographie du tourisme: de l'espace consommé à l'espace maîtrisé.* Paris: Pearson Education.

MRC de Rivière-du-Loup (2012). *Schéma d'aménagement et de développement révisé.* Retrieved March 28, 2015, from http://riviereduloup.ca/mrc/?id=e2796&a=2013#schema

Müller, D. K. & Jansson, B. (2006). The difficult business of making pleasure peripheries prosperous: perspectives on space, place and environment. in D.K. Müller & B. Jansson (Dir.). *Tourism in Peripheries: Perspectives from the Far North and South* (pp. 3–19). Cambridge: CABI.

Nicholls, M. (2014). *Climate Change: Implications for Tourism.* Cambridge: CISL.

Notre-Dame-du-Portage (2010) Info-portage, July 2010. Notre-Dame-du-Portage : municipalité de Notre-Dame-du-Portage.

Ostrom, E. (2009). A general framework for analyzing sustainability of social-ecological systems. *Science*, 325 (5939): 419–22.

Ouranos (2014). *Sommaire de la synthèse des connaissances sur les changements climatiques au Québec.* Montreal : Ouranos

Overton, J. (2010) The consumption of space: land, capital and place in New Zealand wine industry. *Geoforum*, 41, 752–62.

Plante, S. (2011). Les défis de la gestion intégrée des territoires côtiers et riverains du Saint-Laurent. In O. Chouinard, J. Baztan & J.-P. Vanderlinden (eds.) *Zones côtières et changement climatique: le défi de la gestion intégrée* (pp. 100–117). Quebec: Presses de l'Université du Québec.

Plante, S., O. Chouinard & G. Martin (2011). Gouvernance participative par l'engagement citoyen à l'heure des changements climatiques: Études de cas à Le Goulet, Pointe-du-Chêne et Bayshore Drive (Nouveau-Brunswick). *Territoire en mouvement: Revue de géographie et aménagement*, 11. Retrieved from http://tem.revues.org/1234.

Quebec (2012). *Stratégie gouvernementale d'adaptation aux changements climatiques 2013–2020: Un effort collectif pour renforcer la résilience de la société québécoise*, Quebec: Gouvernement du Québec. Retrieved from www.mddelcc.gouv.qc.ca/changements/plan_action/stategie-adaptation2013-2020.pdf.

Saarinen, J. (2014). Nordic perspectives on tourism and climate change issues. *Scandinavian Journal of Hospitality and Tourism*, 14(1), 1–5.

Scheffer, M., S. Carpenter, J. A. Foley, C. Folke et B. Walker. (2001). Catastrophic shifts in ecosystems. *Nature*, 413 (6856), 591–96.

Scott, D., C. de Freitas & A. Matzarakis (2009). Climate change adaptation in the recreation and tourism sector. In K. Ebi, I. Burton & G.R. McGregor (Eds.) *Biometeorology for Adaptation to Climate Variability and Change* (pp. 171–94). New York: Springer.

Shaw, J., R.B. Taylor, D.L. Forbes & S. Solomon (1998). *The Sensitivity of the Coasts of Canada to Sea-level Rise*. Ottawa: National Resources Canada.

Simpson, M.C., S. Gössling, D. Scott, C.M. Hall & E. Gladin (2008). *Climate change adaptation and mitigation in the tourism sector: frameworks, tools and practices*. Paris: UNEP, University of Oxford, UNWTO, WMO.

Smit, B. & J. Wandel. (2006). Adaptation, adaptive capacity and vulnerability. *Global Environmental Change-Human and Policy Dimensions*, 16(3): 282–92.

Smith, N. (1984) *Uneven Development: Nature, Capital, and the Production of Space*. Athens, Georgia: The University of Georgia Press.

Soja,E. (1989). *Postmodern Geographies the Reassertion of Space in Critical Social Theory*. London: Verso.

SRC (2014). *Applying Resilience Thinking. Seven Principles for Building Resilience in Social-Ecological Systems*. Stockholm: Stockholm University, Stockholm Resilience Center.

Tyler, S. R. (2006). *Comanagement of Natural Resources: Local Learning for Poverty Reduction*. Focus, International Development Research Centre, Ottawa.

Walker, B. H. & D. Salt. (2006). *Resilience Thinking: Sustaining ecosystems and people in a changing world*. Washington, DC: Island Press.

10 A resilience approach to collaborative tropical reef conservation on Gili Trawangan, Indonesia

L. Arifin Bakti and Alan A. Lew

Introduction

Resilience thinking uses a systems approach to organizing social-ecological phenomena to understand how and why things change over time. This approach provides the foundation for the concepts of panarchy (a hierarchy of nested systems and subsystems) and the adaptive cycle (modeling the cyclical stages of system changes over time). Fishing is a common livelihood option among island and coastal populations across Southeast Asia, and is often combined with tourism in more accessible destinations. Both fishing and tourism are dependent on healthy coral reef ecosystems, which can be threatened by local human activities and global climate change. Local, regional, national, and international interests (systems) have been involved in coral reef conservation and restoration efforts on the Indonesian island of Gili Trawangan since at least the early 1980s, when fishing was the dominant livelihood and tourism was in its early stages of development. Today, Gili Trawangan is among the most popular scuba diving destinations in Indonesia. The historical interactions among the different interest groups involved makes it a good example of how a panarchy of systems forms and functions in response to threatened coral reefs. The various interest groups involved in this process operate at different scales and moved through their own adaptive cycles as they collaborate and shape a coral reef conservation system for the island. Understanding the historical evolution of the goals, capabilities, and functions of community collaborations on Gili Trawangan from a resilience systems perspective provides valuable insights into the dynamics of tropical coral reef conservation in a developing country context, as well as lessons for other areas facing similar challenges.

Gili Trawangan, Indonesia

Gili Trawangan is the largest in land area (340 ha) and population of three small coral islands that comprise the village of Gili Indah (population 3,586 in 2010; GIV, 2010) (Figure 10.1). The islands are part of the much larger island of Lombok, and is situated off its west coast, facing the island of Bali. Gili Trawangan has a mostly permanent population of about 2,000 people. They are mostly Muslims, although the growth of tourism has brought migrants from around Indonesia and

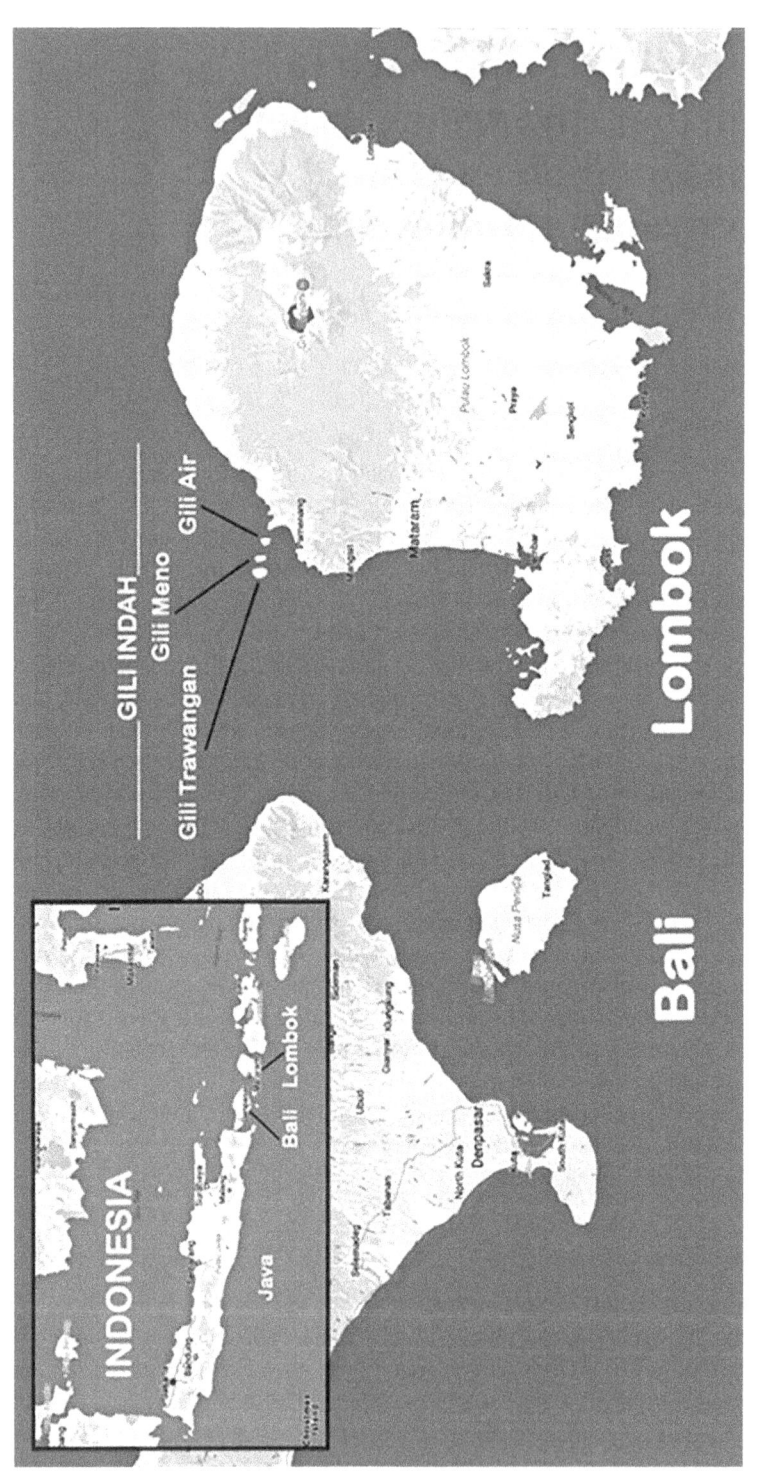

Figure 10.1 Map of Gili Trawangan, Indonesia
Source: Map Data © 2017 by Google, used with permission

abroad, resulting in significant minority populations of Christians, Hindus, and Buddhists. Gili Indah's residents today work in private businesses (53.6%), almost all of which are tourism related, and in fishing (16.3%) (GIV, 2010).

Gili Indah's tourism began on Gili Air, which is closest to the main island of Lombok, but Gili Trawangan is now its most important tourism island. It has experienced many of the characteristics of the Tourism Area Life Cycle (TALC) (Butler, 1980), starting with the TALC stages of exploration (discovery by tourists) and involvement (initial local participation and local organization) from the mid-1970s through the mid-1980s. During this time, the island was governed by traditional rules (*adat*), formulated and agreed upon by the local population, most of whom were ethnic Sasak (from Lombok) and Bugis (mostly from Sulawesi) who worked as farmers and fishermen. These locals were the first to respond to the arrival of backpacker tourists and were the pioneers of small-scale tourism on Gili Trawangan (Hampton, 1998). Their knowledge of the local coral reef ecosystem was very intimate and they maintained a traditional and subsistence lifestyle in the way they developed their businesses, and, generally, the way they maintain the integrity of their social, cultural, and political life in their everyday practices.

The TALC stage of development (equivalent to the growth phase of the adaptive cycle) began in the early 1990s, at which time the economic potential of Gili Trawangan as a diving and beach destination became recognized by outside investors, including businessmen and national and local politicians. In 1992, Indonesia's central government, ruled by the long-time dictator Suharto, sent military forces to evict the residents of Gili Trawangan at the behest of tourism developers (Kamsma & Bras, 2000). The violent act was not successful and residents eventually rebuilt their businesses, while others rented their lands to tourism investors.

In the late 1990s and early 2000s, as Indonesia began its transition into one of the world's largest democracies, Gili Trawangan was entering the consolidation phase of the TALC and adaptive cycle models. Migration to Gili Trawangan increased the island's population from about 400 in 1996 (Hampton, 1998) to 1,089 in 2003 (Satria, Matsuda & Sano, 2006). During that time, many new hotels, bungalows, restaurants, and tourism facilities were quickly built without any planning or regulatory frameworks. This approach was exacerbated by national policies favoring foreign investors' interests over the authority of local communities to manage their development (Silver, 2005). The first fast boat ferries from Bali started arriving in Gili Indah at this time, along with investments in large, upscale resorts and a rapid increase in the number of tourists and outside investors (both Indonesian and foreign). This resulted in a dramatic change in the island's demographics and economy (Vaisutis et al., 2007). Traditional agriculture largely disappeared in the early 2000s, as land was more valuable for tourism and property development speculation. Due to the small size of the island and its growing tourism orientation, the community came to a general agreement, which later become an *awig-awig* regulation (see p. 170 below), to limit transportation to horse-drawn carts and bicycles, and to ban motor vehicles to both ensure safety and reduce air pollution.

By the mid-2010s, most of Gili Trawangan's land base was used for tourism (80%), while tourist numbers grew from 54,957 in 2009 to 503,341 in 2014

(BPS, 2015). Access to Gili Trawangan from Lombok takes 25–30 minutes by regular boat and 10–15 minutes by fast boat. From Bali, about ten fast boats daily make a two- to three-hour trip across the Lombok Strait. For tourists, the fast boat service from Bali has provided a rapid and cheap way for them to visit Gili Trawangan, Lombok, and the province of Nusa Tenggara Barat (West Nusa Tenggara, or NTB). Many tourists come for snorkeling and scuba diving, resulting in 27 certified dive shop operations on Gili Trawangan alone (as of 2015).

Foreign investment in tourism businesses has helped to facilitate the island's economic growth, along with the absence of zoning or other land use controls along the island's coastal areas (Graci, 2013). The Tourism Promotion Agency of North Lombok estimates that, as of 2015, Gili Trawangan had 128 lodging rooms, and as many as 87 homestay rooms (BPS, 2015). During peak periods, additional rooms that are normally not used for tourists are temporarily made into accommodations. Numerous travel agencies operate in Gili Indah to facilitate tours beyond the islands, with Komodo National Park being a popular multi-day boat trip. Other tourist facilities include restaurants, cafes, and bars; spas and salons; fashion and art shops; and banks, money changers, and ATMs.

Decline in the coral reef ecosystems of Gili Trawangan

As a complex, interconnected system, tourism is vulnerable to external disturbances that are both environmental and social in origin (Russell & Faulkner, 2004). If a tourism system has a low level of resilience, it may be vulnerable to having small changes in one factor causing it to cross a threshold or tipping point between a desired state and an undesirable state (Gladwell, 2002). Such a change may occur in an abrupt and unexpected manner, resulting in a surprise response that allows little time for thoughtful action. Resilience thinking suggests that practices that embrace the complexity of multiple, cross-scale interactions can enhance resilience even when change is unanticipated or unplanned for (Berkes, Colding & Folke, 2003; Walker & Salt, 2006).

Consolidation of the tourism industry on Gili Trawangan in the 2000s resulted from a period of dramatic tourism growth which, accompanied by poor planning and land use monitoring, led to significant disruptions to the island's social-ecological system, especially its coral reefs. Several external drivers of change have impacted Gili Trawangan's coral reef ecosystem, including overfishing and illegal fishing practices, increasing numbers of tourists, warming ocean temperatures, and strong tropical storms. The island's fishermen have seen declining fish catches since at least the early 2000s, caused by their own overfishing and illegal fishing practices that have also damaged the coral reefs. Tourism-related impacts on coral reefs come from boats that drop their anchors carelessly and the release of sewage from the island into the surrounding seas. These impacts are more likely to affect coral populations that are close to the island's shoreline, where large numbers of boats are coming and leaving on a regular basis. The areas of Gili Trawangan that are predominantly visited by tourists for sunbathing and swimming are the eastern and southern coasts. Some parts of the eastern coast are often crowded with swimmers and sunbathers, with densities reaching 100 people per 100 m². It is also in this area where snorkelers are in large numbers and are more likely to step on shallow corals and kill them. It is these near-shore coral reefs that have been the most damaged in Gili Trawangan (and also Gili Meno and Gili Air). Further out from the shoreline, unskilled scuba divers may also step on coral and kill it, which is made worse by dive groups that exceed suggested capacities during the high summer and winter seasons.

Global warming is another driver of change that has caused a loss of the island's coral reefs. Rising ocean temperatures cause the health of the coral ecosystem to decrease slowly, until a threshold is reached where coral bleaching occurs across large portions of the reef. This condition transforms the coral-dominated system into an algae-dominated system, because algae, which has always been present, is more resilient and stable under a higher temperature regime (Walker et al., 2012). The most serious coral bleaching in recent decades occurred during the intense El Nino of 1997–8 (GET, 2010; 2012; Kartawijaya et al., 2013). Strong tropical storms also damage coral reefs, although the recovery period is usually shorter in those occurrences.

Gili Trawangan's community, which today consists of both foreigners and local people, has become increasingly aware of the environmental damage that has accompanied increasing numbers of tourists and levels of tourism development (Bakti, Robb, Virgota & Goreau, 2008). Because the island is small enough (7.5 km in circumference), residents know one another and can discuss their shared problems and needs face to face. For that reason, they may be more ready to take responsibility and participate in efforts to address those challenges. Collaboration across many different scales and involving different stakeholders has been integral to the island's efforts to conserve and restore its coral reef resources. This collaboration has also demonstrated the resilience of the community of Gili Indah overall, but especially Gili Trawangan. To understand this process requires an understanding of resilience theory, with particular attention to the adaptive cycle model and the panarchy model.

Resilience and the adaptive cycle

The resilience thinking framework allows for community leaders and policy makers to develop and assess measures that point to the level of resilience within a social system, and to adopt policies that seek to enhance the resilience of those variables. Holling and Gunderson (2002, p. 28) define resilience as "the magnitude of disturbance that can be absorbed before the system changes its structure by changing the variables and processes that control behavior." The resilience conceptual model comprises numerous distinct scales with cross-scale connections that form a multilevel and hierarchical social-ecological system (SES) structure. An SES, therefore, exists across multiple levels and time frames (Walker & Salt, 2006) and is able to absorb a spectrum of shocks and disturbances, as it maintains its capacity for renewal, reorganization, and development (Walker et al., 2002; Folke, 2006). Resilient systems also have the potential for multiple metastable regimes, rather than a single equilibrium, and even though change may occur within any one regime, the set of dynamically important variables and interactions remains fixed within each regime (Holling & Gunderson, 2002).

The essence of resilience theory includes the following components and assumptions about an SES (Walker & Salt, 2006):

- External change drivers are disturbances that are exogenous to the SES, such as climate conditions, non-local governance systems, and external economic factors.
- Fast moving internal variables are those that change quickly when stressed, such as air quality, water supply, and gasoline prices, and are often referred to as "disasters" or "crises" when they change too quickly.
- Slow moving internal variables, also known as controlling variables, are those that remain more stable under pressure, such as soil fertility conditions (for agriculture), lower food chain organisms, religious and cultural traditions, and monetary systems, and are usually seen as "annoyances" when they change because their shifts occur over a long period of time.
- Both fast and slow variable changes can occur in non-linear patterns, making them difficult to predict (Walker et al., 2002).

Based on these fundamental building blocks and assumptions, Walker and Salt (2006) developed a framework to provide a way of thinking in which a system can sustain itself by:

- retaining control over its core essential functions and structures;
- being capable of self-organization under changing conditions; and
- building and enhancing its future capacity for learning and adaptation.

The aim of resilience thinking is to understand how nature and humans operate together in complex adaptive systems to achieve these goals (Walker et al., 2002; Folke et al., 2002; Allen et al., 2011; Folke, 2016).

Another key consideration in resilience theory is that changes in an SES reflect movement through the "Adaptive Cycle" (Figure 10.2). Different systems move through this cycle at different scales and time. Holling and Gunderson (2002) suggest that most, although not all, systems follow a four-phase cycle:

- *a rapid growth or exploitation phase* [r] wherein new resources and opportunities are discovered and exploited;
- *a consolidation and conservation phase* [K] in which the system matures, consolidates, and become rigid against change;
- *an energy release or collapse phase* [Ω] where the stress of conservation against increasing levels of change leads to a collapse of the system's structures; and
- *a reorganization and renewal phase* [α] during which a less structured system context allows innovations, adaptations, and eventually new opportunities for exploitation and development to emerge (Holling & Gunderson, 2002; Walker & Salt, 2006).

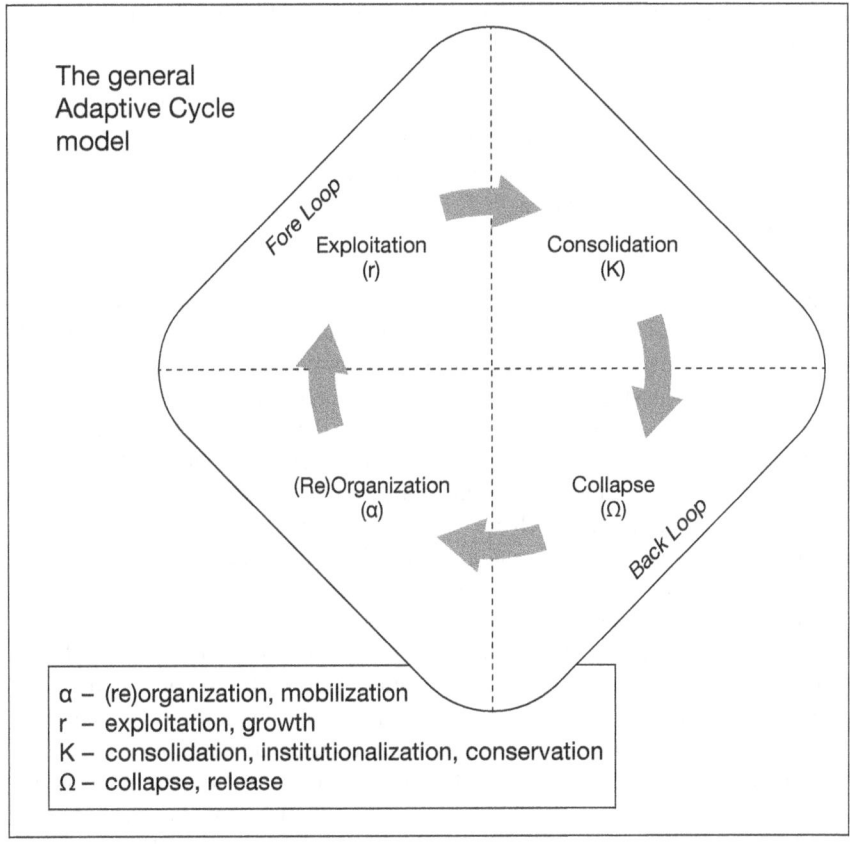

The general
Adaptive Cycle
model

Fore Loop

Exploitation
(r)

Consolidation
(K)

(Re)Organization
(α)

Collapse
(Ω)

Back Loop

α – (re)organization, mobilization
r – exploitation, growth
K – consolidation, institutionalization, conservation
Ω – collapse, release

Figure 10.2 The general Adaptive Cycle model
Source: Lew, Ng, Wu & Ni, 2016

The growth [r] and consolidation [K] phases emerge from standard ecological theory, in which an ecosystem's growth [r] stage is controlled by colonizing species tolerant of environmental variation which then transitions to the consolidation [K] stage, which is dominated by the species adapted to a narrower range of variation (Holling & Gunderson, 2002). The other phases consist of the collapse (or energy release) stage [Ω] and the reorganization stage [α] and are typically briefer events. For example, in the context of a forest, a collapse [Ω] could be a wildfire or insect outbreak that releases biomass nutrients, while the reorganization [α] phase could include an adjustment to a loss or change in soil nutrients. The phase that leads into a new adaptive cycle, begins with completion of the reorganization [α] and the start of the growth [r] state. Not all systems move through all stages of the adaptive cycle, with the collapse phase often being skipped though planned reorganization (Walker, et al, 2006). On the other hand, some systems may be "trapped" in a collapsed cycle, never being able to escape degraded or impoverished conditions (Allison & Hobbs, 2004).

Through this adaptive cycle, three main variables oscillate in influence. Holling (2001) referred to these as:

- *resilience* (characterized by a high capacity for innovation and adaptation);
- *potential* (having a high capacity for change due to accumulated resources, such as biomass or human social capital); and
- *connectedness* (providing a high capacity for control and management to guide future directions) (Table 10.1).

Innovative resilience decreases in the consolidation stage, due to its high degree of control. It reaches its lowest point in the collapse phase, but then increases through reorganization as new opportunities for innovation emerge, and is highest in the growth phase. Connectedness moves in a roughly opposite direction to resilience, being highest in the consolidation stage, decreasing through collapse, and reaching its lowest point in the reorganization phase before increasing again in the growth phase. Potential is highest when an accumulated wealth in term of natural or human resources is most available, which is in the reorganization and consolidation stages; potential is lowest in the transitional growth and collapse stages.

Collaboration is strongest in the two high potential phases (see Table 10.1), although it is more likely to be associated with resilience in the (re)organization stage when collaborators are working to find innovative solutions to a clearly defined problem (a collapse). In the consolidation stage, collaboration is undertaken to conserve existing institutional structures, leading to system rigidity and increasing vulnerability. Collapse, however, can be avoided if the collaborators recognize their vulnerability and the system is adequately reorganized to return to a growth scenario.

The adaptive cycle is a useful metaphor to understand a complex world of rapid change in terms of resilience, potential, and connectedness within a system (Holling & Gunderson, 2002). These insights suggest potential management interventions that can enhance the system capacities to bounce back following a disturbance, to better manage the impacts of external drivers of change on the

Table 10.1 The four stages and three variables of the Adaptive Cycle.

Adaptive Cycle		*VARIABLE*		
STAGE		*Resilience*	*Potential*	*Connectedness*
1. (Re)Organization	α	Increasing	High	Low
2. Exploitation; Growth	r	High	Low	Increasing
3. Consolidation; Conservation	K	Decreasing	High	High
4. Decline; Collapse	Ω	Low	Low	Decreasing

Notes: *resilience* = capacity to innovate and adapt
       potential = capacity to change using accumulated resources
       connectedness = capacity to control and manage
Source: Lew, 2017; based on Holling (2001) and Allison & Hobbs (2004)

system, and to work to avoid a system collapse by periodic and planned structural reorganizations (Folke 2006; Walker & Salt, 2006). In these ways, institutions and social networks can better manage both the human and the biophysical domains of the system, and focus on drivers of change and the slow changing variables that may create vulnerability to the system in the future.

Spatial hierarchy in the adaptive cycle consists of multiple SES relationships and influences that extend across scales, and is also known as the panarchy model of resilience (Holling 2001; Gunderson & Holling, 2002). Ecological and social-ecological systems form nested sets of systems and subsystems, each with its own adaptive cycle responses to varying drivers of change (Gotts, 2007). In most cases, the larger, slower cycles are associated with slow, controlling variables and restrict the smaller, faster cycles, thereby helping to maintain system integrity. Critical cross-scale interactions can operate throughout the phases of collapse and reorganization in the form of "revolt" connections, where a collapse on one level can trigger a crisis or collapse in a higher- or lower-level system, and "remember" connections, where the reorganization phase of one system is regulated or moderated by the conservative phase of a higher-level system. Holling et al. (2002) outline three types of changes that can occur within panarchies:

- "incremental" change in the growth and consolidation phases as the system moves toward maturity;
- "abrupt" changes in often unanticipated transition from consolidation through the collapse and reorganization phases; and
- "transformational learning," which refers to changes in the interaction among various sets of unstable variables, and which may affect some systems more than others in unpredictable ways.

Tourism applications

Farrell and Twining-Ward (2004) outline a tourism panarchy model following Gunderson and Holling's (2002) hierarchical nesting of one system level within another. In a tourism panarchy model, the core of a local tourism system is shown as being part of the larger regional tourism system that contains other, parallel local tourism systems. The regional tourism system is similarly embedded in a hierarchy of a larger tourism system leading to the global tourism system. It is also important to recognize that a tourism system, no matter the hierarchical level, is deeply embedded in a social-ecological system that includes many non-tourism elements, along with many other parallel systems. This model creates a comprehensive and complex tourism system (Baggio, 2008), covering social, economic, geopolitical, and ecological components, along with the processes and functions that complement the totality and are essential for sustainability (Lew & Hall, 1998). These systems and processes are related to one another, although some linkages are much stronger than others. We usually use our recognition of these stronger connections to define the core and periphery of different systems, including the tourism system. Debates over how to define the tourism system (Leiper, 1990), and even whether such a system exists (Leiper, 1982), demonstrates the challenges and flexibility of a systems approach to social science research. Even the Tourism Satellite Accounting System, which is used by many countries to monitor their tourism economies, is not fixed, but must be redefined by each country separately to meet the nuances (and politics) of their tourism industry (Hall & Lew, 2009).

As noted in Table 10.1, the level of resilience in a system changes over time through the adaptive cycle, which is an integral part of the panarchy model (Holling & Gunderson, 2002; Walker & Salt, 2006). As others have pointed out (Farrell & Twining-Ward, 2004; Hamzah & Hampton, 2012), the four-phase adaptive cycle, equates to different stages in Butler's (1980) Tourism Area Life-Cycle (TALC) model (Figure 10.3). A young, newly emerging, tourism system [α] enters a phase of growth [r] (equated to the exploration, discovery, and early development stages in TALC), as it responds to the potential for new kinds of relationships that have a higher function. As tourism investments increase, the system grows, more energy is created, and relationships become more permanent [K] (equated to the advanced development and consolidation stages in TALC). The rigidity of the consolidation stage causes a drop in the resilience of the system as a level of complacency sets in based on assumptions of past momentum. The system's capacity to respond to change and absorb shocks, therefore, may not keep pace with changes in the tourism market or possibly in its tourism product. This can lead to a phase that includes the release of suppressed and pent-up

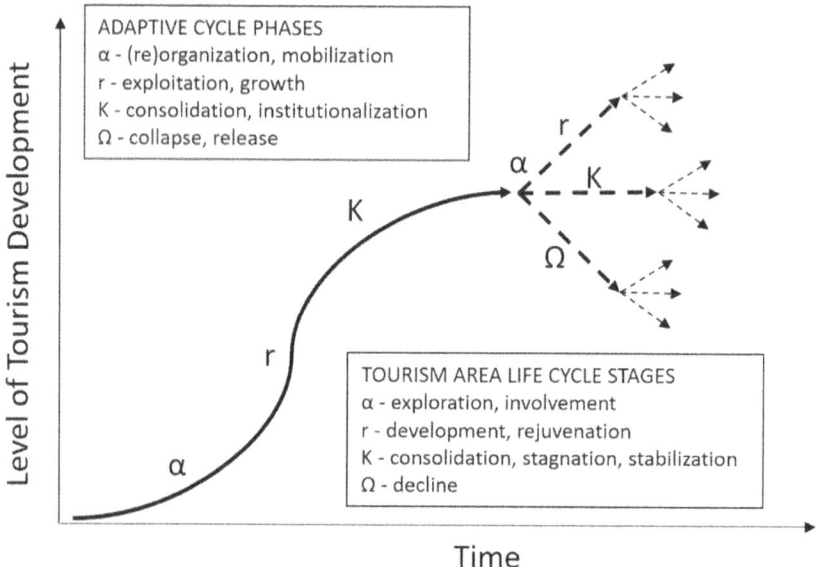

Figure 10.3 The Tourism Area Life Cycle (TALC) model showing comparable stages in
the Adaptive Cycle model
Source: Lew, 2017 (based on Butler, 1980, and Holling, 2001)

energies (possible social instabilities) and a collapse in tourist arrivals [Ω] (referred to as stagnation or decline in TALC). This collapse is a "destabilizing event" and the dissipated energy is then recycled back into a reorganization phase (post-stagnation in TALC), which is hopefully followed by a new phase of change (exploration, discovery in TALC) and growth (development in TALC).

McKercher's (2005) reflection on the TALC recognized that destinations behave like living ecosystems rather than discrete products. Tourism destinations are not distinct entities but multi-agent, multilevel, multi-dimensional social-ecological systems that are in constant states of evolution (Davoudi, 2012). From this perspective, the resilience of a tourism system would be its capacity to effectively evolve to maintain its core functions at a desired level, while also recognizing its relationships in a panarchy of parallel and hierarchical subsystems. The adaptive cycle "loop," along with the "revolt" and "remember" influences between subsystems, captures this complexity and goes further than the TALC in recognizing the agency of individuals where significant change to one part of the tourism panarchy system can have unpredicted consequences on other parts of the system (Folke, 2016). Panarchy also shows how diverse subsystems may be in different adaptive cycle phases at the same time, and could consequently influence each other in unpredictable and non-linear ways.

Systems and subsystems on Gili Indah

A basic principle of resilience theory is that change occurs at different spatial and temporal scales (Carpenter & Gunderson, 2001; Lew, 2014). Spatial scale refers to the location of a specific system (or subsystem) within the nested hierarchy of larger and smaller systems. The Gili Trawangan community is a subsystem (subvillage) within the Gili Indah Village, which itself is a political subsystem that is nested within a larger district of North Lombok, which is part of the West Nusa Tenggara provincial system. The community of Gili Trawangan contain several smaller subsystems, including a dive operator system, hotel and restaurant system, educational system, and local transportation system. Parallel systems for each of these exist on each of the three islands of Gili Indah. A community tourism system in Gili Trawangan could consist of a tourism services subsystem (accommodation, restaurants, diving operator, banking, and travel services), a tourist attraction subsystem (commercial and non-commercial attractions), a tourism marketing subsystem (promotional and social media), and a tourism governance subsystem (policies and regulations impacting tourism development).

The specific resilience of any one of these subsystems can be assessed in relation to a specific external driver of change (Folke, 2016). In contrast, general resilience is a broader concept that seeks to assess the overall resilience of a system (which also implies the panarchy of its subsystems). Thus, we can seek to understand the general resilience of the small island system of Gili Indah (Figure 10.4), which includes its three island subsystems and all social and environmental subsystems that cut across the three islands. If the focus is on Gili Indah, then the external driver(s) of change that is examined should cut across the entire island

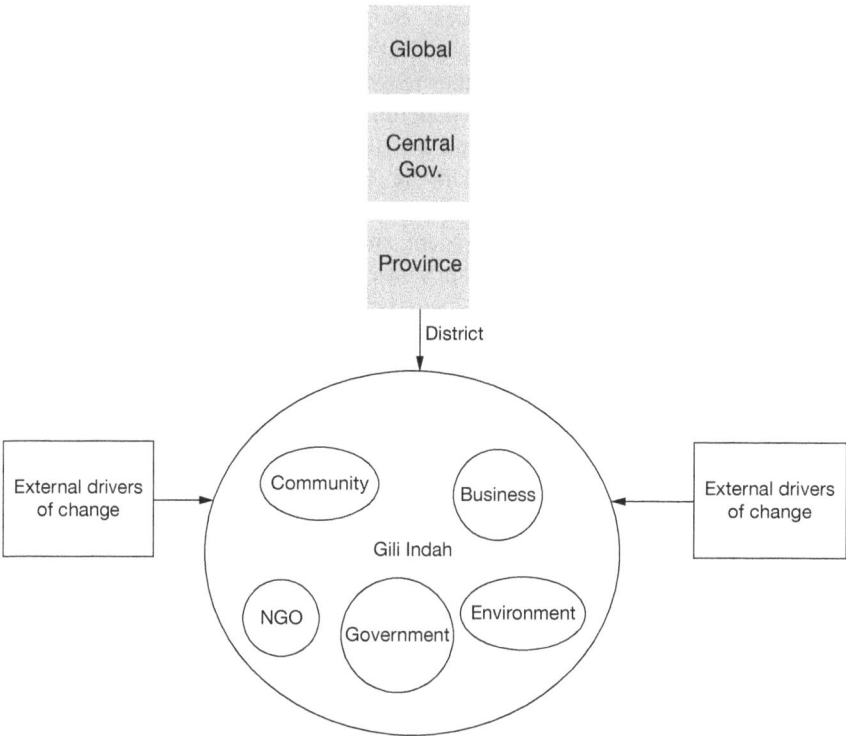

Figure 10.4 A small island tourism panarchy of systems and subsystems, and external
influences
Note: The shaded boxes are higher-level systems that are also external drivers of change.

group, and the internal variables that respond through the adaptive cycle should be ones that can be assessed at the island group level. If an external driver or internal response variable only applies to one of the three islands, then it is not appropriate to use at the Gili Indah archipelago level. Similarly, if the island of Gili Trawangan is the system of interest, then the external and internal variables must also be appropriate for that level of analysis. Of course, the system is an open one that also undergoes cross-scale interactions, with influences from both higher and lower systems in the panarchy hierarchy, so interactions with other systems cannot be ignored (Gunderson & Holling, 2002). But the variables of analysis must be properly recognized and applied at the correct scale.

The core of an island tourism panarchy model (Farrell & Twining-Ward, 2004) consists of an assemblage of structures, services and resources directly contributing to the tourism sector, as well as significant social, economic, geopolitical, and ecological components, processes and functions that shape development paths, resource use, and quality of life. Despite their modest size, small islands are characteristically complex social-ecological systems due to their diverse and often unpredictable institutional arrangements, their resource or economic dependency, and the tension between managing the conservation of marine ecosystems and developing them for tourism. The local complexity in the Gili Indah system in these areas is further impacted by external influences from regional, national, and even global governance and economic systems. The global system, for example, brings international tourists and international investors and business people to the islands, as well as crises such as the Asian economic crisis (1997), the Bali terrorist bombing (2002), and global warming (resulting in coral bleaching).

In the panarchy model, the lower levels (such as Gili Indah) are only semi-autonomous, being impacted more by higher levels than vice versa. Higher-level systems are slower moving and usually unaffected by many lower level disturbances. Change occurs at different spatial scales and across temporal scales (Lew, 2014), with change drivers that impact higher levels in the hierarchy forcing their impacts to lower levels (Carpenter & Gunderson, 2001). A regional economic crisis, for example, can influence the spatial systems ranging from the national to the provincial, and from an entire city to an individual tourism business (Hillmer-Pegram, 2014). These impacts can potentially move in all directions both within and across spatial and temporal scales. Because of these influences, it is necessary to look at the larger context of a tourist destination system (such as, Gili Indah or Gili Trawangan) to fully understand its social, economic, and environmental changes (Hall & Lew, 2009).

Collaboration for social-ecological resilience on Gili Trawangan

For Gili Trawangan, the island system and the tourism system are essentially identical. While not all residents are directly involved in tourism activities, they are all impacted by tourism due to its overwhelming presence. One could argue that coral reef degradation from over use is an obvious potential impact of excessive tourism development on a small tropical island, which should have been anticipated and planned for. Such an approach (planning forethought) would be seen in a continuous learning and adaptation strategy that is deeply integrated into the social structure and continually evolving (evolutionary resilience) (Davoudi, 2012). While this is more evident in developed economies with more extensive resources, even there it can sometimes fail, as was seen when Hurricane Katrina hit New Orleans in 2005.

Threats to the coral reef system are examples of slow changes that have built over time, and were not so evident to the local population in this developing society with its very limited financial and technical resources. Modest attention was paid to these issues until various external pressures pushed them to a crisis threshold (Ω collapse) that was visibly apparent to enough residents and community leaders to force them into action (α reorganization).

Gili Trawangan's island system has attempted to self-organize itself to address these challenges. Because of the relatively weak local government, the island's response has mostly been through collaborative efforts organized by local leaders and involving multiple stakeholders, including owners of dive shops, hotels, and restaurants, as well as the *Kadus* (subvillage head), the *Kades* (village head), and local fishermen. Such collaboration is a direct reflection of the self-organizing characteristic of a well-integrated system. On Gili Trawangan, this is possible because:

- there is a broad awareness of the huge impacts, mostly seen as beneficial, that tourism growth has brought to the island;
- the island is small enough that people know each other and communicate regularly; and
- there is a mix of Indonesian and immigrant populations from around the world who bring different skills to address the challenges they face.

Several strategies have been employed to meet the challenges that Gili Trawangan has faced since the island entered the early consolidation phase of development in the early 2000s. These have ranged from protection and conservation programs to recovery and restoration efforts. They have been largely self-organized initiatives that have included significant community participation, including both indigenous and foreign residents and business people (Bakti et al., 2008; GET 2010, 2012; Robbe et al., 2011; Bakti, Goreau, Robbe & Virgota, 2010). The initiatives below have been undertaken with almost no government financial support (extremely limited involvement of higher level national systems), having relied mostly on local volunteers (lower-level subsystems), some international aid sponsors (global-level systems) and some local government (including university) involvement (higher level provincial systems).

Collaborating to regulate coral reef activities

Upheavals of the reform era (*reformasi*) in Indonesian politics lasted from 1998 through much of the 2000s, and resulted in a series of crises, including financial and political, that stressed the ability of local governments to properly provide needed services to communities. One of the major problems before and during this period was a lack of any regulations governing the protection of marine resources. Under the Suharto regime, fisheries and marine resources were a small division under the Ministry of Agriculture. It were underfunded and lacked authority to regulate the large and complicated seaways of Indonesia. As a result, provincial and local authorities were also unable to regulate the exploitation of marine resources. The Coral Reef Rehabilitation and Management Program (COREMAP) Indonesia is a program funded by the World Bank, the Asian Development Bank, and USAID since the mid-1980s to assist Indonesia in managing its coral reef ecosystems. In 2000, COREMAP was working to zone the waters of Gili Indah into areas for conservation, recreation/tourism use, and fishing. As part of this process, they suggested that the village adopt a coral reef *awig-awig* (Satria, Matsuda & Sano, 2006). *Awig-awig* (or awik-awik) is a traditional system of rules that originated in the cultural practices of Bali and was later introduced to Lombok (MCA-Indonesia, 2013). It had never actually been part of Gili Indah's traditional culture, but with many Sasak immigrants from Lombok, it was considered culturally appropriate, and was therefore adopted by the local Gili Indah leadership.

The *awig-awig* for coral reef management on the Gili Indah islands was adopted in 2001 by Gili Indah Village Decree (Surat Keputusan Desa No. 12/ Pem.1.1/06/2001) specifically to protect traditional fisheries and live coral reefs from the increasing use of destructive fishing practices, and to maintain the fishing culture of the islands. It was created through consultation with the elected *Kades* and *Kadus*, without involvement of the fishing community, as was the normal practice among government authorities at that time (Schellhorn, 2010). Fishing activities, however, were the main target of the regulations, which were locally seen to benefit tourism more than fishing (Satria et al., 2006). Net fishing and seaweed collection were forbidden in areas where coral reefs were abundant. In these protection zones, snorkeling and diving were (and are still) the only permitted activities. Buffer zones surrounding the core coral reef areas allowed diving and snorkeling, along with line fishing (angling). Exploitation zones were areas where coral reefs were limited and in which most activities are allowed, including snorkeling, diving, seaweed cultures, pearl cultures, net fishing, angling, and spear fishing. Destructive fishing practices, such as bombing and the use of poisons, were highly prohibited everywhere, even though the local fishers had practiced these methods in the past. Using the *awig-awig* regulations, the local community in Gili Indah demonstrated a degree of self-organization and self-governance over their marine resources, through the political instability of Indonesia's reform era.

The reorganization of Indonesia's governance system since the mid-2000s has been accompanied by new initiatives that have impacted marine protection in Gili Indah's waters, eventually replacing the *awig-awig* with more formal regulations. In 2007 the national government revised and approved its laws on the management of coastal areas and small islands (Undang-Undang Republik Indonesia Nomor 1 / 2014 Tentang Perubahan Atas Undang-Undang Nomor 27 / 2007 Tentang Pengelolaan Wilayah Pesisir dan Pulau-Pulau Kecil). Based on this, the Province of West Nusa Tenggara (NTB) approved regional regulations on the management of coastal and marine areas in small islands in 2008 (Peraturan Daerah Nomor 2 / 2008 tentang Pengelolaan Wilayah Pesisir dan Puau-Pulau Kecil). This legislation allowed district level actions, which saw the North Lombok District (which Gili Indah is part of) enacting local regulation on spatial planning in 2011 (Peraturan Daerah Kabupaten Lombok Utara Nomor 9 / 2011 Tentang Rencana Tata Ruang Wilayah Kabupaten Lombok Utara Tahun 2011–2031), which gave greater legal planning authority to village level governments.

At the national level, the new Ministry of Maritime Affairs and Fisheries issued regulations in 2014 for a management plan and zoning specifically for the Gili Indah islands (Keputusan Menteri Kelautan dan Perikanan Republik Indonesia Nomor 57 / KEPMEN-KP/2014 Tentang Rencana Pengelolaan dan Zonasi Taman Wisata Perairan Pulau Gili Ayer, Gili Meno dan Gili Trawangan di Provinsi Nusa Tenggara Barat Tahun 2014–2034). In that same year, the Gili Indah Village community replaced the *awig-awig* regulations on marine resources zoning with a formal village regulation on the management of coastal and marine zones (Peraturan Desa Gili Indah No: 03 / 2014 tentang Pengelolaan Kawasan Pesisir dan Laut Desa Gili Indah) that conformed to the policies adopted by the Ministry of Maritime

Affairs and Fisheries. This village regulation revised the zoning usage map for the Gili Indah waters, based on an extensive public participation process.

Collaborating to enforce the coral reef regulations

Several local leaders (who were former fishermen now involved in the Gili Trawangan tourism industry) created the Gili Youth Task Force (Yayasan Front Pemuda Satgas Gili) in 2000 at the same time that COREMAP was working with the Gili Indah Village government to create the *awig-awig* for coral reef protection. The Gili Youth Task Force (known locally as SATGAS, which is an abbreviation of "task force" in Bahasa Indonesia), was a non-governmental organization (NGO) made up of local, high-school age youths who were appointed by the *Kades* and *Kadus* to enforce the *awig-awig* regulations. SATGAS youths would patrol the waters of Gili Indah two or three times a day in a power boat provided for the organization. They reported violations to the Lombok police and the Natural Resources Conservation Center (KSDA, part of the Ministry of Forestry). Under the *awig-awig*, the local community determined penalties for violators on a case-by-case basis, which could range from warnings to public humiliation and banishment from the islands.

SATGAS, however, lacked sufficient resources, such as speedboats, communication facilities, and operating funds to effectively monitor the *awig-awig*. To address this problem, SATGAS leaders sought support from the tourism business community on Gili Trawangan, which included dive operators, hotels, shops, and restaurants. In response, several leaders among the tourism entrepreneurs (which included both Indonesians and Europeans) established the Gili EcoTrust (GET) in 2001. GET created a voluntary fee that all tourism businesses on Gili Trawangan pay (through peer pressure) to the Trust to support its conservation programs. The rate of this fee was set annually by GET and varied depending on the number of tourists that the island received. The voluntary fee was ended in 2017.

Collaboration between the Gili Indah Village government (who created the coral reef *awig-awig* regulations), SATGAS (who were given authority to enforce the regulations), and GET (which provided funds for enforcement operations) enabled the mostly effective enforcement of the coral reef *awig-awig*. The *awig-awig* regulations applied to all three of the Gili Indah islands, whereas SATGAS and the Gili EcoTrust were based solely on the island of Gili Trawangan, which is the most developed in terms of tourism activities. Fishing is still a primary occupation of a few small communities on the lesser-developed islands of Gili Meno and Gili Air. These fishermen were paid a subsidy by GET to prevent them from fishing in the protected zones that were originally designated by the *awig-awig*, and later came under the more formal marine protection policies. Unfortunately, destructive fishing techniques are still occasionally used in the less protected areas.

Since 2009, Gili Indah has been designated as a National Water Conservation Area based on a Ministry of Maritime and Fisheries Decree (Keputusan Menteri Kelautan dan Perikanan No. Kep. 67/MEN/2009), and through which the central government formed a technical unit called the National Marine Waters

Conservation Agency (Balai Konservasi Kelautan Perairan Nasional, BKKPN). BKKPN replaced the Natural Resources Conservation Center (KSDA), which SATGAS would report fishing violators to. In 2014, BKKPN adopted a new zoning map for Gili Indah (KEPMEN-KP No. 57 of 2014), to which was created through a participatory process that sought to reflect sustainable resource management practices. The GET is now working with the BKKPN, representing the Ministry of Maritime Affairs and Fisheries, to enforce the protective coral reef zoning system, although resources for this are still somewhat limited (GET, 2017).

The *awig-awig*, SATGAS, GET, and BKKPN represent successive stages that the local community of Gili Trawangan entered upon as it self-organized and reorganized activities to protect the coral reefs that were vital to their tourism industry. These functions might have been undertaken by a single government agency in another context, but this was not possible in Indonesia in the late 1990s and early 2000s when maritime regulations were non-existent and a series of national crises had further weakened the government's ability to manage these resources to the degree desired by Gili Indah's residents and international development agencies. Over time, as the country's governance system was restructured, more formal institutional arrangements came to replace the traditional and local-based systems to regulate human activities in coral reef areas, as well as to enforce those regulations.

Collaborating for coral reef restoration

With the rapid expansion of unregulated tourism development in the 1990s, coral reef areas adjacent to the most heavily settled parts of the islands of Gili Indah had mostly died due to human impacts. The main causes of these losses were blast fishing (bombing) and *muro-ami* net fishing (in which corals are smashed to scare the fish into nets). Fish poisoning (which kills coral organisms), tourist boat anchors, and tourist snorkeling and diving activities also caused damage in the high use areas. Climate change-caused warming of sea waters was the primary non-local cause of coral death, with the 1997 El Nino event causing widespread coral bleaching across Southeast Asia and in other tropical waters around the globe (GET, 2010; 2012; Goreau, McClanahan, Hayes & Strong, 2000).

As part of its coral reef conservation efforts, the Gili EcoTrust initiated a collaboration in 2006 with the Global Coral Alliance (a nonprofit organization based in the US that is the license holders of Biorock technology; www.globalcoral. org) to use their technology to restore damaged reefs in the heaviest used shoreline areas adjacent to the three Gili Indah islands. This collaboration also included researchers from the NTB province's University of Mataram and representatives of smaller community-based NGOs in Gili Indah. Biorock technology (www.biorock. org) uses low-voltage structures to stimulate coral growth at rates that were found to be 2–6 times faster than the natural growth of corals in tests conducted in Gili Indah's waters (Bakti et al., 2012). Over 120 Biorock structures were installed around the Gili Indah islands between 2006 and 2015 (GET, 2017). GET hosted numerous Biorock workshops, in their collaboration with the Global Coral Alliance, attracting participants from around the world seeking to restore and strengthen their coral reef ecosystems. In each workshop participants build Biorock structures that further expand the network in the more damaged reefs of Gili Indah. In recent years, the coral reef restoration efforts of GET have been supported by Indonesia's National Marine Waters Conservation Agency (Balai Konservasi Kelautan Perairan Nasional, BKKPN), which was created in 2010 by the Ministry of Maritime Activities and Fisheries. BKKPN also worked with GET to place mooring buoys to reduce boat anchor damage to coral reefs, and worked closely with the Gili Indah Village government in its adoption of the 2014 marine waters zoning regulations that replaced the less formal *awig-awig*.

Discussion

An island panarchy system

Gili Trawangan and the other islands of Gili Indah are heavily dependent on tourism activities that are associated with their beaches and surrounding coral waters. Although fishing is not as significant to the local economy as it was in the past, it too is dependent on healthy reef habitats. There are many ways in which the Gili Indah system can be subdivided from a system perspective. Table 10.2 summarizes the systems and subsystems that share an interest in the management and use of the coral reefs of Gili Indah. These systems consist of three types:

- geographic administrative systems, most of which are directly associated with administrative governance jurisdictions;
- marine biological ecosystems, of which the coral ecosystems are the primary focus for conservation and restoration; and
- social systems and subsystems that operate and are embedded within corresponding levels of the geographic administrative boundaries.

The social systems are primarily composed of individual actors who work either individually or collectively toward achieving their goals. (Note that the terms "system" and "subsystem" are used interchangeably here depending on their relative context to one another, with an awareness that all systems are subsystems of the global earth system.)

In resilience theory, the many systems and subsystems identified in Table 10.2 influence one another by either constraining change (remember) or encouraging it (revolt). The experience of working to conserve and restore coral reef ecosystems in the waters of Gili Indah show how the collapse of authority at higher levels of geographic administration result in a high degree of self-organization at the lowest levels. Very limited remember and revolt influences occured between the different levels within the panarchy hierarchy, as each was largely left to its own accord. It is more likely that parallel systems at the lowest levels were influencing one another. Thus, the NGO groups often collaborated toward shared goals, while also pushing for change in behavior and practice in the Fishermen Subsystem, the Tourism Business Owners Subsystem, and even the Local Governance Subsystem of the Gili Indah Village System.

Table 10.2 Systems and subsystems related to coral reef protection in Gili Indah Village

I. Geographic Administrative Systems associated with Gili Indah Village (nested hierarchy)			
Level 1	The Indonesia Country System, which includes:		
Level 2	The Nusa Tenggara Barat (West Nusa Tenggara) Province System, which includes:		
Level 3		The Lombok Island System, which includes:	
Level 4		The North Lombok District System, which includes:	
Level 5			The Gili Indah Island Village System, which is comprised of:
Level 6			The Subvillage Systems of Gili Trawangan Island, Gili Meno Island, and Gili Air Island; and
			The Gili Indah Waterways System that lies between and around these three islands

II. Biological Ecosystems in the Gili Indah Village Waterways System			
	Non-Coral Reef Ecosystems (not a focus in this study)		
	Coral Reef Ecosystems, including:		
		Dead Coral Reef Ecosystems – These are targets for coral restoration efforts	
		Living Coral Reef Ecosystems – These are targets for conservation efforts, and include:	
			Coral Organisms – These are the primary indicators of conservation success
			Fish and other marine animals that inhabit the coral structure and feed on its organisms – These are secondary indicators of conservation success

III. Social Systems and Subsystems in the Gili Indah Village (Island) System relevant to coral reef conservation. (Some are parallel to one another and many have overlapping memberships.)		
	The Tourism System, including subsystems of:	
		Tourism businesses owners (travel, transportation, dive/snorkel, accommodations, food, shops, support services)
		Tourists (domestic and international; scuba divers, snorkelers, others)
	The Fishing System, including subsystems of:	
		Fishermen
		Fishing support service providers (fish processing, fish sales, boat makers and repairs)
		Fish consumers
	The Governance System, including subsystems of:	
		Local village government officials
		Regional and central government officials involved in Gili Indah activities
		University and scientific agency staff
		Non-governmental and non-profit organization representatives
		Residents and others for whom government and other representatives speak (alternatively, this might be considered a separate subsystem)

Source: Authors

At the lowest levels, the social subsystems interact in complex and changing ways. In terms of coral reef conservation efforts on Gili Indah, initiatives in the early years of *reformasi* began with the local village government's coral reef *awig-awig* regulations, then moved to the NGO realm, first with SATGAS being designated to enforce the regulations, then to Gili EcoTrust (GET) to fund the enforcement. The local NGO subsystem, as might be expected, overlaps considerably with other social subsystems of the community. SATGAS draws on local Indonesian residents for its membership, while the more diverse GET NGO comes out of the Tourism Business Owners Subsystem, which is why they were successful in receiving funds from that subsystem.

The protection of living coral reef ecosystems was shown to be organized and administered almost entirely among subsystems within the Gili Indah Village administrative system. The restoration of dead coral reef ecosystems, on the other hand, while initiated at the local village-level NGO subsystem, entailed higher levels of administrative systems, including provincial and national, as well as international in the form of the Global Coral Alliance NGO. This effort drew upon the distinct global experience and perspectives offered by the foreign entrepreneur component of the Tourism Business Owners Subsystem in Gili Indah, especially through their involvement in the Gili EcoTrust.

Adaptive cycles in coral reef conservation

The adaptive cycle in resilience theory models how systems move through stages of crisis to recovery in systems in a cyclical manner. In the case of coral reef health in Gili Indah, there are several systems whose adaptive cycles play a significant role.

The Indonesia Country System experienced a collapse [Ω] in the late 1990s, which started with the Asian Economic Crisis in 1997, leading to the overthrow of the Surharto regime in 1998, which was followed by several years of economic, social, and political instability as the country moved through a long reorganization phase [α] (Satria & Matsuda, 2004). Indonesia in recent years has entered a growth [r] and perhaps an early consolidation [K] phase, with central and regional governments slowly beginning to assert authority on issues such as marine conservation. At the same time, national government policies have intentionally decentralized political decision making. Within this context, it is still necessary for local interests to take initiatives to achieve their desired goals.

As noted on pages 160--161, the living coral reef ecosystems around Gili Indah have experienced stresses that have resulted in varying degrees of system collapse [Ω] for different reefs. Some of the 'living reef ecosystems' have collapsed and not recovered, becoming 'dead reef ecosystems.' Others have only partially experienced this, while still others continue to be healthy. The natural recovery [α] of coral reefs is a much longer-term process, and one that would probably not be successful if the local human-induced stresses experienced were to continue. The result could be a total transformation of the living coral reef ecosystems into dead coral reef ecosystems.

To this point, we have considered all the coral reefs in the Gili Indah Waterways System to comprise a single system. However, it is evident that the various parts of this system are exposed to different types and degrees of stress, which accounts for the wide variation in degrees of health. For a more in-depth analysis, it would, therefore, be appropriate to treat each reef site as a distinct subsystem in the larger coral reef ecosystem. An adaptive cycle assessment of the state of each coral reef system could then be undertaken to assess its current and possible future health from a biological perspective.

What is of greater interest here is how the actions of the human social systems have impacted the health of the reef systems. Global warming aside, different local social systems have had variable impacts on some coral reefs over the years. In terms of anthropogenic-based declines in coral health, the Fishermen Subsystem has killed some or all of some reef systems. The transportation portion of the Tourism Business Owners Subsystem, and the divers and snorkelers of the Tourists Subsystem have also damaged coral, especially near the core tourist areas. Water pollution from other parts of the Tourism Business Owners Subsystem has further eroded coral health.

However, social systems have also contributed to the recovery and possible expansion of some unhealthy and dead coral reefs, and to maintaining the integrity of healthy coral reefs. This has been done by reducing the levels of human impacts through the regulation and zoning of fishing and recreation activities to allow the reefs to recover and grow through natural means. For those reef systems that are largely, if not completely, dead, such recovery and growth might not be feasible in human time scales, leading to the introduction of reef restoration technology projects, as described above (p. 173).

The Fish and Other Marine Animals Subsystem has also benefited from actions to protect the Living Coral Reef Ecosystem, although not to the same degree since fishing is still a legal occupation in the Gili Indah waterways system. Because of this, fish populations remain modest in comparison to historically high levels, as is the situation in most tropical seas due to the increasing human consumption of seafood (Allendorf, Berry & Ryman, 2014). In comparison to the past, the Fish Subsystem today is in an impaired state, although one that is stable and representative of its balance in relation to the social systems that manage it. From an adaptive cycle perspective, the Fish Subsystem has moved from a high level [K], through a decline [Ω], and then a reorganization [α]. The reorganization moved it into a new state which did not include a growth [r] stage, but instead maintained [K] the Fish Subsystem at a lower level of productivity.

Both the Living Coral Reef Ecosystem and the Fish Subsystem that exist today are not independent of the many human systems and subsystems that manage them for both exploitation (consumption) and conservation (protection). This shows the difficulty of defining system boundaries. We use a systems approach to better understand the forces of change and variable responses to those forces. However, because everything is related to everything else, we need to be aware of the limitations of our artificial constructs, and not let them blind us to the true complexities of the world.

Conclusions

This case study shows how one community (Gili Indah/Trawangan) has addressed a decline in an economic resource (coral reef ecosystems). The complexity of this example is made evident using a systems approach that is theoretically grounded in resilience theory and resilience thinking (Holling, 1973; 2001; Gunderson & Holling, 2002). The complexity comes from two general directions:

- First are the external forces of change that impact the systems involved. For the Gili Indah/Trawangan case, the major forces of change included:
  1   national- and regional-level political upheavals in the post-Suharto era;
  2   rapid increases in the numbers of tourists, tourism entrepreneurs, and tourism businesses;
  3   local fishing practices to meet growing consumption demands; and
  4   global climate and ocean warming causing ecosystem changes.
- Second are the many systems and subsystems that came into play over the single issue of coral reef protection in the waters of Gili Indah (Table 10.2). Each of these were impacted to some degree by the many forces surrounding the Gili Indah island system, and responded in different ways to them. Responses in the various systems then served as forces of change in other systems. Actions taken by the NGO system, especially on Gili Trawangan, were made in direct response to how the living coral reef systems were responding (in a bad way) to forces of change brought on by practices of the local fishing and tourism systems, as well as to global warming. Actions by the NGO system, however, were also affected by national and regional political transformations that impacted local governance systems.

In these ways, not only are systems responding to generic external forces of change, but as each system (or subsystem) responds, it has the potential to become an external force of change on other systems. In the case study described here, most of these forces of change were of the "revolt" type, with one system trying to force a change in behavior (or path divergence) in another system. It is likely that "remember" (or path dependency) influences were also present, but were overlooked because they were not within the focus of the present study and may have been subtle in the context of Indonesia's *reformasi*. Now that the national government has established its authority over the management of maritime resources, it is likely to have a greater "remember" role in influencing local policies and practices.

In addition to the complex panarchy of interrelated systems being impacted by external forces of change, and becoming external forces of change themselves, the study examined how the adaptive cycle from resilience theory was evident in the behavior and evolutionary path of some of the major systems involved. This provided additional insight into how different systems move through different cycles, although there were some shortcomings due to imprecision in identifying where a system might currently be in a cycle. This was seen in the current state of politics in Indonesia (as a system), as well as in generalizing the current health of the system of fish and other marine life that inhabit live coral reef ecosystems.

In the context of the geographic, social, economic, and political setting of Gili Trawangan, collaboration among various stakeholders was shown to have a significant impact on the legitimacy and effectiveness of the programs adopted to manage the coral reef ecosystems. Bottom-up participatory structures, as was evident in this case, can be key components in large-scale social change efforts (Light, 2002). These are modeled as a "revolt" influence in the resilience adaptive cycle model, with lower-level systems (like Gili Trawangan) taking initiatives that trigger change in higher-level systems.

The lack of an effective top-down management structure was, perhaps, an advantage in the Gili Trawangan case. It placed the island in a (re)organizational stage of the adaptive cycle, where resilience approaches (e.g., adaptation and innovation) are strongest because of the urgency of the coral degradation. Through various environmental management efforts, the community of Gili Trawangan demonstrated this resilience by engaging in collaborative self-organization and self-education in the face of threats to their economic livelihood. Stronger central government involvement, on the other hand, might have imposed institutionalized structures, representative of a more rigid consolidation stage of the adaptive cycle, which may also be more vulnerable to failure.

This could provide lessons for other communities that need to 'boot-strap' themselves to respond to the threats they face. These lessons include:

- *The importance of collaborative inclusion.* All the major, and perhaps minor, social systems (interest groups, stakeholders) must be involved in planning processes and be supportive of system-wide initiatives, because all are part of an integrated system that includes both humans and the environment in which they are situated. While this was not always evident in the local practices on Gili Indah (especially by the government), it was more evident in the possibly more effective collaborative approaches by the NGO community.
- *The importance of leadership.* It is difficult to predict where leadership will come from on an issue, but it can be expressed in the form of individuals and through key organizations that have earned the respect of others within the larger community system, both of which can change over time. This was clearly seen in the NGO community of Gili Trawangan, especially when government leadership was seriously lacking.
- *The importance of different levels of support.* Systems (communities, organizations, stakeholder groups) do not operate in isolation, but need to be aware of their relationship to higher-level systems and lower-level systems, both of which can be drawn upon to enhance resources toward shared goals. When a crisis problem or issue arises, key stakeholders will seek the support they need. If structures to facilitate that support are already in place, then responses can be more timely and effective.

Gili Indah is typical of many of the more popular tropical island tourism destinations in Southeast Asia. The growth in tourism worldwide, but especially within Asia (Hampton, 1998; Daldeniz & Hampton, 2013), has placed tremendous pressure on these island systems (both human and environmental) that must be managed to maintain their natural beauty and value. In such a context, system collapse is a constant threat, which from a resilience perspective may be a good thing because it can foster self-organizing collaboration and innovation, which can then lead to new opportunities for community development and growth.

References

Allen, C. R., Cumming, G. S., Garmestani, A. S., Taylor, P. D. & Walker, B. H. (2011). Managing for resilience. *Wildlife Biology*, 17: 337–49.

Allendorf, F.W., Berry, O. & Ryman, N. (2014). So long to genetic diversity, and thanks for all the fish. *Molecular ecology*, 23(1): 23–25.

Allison, H.E. & Hobbs, R.J. (2004). Resilience, adaptive capacity, and the "Lock-in Trap" of the Western Australian agricultural region. *Ecology and Society* 9(1): 3. Retrieved from http://www.ecologyandsociety.org/vol9/iss1/art3

Bakti, L.A.A., Goreau, T., Robbe, D. & Virgota, A. (2010). Using Community Action Planning (CAP) tools for sustainable ecotourism development in Gili Trawangan, Lombok. Paper presented in The International Seminar on Harmonization of Tourism Development, 27 April 2010, Udayana University, Bali, Indonesia.

Bakti, L.A.A., Robb, D., Virgota, A. & Goreau, T.J. (2008). *Biorock Reef Restoration for Sustainable Ecotourism in Gili Trawangan. Global Coral Reef Alliance*. Retrieved from http://www.globalcoral.org/biorock-reef-restoration-for-sustainable-ecotourism-in-gili-trawangan/

Bakti, L.A.A., Virgota, A., Damayanti, L.P.A., Radiman, T.H.U., Retnowulan, A., Hernawati, A.S. & Robbe, D. (2012). Biorock reef restoration in Gili Trawangan, North Lombok, Indonesia. In Goreau, T.J. & Trench, R.K. (2012). *Innovative Methods of Marine Ecosystem Restoration*. London: Taylor & Francis.

Baggio, R. (2008). Symptoms of complexity in a tourism system. *Tourism Analysis*, 13(1), 1–20.

Berkes, F., Colding, J. & Folke, C. (2003). *Navigating Social-Ecological System: Building Resilience for Complexity and Change*. Cambridge: Cambridge University Press.

Badan Pusat Statistik Kabupaten Lombok Utara (BPS). (2015). *Lombok Utara In Figures 2015*. Retrieved from http://lombokutarakab.go.id/v1/images/doc/2014/struktur2014/Lombok-Utara-Dalam-Angka-2015.pdf

Butler, R. (1980). The concept of a tourist area cycle of evolution: implications for management of resources. *Canadian Geographer*, 24(1), 5–12.

Carpenter, S.R. & Gunderson, L.H. (2001). Coping with collapse: ecological and social dynamics in ecosystem management. *BioScience* 6:451–57.

Daldeniz, B. & Hampton, M.P. (2013). Dive tourism and local communities: active participation or subject to impacts? Case studies from Malaysian dive tourism and local communities. *The International Journal of Tourism Research*, 15(5), 507–20.

Davoudi, S. (2012). Resilience: a bridging concept of a dead end? *Planning Theory and Practice*, 13(2), 299–333, http://dx.doi.org/10.1080/14649357.2012.677124

Farrell, B. H. & Twining-Ward, L. (2004). Reconceptualizing tourism. *Annals of Tourism Research*, 31(2), 274–95.

Folke, C. (2006). The emergence of a perspective for social-ecological systems analyses. *GLobal Environmental Change*, 253–67.

Folke, C. (2016). Resilience. In *Oxford Research Encyclopedia of Environmental Science*, pp. 1–68. Oxford University Press USA DOI: 10.1093/acrefore/9780199389414.013.8

Folke, C., Carpenter, S., Elmqvist, L., Gunderson, L., Holling, C. S. & Walker, B. (2002). Resilience and sustainable development: building adaptive capacity in a world of transformations. *Ambio,* 31(5), 437–40.

GET (2010). Seventh Indonesian reef restoration training/workshop. Gili EcoTrust, Gili Islands, Lombok NTB, Indonesia, November 2010, Hotel Vila Ombak, Gili Trawangan, Indonesia.

GET (2012). Eighth Indonesian reef restoration training/workshop. Gili EcoTrust, Gili Islands, Lombok NTB, Indonesia, November 2012, Hotel Vila Ombak, Gili Trawangan, Indonesia.

GET (2017). *Situation on Gili and Objectives.* Gili EcoTrust. Retrieved from http:// giliecotrust.com/about-biorock-project-on-gili-islands-in-indonesia.html

Gili Indah Village (GIV) (2010). *Medium Term Development Plan for Gili Indah Village 2010–14* / Rencana Pembangunan Jangka Menengah: RPJM Desa Gili Indah 2010–2014. Mataram, Indonesia: Nusa Tenggara Barat Provincial Government.

Gladwell, M. (2002). *The Tipping Point: How Little Things Can Make a Big Difference.* Boston: Little, Brown.

Goreau, T., McClanahan, T., Hayes, R. & Strong, A., (2000). Conservation of coral reefs after the 1998 global bleaching event. *Issues in International Conservation,* 14(1), 5–15.

Gotts, N. M. (2007). Resilience, panarchy, and world-systems analysis. *Ecology and Society,* 12(1), 24.

Graci, S. (2013). Collaboration and partnership development for sustainable tourism. *Tourism Geographies: An International Journal of Tourism Space, Place and Environment,* 15(1), 25–42.

Gunderson, L. & Holling, C. S. (2002). *Panarchy: Understanding Transformations in Human and Natural Systems.* Washington DC: Island Press.

Hall, C.M. & Lew, A.A. (2009). *Understanding and Managing Tourism Impacts: An Integrated Approach.* London: Routledge.

Hampton, M.P. (1998). Backpacker tourism and economic development. *Annals of Tourism Research,* 25(3), 639–60.

Hamzah, A. & Hampton, M. P. (2012). Resilience and non-linear change in island tourism. *Tourism Geographies,* 15, 43–67.

Hillmer-Pegram, K.C. (2014). Understanding the resilience of dive tourism to complex change. *Tourism Geographies,* 16(4): 598–614.

Holling, C. S. (1973). Resilience and stability of ecological systems. *Annual Review of Ecology and Systematics,* 1–23.

Holling, C. S. (2001). Understand the complexity of economic, ecological, and social systems. *Ecosystem,* 4, 390–405.

Holling, C.S. & Gunderson, L. (2002). Resilience and adaptive cycles. In L. Gunderson & C.S. Holling, *Panarchy: Understanding Transformations in Human and Natural Systems* (pp. 25–62). Washington DC: Island Press.

Kamsma, T. & Bras, K. (2000). Gili Trawangan: from desert island to "marginal paradise." In D.R. Hall & G. Richards (Eds.), *Tourism and sustainable community development* (pp. 170–84). London: Routledge.

Kartawijaya, T., Tarigan, S.A., Ningtias, P., Herdiana, Y., Hasbi, K.M., Muttaqin, E. (2013). *Kajian Dampak dan Daya Dukung Kegiatan Wisata di Taman Wisata Perairan Gili Matra.* Bogor, Indonesia: Wildlife Conservation Society.

Leiper, N. (1982). Why "the tourism industry" is misleading as a generic expression: The case for the plural variation, "tourism industries". *Tourism management,* 29(2), 237–51.

Leiper, N. (1990). Tourist attraction systems. *Annals of Tourism Research,* 17(3), 367–84.

Lew, A.A. (2014). Scale, change and resilience in community tourism planning. *Tourism Geographies,* 16(1), 14–22.

Lew, A. A. (2017). Modeling the resilience adaptive cycle: collaborative for sustainable tourism and resilient communities blog, (21 January). Retrieved from http://www.tourismcommunities.com/blog/modeling-the-resilience-adaptive-cycle.

Lew, A.A. & Hall, C.M. (1998). The geography of sustainable tourism: lessons and prospects. In Hall, C.M. & Lew, A.A., eds., *Sustainable Tourism: A Geographical Perspective* (pp. 199–203). London: Addison Wesley Longman.

Lew, A.A., Ng, P.T., Wu, T-C. & Ni, C-C. (2016). Some new resilience figures and diagrams: collaborative for sustainable tourism and resilient communities blog (30 September). Retrieved from http://www.tourismcommunities.com/blog/some-new-resilience-figures-and-diagrams.

Light, A. (2002). Restoring ecological citizenship. In B. Minteer & B. PeppermanTaylor, eds., *Democracy and the claims of nature* (pp. 153–72). Lanham, MD: Rowman and Littlefield.

McKercher, B. (2005). Destinations as products? A reflection on Butler's Life Cycle. *Tourism Recreation Research*, 20(3), 97–102.

MCA-Indonesia (2013). *Awik-awik: Local Wisdom for Natural Resource Management.* Millenium Challenge Account-Indonesia. Retrieved from http://mca-indonesia.go.id/en/kabar-kami/awik-awik-kearifan-lokal-pengelolaan-sumber-daya-alam/.

Robbe, D.R., Purnawadi, Ali U. & Bakti, A. (2011). Gili Matra, marine protected area: ecotourism and community management. In *Proceedings of the 2nd Coral Reef Management Symposium on Coral Triangle Areas*, September 28–30, Kendari, Indonesia.

Russell, R. & Faulkner, B. (2004). Entrepreneurship, chaos and the tourism area lifecycle. *Annals of Tourism Research*, 31(3), 556–79.

Satria, A. & Matsuda, Y. (2004). Decentralization of fisheries management in Indonesia. *Marine policy*, 28(5), 437–50.

Satria, A., Matsuda, Y. & Sano, M. (2006). Questioning community-based coral reef management systems: case study of Awig-Awig in Gili Indah, Indonesia. *Environment, development and sustainability*, 8(1), 99–118.

Schellhorn, M. (2010). Development for whom? Social justice and the business of ecotourism. *Journal of Sustainable Tourism*, 18 (1), 115–35.

Silver, C. (2005). Do the donors have it right? Decentralization and changing local governance in Indonesia. In H. Richardson & C.H. Bae, eds., *Globalization and Urban Development* (pp. 95–108) (Advances in Spatial Science Series), Heidelberg: Springer Berlin.

Vaisutis, J., Bedford, N., Elliot, M., Ray, N., Stewart, I., Ver Berkmoes, R., William, C., Witton, P. & Yanagihara, W. (2007). *Indonesia* (8th edn). Melbourne: Lonely Planet.

Walker, B.H. & Salt, D. (2006). *Resilience Thinking: Sustaining Ecosystems and People in a Changing World.* Washington, DC: Island Press.

Walker, B., Carpenter, S., Anderies, J., Abel, N., Cumming, G., Janssen, M., … Pritchard, R. (2002). Resilience management in social-ecological systems: a working hypothesis for a participatory approach. *Conservation Ecology*, 6(1), 14.

Walker, B. H., Carpenter, S. R., Rockstrom, J., Crépin, A.-S. & Peterson, G. D. (2012). Drivers, "slow" variables, "fast" variables, shocks, and resilience. *Ecology and Society*, 17(3), 30, doi: 10.5751/ES-05063–170330

Walker, B.H., Gunderson, L.H., Kinzig, A.P., Folke, C., Carpenter, S.R. & Schultz, L. (2006). A handful of heuristics and some propositions for understanding resilience in social-ecological systems. *Ecology and Society* 11(1): 13. online http://www.ecologyandsociety.org/vol11/iss1/art13/

Part III
Disaster events and tourism

11 Disaster resilience of small businesses in Guanxian Ancient Town, Sichuan, China

Honggang Xu, Fangfang Chen and Shanshan Dai

Introduction

As one of the most essential components in a destination, small businesses provide livelihoods to local residents and are the economic backbone of many communities (Prasad et al., 2015). Because of their flexibility and innovation ability, small businesses can absorb disturbances and respond effectively in the face of exogenous changes, benefiting the whole economic system (Williams & Vorley, 2014). However, small businesses are also exposed to many external disturbances, and their survival is always an issue. Thus, small tourism business resilience is vital to the sustainable development of tourism destinations.

At present, research on small tourism business resilience is concentrated on coastal destinations, and little attention has been paid to other tourism destinations. However, resilience is context-dependent (Biggs, Hicks, Cinner & Hall, 2015), and levels of resilience and factors influencing resilience may differ in social and economic settings. With this in mind, this chapter takes Guanxian Ancient Town, a cultural tourism destination in Sichuan Province, China, which was struck by the Wenchuan Earthquake in 2008, as the study site to investigate the disturbances suffered by small enterprises and their responses. The resilience of small enterprises in Guanxian Ancient Town is analyzed to contribute to the understanding of small tourism business resilience.

Literature review

Resilience and tourism research

The concept of resilience emerged from the ecological sciences and referred to an ecological system's ability to absorb changes and disturbance and still maintain the same internal and external relationships (Holling, 1973). This was then expanded to include social and social-ecological systems (Adger, 2000; Holling, 2001). Tourism scholars began engaging in the concept of resilience after it was introduced to social and social-ecological systems, and the various threats faced by tourism and the way tourism businesses have responded has aroused more and more attention over the past decade (Ruiz-Ballesteros, 2011; Calgaro, Dominey-Howes & Lloyd, 2013; Becken, 2013). Currently, resilience studies in tourism have been undertaken at multiple scales, across different sectors, and center on different tourism stakeholders (Hillmer-Pegram, 2013). Destination resilience, community resilience and small business resilience have been the main study themes.

Small tourism business resilience

Conceptualizing small tourism business resilience

Though centered around the ability to respond effectively to unpredictable crises, the key indicators of business resilience vary among researchers. Biggs et al. (2015) focused on the operational states before and after a crisis, and he defined enterprise resilience as the ability to maintain and adapt its essential structure, identity, and operations (Biggs et al., 2015). In the face of crises or shocks, a resilient tourism enterprise is able to maintain or grow its existing level of employment and income and stay operational (Biggs, Hall & Stoeckl, 2012); running into debt, shutting down, downsizing, or switching to other sectors are all seen as lack of resilience (Biggs et al., 2015). To assess enterprise resilience, Biggs et al. (2015) did not look at the specific criteria for different operating states, but instead adopted "perceived resilience," where four measures of perceived resilience and four measures of a perceived lack of resilience were introduced to evaluate entrepreneurs' perceptions of business resilience.

Some scholars are concerned with the responses of small businesses to crises. Enterprise resilience is defined as "the ability to take situation-specific, robust and transformative actions" in unexpected changes (Amann & Jaussaud, 2012), and has three different forms:

- surviving;
- adapting; and
- growing (Dahles & Susilowati, 2015; Prasad et al., 2015).

The survival or resilience of a business is reflected in its ability to return to the previous state of normality, while adapting a business may involve shifting to a new business concept; growing a business involves seizing new opportunities in a crisis. Dahles and Susilowati (2015) suggest that business resilience is an embedded concept and should never be viewed separately from other social and economic sectors. In the face of a crisis, downsizing a tourism business, seeking side businesses in other economic sectors, and switching to more promising markets are all characteristics of resilience. Diversity and flexibility are closely related to resilience (Dahles & Susilowati, 2015; Ruiz-Ballesteros, 2011).

In this chapter, how small businesses respond to exogenous disturbances is adopted as the main criterion of resilience. In addition, the "essential structure, identity, and functioning" of small businesses are also taken into consideration, and they are identified in this study as persisting with the business operations in Guanxian Ancient Town. A resilient small business in Guanxian Ancient Town is defined as one that is able to survive, adapt, grow, or otherwise keep operating through a crisis.

Factors that influence small tourism business resilience

Although small enterprises may be vulnerable to exogenous disturbances over which they have no control, their flexibility, embeddedness and practical abilities provide resources for resilience (Williams & Vorley, 2014, Cioccio & Michael, 2007, Williams & Vorley, 2014). In the face of external changes, small enterprises tend to find alternative resources or innovative products (Williams & Vorley, 2014), and though they have no formal plans, they learn in practice to cope with crisis (Cioccio & Michael, 2007). Embedded in the social and economic contexts of destinations, small enterprises can seek alternative livelihoods in other economic sectors if necessary (Dahles & Susilowati, 2015).

Overall, there are five factors that may influence small business resilience:

- lifestyle values;
- social capital;
- human capital;
- financial capital; and
- natural capital (Biggs et al., 2015; Sydnor-Bousso, 2009; Baker & Coulter, 2007; Danes, Lee & Amarapurkar, 2009).

Lifestyle factors refer to the desire of enterprise owners and staff to live in a particular location because of its amenity values and associated quality of life (Biggs, Ban & Hall, 2012), and is closely related to sense of place and personal identity (Biggs, 2011). Biggs et al. (2015) further divided it into identity, love and shared knowledge. Lifestyle factors influence an entrepreneur's attitude towards profit; for lifestyle entrepreneurs, profit is not the main concern as they are more willing to tolerate poor financial performance and are reluctant to quit in difficult times (Biggs et al., 2012).

Besides amenity lifestyle concerns, there may be other factors that can influence an entrepreneur's business attitude, such as economic livelihood. Livelihood businesses comprise the main group of small enterprises in most tourism destinations, especially destinations in developing or low-income areas. For livelihood entrepreneurs their business is the main source of their family's income (Liang, Xu & Thomas, 2010). For these entrepreneurs, profit is their main concern in operating the business, but they are different from profit-oriented entrepreneurs who require high levels of return on investment. Although Biggs conducted some comparative studies between low- and high-income areas, those studies did not distinguish between the meanings of "lifestyle factor" for amenity and livelihood entrepreneurs.

Social capital is defined as the actual and potential resources embedded in the relationships possessed by the enterprise (Nahapiet & Ghoshal, 1998) and is positively related to business resilience (Prasad et al., 2015; Ledogar & Fleming, 2008). The social capital of small enterprises is embedded in their relationships with government, family members and friends, local community groups, and other enterprises (Biggs et al., 2012). In the face of crisis, social capital provides information (business information, operating advices, etc.), resources (Prasad et al., 2015), cooperative opportunities (Baker & Coulter, 2007), government support (Biggs et al., 2015) and other needs.

Human capital refers to the skill and coping capacity of the entrepreneurs and their employees, including business experience and the essential skills required in a specific sector (Biggs et al., 2012). Human capital benefits business resilience (Kizos et al., 2014; Prasad et al., 2015). Financial capital includes financial conditions (e.g., profits, income, and investments) and access to finance. It is of great importance to an enterprise's survival and success (Biggs et al., 2015). For a small tourism enterprise, natural capital depends on the types of main attractions in the destination (Biggs et al., 2012).

Embedded in the context of China, this chapter investigates:

- what were the disturbances and crises from the 2008 Wenchuan Earthquake and the reconstruction that followed that impacted small businesses in Guanxian Ancient Town;
- how did the small businesses respond to those crises;
- what levels of resilience did their responses reflect; and
- what influenced their resilience.

Methodology

Study site

Guanxian Ancient Town, located in the core area of Dujiangyan City, Sichuan Province, has an area of 1.01 sq. km and a history of more than 1500 years (Figure 11.1). Tourism development in Guanxian Ancient Town is closely dependent on the adjacent UNESCO World Cultural Heritage site of the Dujiangyan Irrigation Project.

Being only 38 km from downtown Chengdu (one of the most popular tourism cities in China) and possessing exclusive attraction resources such as Qingcheng Mountain (a famous Daoist religious site) and the Dujiangyan Irrigation Project, Dujiangyan City's tourism industry was flourishing well before the earthquake in 2008. However, because it was only 15 km from the epicenter of the Wenchuan earthquake, it was damaged heavily and all industries, including tourism, collapsed. An extensive effort was made to help the city recover from the disaster, and, as the pillar industry, tourism's recovery and development was a high priority. On 29 September 2008 (4 months after the earthquake), Qingcheng Mountain and the Dujiangyan Irrigation Project reopened, after which receipts and revenues from tourism in Dujiangyan City began to rise steadily. Tourist receipts for the National Day (10 October 2008) and the Spring Festival (February 2009) Golden Week holidays exceeded those from 2007, as did total tourist receipts for 2009 (Figure 11.2).

Before the earthquake, Guanxian Ancient Town, in downtown Dujiangyan, had the dual function of tourism and being the city's commercial center. Tourism businesses were mainly operated by local families, providing accommodations, catering, souvenirs, and local specialties. In addition, the town was where the government, hospital, and school were located, and non-tourism urban commercial services were an integral part of business activities there.

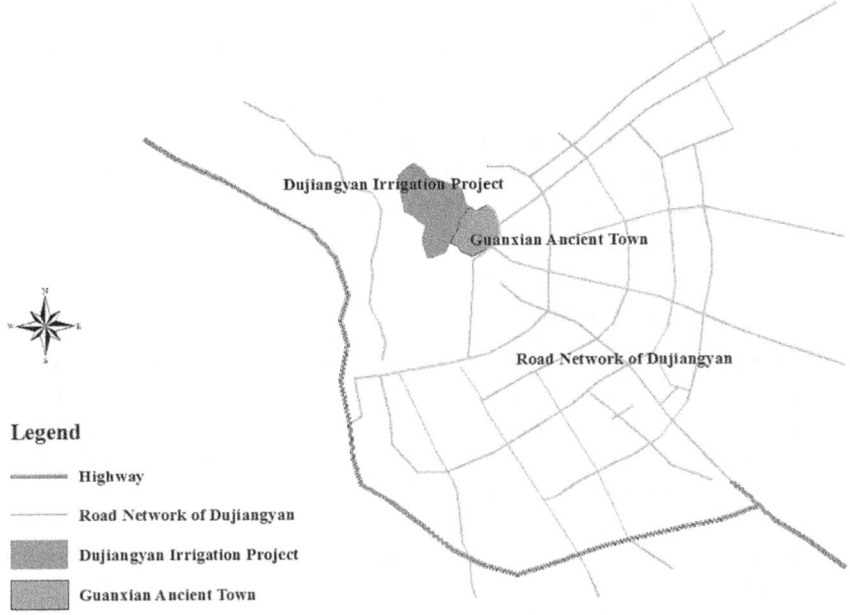

Figure 11.1 Guanxian Ancient Town in Dujiangyan City, Sichuan, China
Source: Authors

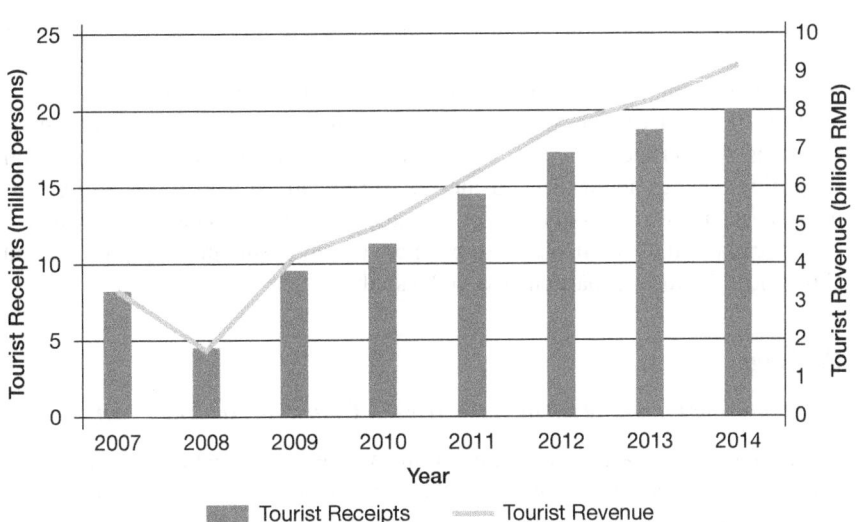

Figure 11.2 Tourist receipts and tourism revenue of Dujiangyan City, 2007–14
Source: Bureau of Statistics of Dujiangyan, 2007–14

The earthquake damaged historical buildings on West Street, North Street, South Street, and Baoping Alley, crippling the businesses in the town center. Luckily, however, only a few buildings collapsed completely, and thus the layout and overall features of the town were maintained. After the earthquake, the town was reshaped into a typical Chinese-style ancient town, with the tourism functions being reinforced and non-tourism urban commercial services removed. This created an unanticipated post-earthquake change on the character of small businesses in the town.

Data collection and analysis

Semi-structured interviews were adopted as the main method to collect data. A preliminary visit was carried out in February 2015, and fieldwork was conducted in August 2015. Interviews were carried out with 43 small business owners, of whom 25 had started their businesses before the Wenchuan earthquake, with the other 18 starting afterwards. Questions were asked regarding:

- basic information about the business and business owner, and motivation for starting the business;
- influences the earthquake had on their business;
- the process of restarting or starting the business after the earthquake; and
- business conditions after the restart or start of the business (e.g., other disturbances that may have arisen).

After the fieldwork was conducted, the authors analyzed the interview transcriptions and found that the enterprises that started before the earthquake should be the main study objects. Therefore, additional interviews were carried out from 28 January to 2 February 2016. Complementary interviews regarding the resources the business owners applied in their recovery strategies and their subjective feelings about their businesses were included with some participants interviewed in the first set of interviews. At the same time, another 13 interviewees were approached. Overall, 32 small business entrepreneurs in Guanxian Ancient City (coded S1 to S32) participated in the study.

Findings

Basic information of the small enterprises in Guanxian Ancient Town

Table 11.1 shows basic information from the businesses samples. Most of the participants were local residents (with three exceptions); and they had started their business for economic livelihood purposes, with no specific business plans. Nearly half of the enterprises had a history of more than 20 years, and these small entrepreneurs were quite familiar with the history and recent development of the town. The majority of the business owners were self-employed, with the most common business forms being local snack shops, restaurants, and souvenir shops.

Table 11.1 Basic information for the sample businesses (n = 32)

Years of Operation			Employees		
>20	15	46.88%	Has at least one employee	11	34.38%
10–20	9	28.13%	Has no employees	21	65.63%
<10	6	18.75%	*Business Form*		
Shut down for several years and restarted lately	2	6.25%	Local snack shop	9	28.13%
Property Ownership of the Shop			Restaurant	8	25.00%
Rent	15	46.88%	Souvenir shop	5	15.63%
Rent (Muslims)*	2	6.25%	Inn	3	9.38%
Self-owned	13	40.63%	Tea house	2	6.25%
Self-owned & Rent	2	6.25%	Clothing shop	2	6.25%

* Rents are lower in price for Muslims
Source: Authors

Table 11.2 Length of business closing after the earthquake

Length of Closing	Number	Percentage	Cumulative Percentage
<1 month	5	15.63%	15.63%
1 to 2 months	6	18.75%	34.38%
3 to 4 months	4	12.50%	46.88%
5 to 6 months	7	21.88%	68.75%
1 year	6	18.75%	87.50%
other	4	12.50%	100.00%
Total	**32**		

Source: Authors

The earthquake and small enterprise coping strategies

The Wenchuan earthquake caused a temporary paralysis of the town. All of the businesses had to shut down for a certain period of time, which was the major challenge faced by the small businesses (see Table 11.2). Besides the loss of income, the raw materials, goods and fixed assets of some small entrepreneurs were also damaged; for example, shop walls or roofs were shattered, restaurant chairs and tables were crushed, and food and goods were buried. Luckily, however, none of the participants suffered the loss of a close family member, and the core assets such as financial capital were retained; thus the damage was not insurmountable.

Some six months later, life in the town was back to normal, in terms of overall business activity. As noted above, the Dujiangyan Irrigation Project was quickly reopened and tourism activities were soon thriving. In addition, a large number of reconstruction workers came to the town, further driving the recovery of the consumer market. Influenced by all these factors, business in the town soon recovered, with 70 percent of the sampled enterprises having restarted within six months of the earthquake. Most of the businesses (87.5 percent) restarted within one year, and returned to their former business levels. However, some entrepreneurs chose to move elsewhere or shift their focus to other sectors temporarily, thus keeping their businesses closed for a longer time.

The strategies the enterprises adopted during the period of closing included:

- restarting the business as quickly as possible by employing alternative sites and materials (S1, S5, S15, S16, S17, S29);
- accessing government assistance, support from friends and personal savings to sustain their livelihoods, and waiting to restart the business (S7, S10, S11, S19, S21, S22, S30); and
- temporarily switching to other places or sectors (S2, S8, S9).

Reconstruction and small enterprises' coping strategies

A large number of people chose to visit the stricken areas after the Wenchuan earthquake, either to help or just out of curiosity. The visitors and reconstruction workers rejuvenated the consumer market in Guanxian Ancient Town, bringing business in the town to a peak in 2009.

However, this prosperity did not last for long. The reconstruction project beginning in 2010 became the second crisis for many small enterprises in Guanxian Ancient Town. Although historical buildings were repaired, old streets renovated, and the number of tourists increased, the reconstruction caused major disturbance for the town's small businesses. These disturbances included:

- road construction impacted the accessibility of some businesses for several months;
- government offices, school, and hospital, along with some residents, were relocated to a new town, causing a sudden decrease of demand in the old town;
- the government attracted new tourism business investments to the town, resulting in increased competition;
- motor vehicles were prohibited from entering the ancient town by the main gate during the daytime, causing some tourists to choose accommodations and amenities outside the old town area;
- stalls, signboards, and tea tables along the riverbank, along with other business strategies commonly used by small businesses, became strictly forbidden or regulated by the government; and
- the proportion of tourism revenue concentrated in the ancient town increased, causing more distinct seasonal fluctuation which businesses needed to plan for.

Some small entrepreneurs made adjustments to their product categories, sales methods, and operating strategies. These included, for example:

- adding or changing their types of products (S5, S17);
- switching to another business (e.g., from souvenir shop to noodle restaurant) (S3);
- boosting sales by using online bookings, advertising and supplying takeouts;
- improving facilities to try to catch up with chain enterprises (however, with far from satisfactory outcomes) (S24, S13);
- laying off some staff or switching to side businesses in the off-season (S26);
- informal cooperation with other businesses (S1); and
- using word of mouth, renting, and employment strategies (S16, S21).

However, with the small entrepreneurs' limited ability and financial capital, these strategies had little effect. Most of the actions could not effectively address changes in market demand. Most entrepreneurs lacked the resources required to switch to new businesses, and one of them even shut down permanently (S15).

Compared with the earthquake, the reconstruction of the ancient town had a larger and longer-lasting influence on the small businesses, and the following strategies were used by small entrepreneurs to cope with the changes:

Strategy 1: Adjusting their business attitudes, which meant reducing their expectations and requirements for profit (S11, S14, S18, S19, S20, S23, S24);
Strategy 2: Keeping operations in the existing sectors and markets by making some adjustments in product categories, sales methods, and operating strategies (S1, S5, S6, S7, S12, S13, S22); and
Strategy 3: Seizing new opportunities, switching to more promising sectors, or expanding their business scale (S2, S3, S4, S25, S31, S32).

Typologies of small business resilience

According to Dahles and Susilowati (2015), there are three levels of resilience for small businesses:

- to survive by adopting coping strategies in the readily available market, where resilience is demonstrated in the capacity to wait for a return to the former state of normalcy;
- to adapt by effectively coping with the changes, where new markets may be explored; and
- to seize the opportunities in the crisis and grow.

Using the responses of small entrepreneurs to the earthquake and reconstruction as the main criterion, with their business conditions before and after the crises as a supplement, two of the participants were categorized as "lacking resilience," while another 20 survived, eight adapted, and two grew. Overall, in the face of the earthquake and the reconstruction that followed, small businesses in Guanxian Ancient Town showed relatively strong resilience.

Lack of resilience

Only two businesses were found to have a lack of resilience: one closed down permanently (S15); the owner of the other expressed a strong desire to close the business and move to a new town (S8).

Sample 8 is a restaurant which became famous in the 1990s, and the business was very successful before the earthquake. However, in the pursuit of more profit, when the earthquake and reconstruction reduced business in the old town, the owner decided to leave and relocate to a more promising market. In the process of moving repeatedly, Sample 8 lost its social capital (fame and old customers), financial capital (due to investment losses and reduced profit), and natural capital (the prior relative location of its site to the Dujiangyan Irrigation Project), and thus lost its resilience. Sample 15, however, survived the earthquake by switching to stalls and alternative materials. However, because of financial constraints, the owner could not afford the soaring rent after reconstruction and eventually had to close the business.

Survival

Of the 33 samples studied, 20 survived the crises caused by the earthquake in 2008 and the reconstruction in 2010. In the face of the Wenchuan Earthquake, they adopted Strategy 1 (restarting the business as quickly as possible by employing alternative sites and materials) and Strategy 2 (accessing government assistance, friends' support and personal savings to sustain their livelihoods, and waiting for a proper time to restart their business). Strategy 1 reflected their flexibility in finding alternative resources, but also their economic weakness. For them, it was urgent to restart the business because of family economic needs:

> We had just a few hundred yuan in hand when the earthquake struck; and they ran out very quickly. So we need to restart the noodles stall, even though the income was low, we needed it for basic living. (S5)

The entrepreneurs who chose Strategy 2 were in a better economic condition than those who adopted Strategy 1. They faced their businesses' temporary closure more easily, and some of them even took it as an "excuse" to have a "vacation." But in general, because of the limitations in the operators' ability and financial capital, the "survival" businesses all stuck to their pre-existing business activities, and none of them used Strategy 3 (switching to other sectors).

When the reconstruction changed the market conditions of the town, the "survival" businesses showed a sense of helplessness; Strategy 1 (adjusting business attitudes) and Strategy 2 (keep operating in the existing markets by making some adjustments) were their main responses. "Adjusting business attitudes" is a reflection of lifestyle values. Some small entrepreneurs persisted in operating because of the climate and lifestyle in Guanxian Ancient Town (S19), while others started their business for an economic livelihood, but became lifestyle entrepreneurs when their children started working, so they no longer needed the money from the business. The lifestyle entrepreneurs did not quit simply because of decreased profits: "[We just] make some money when the situation is good and have a rest when it's not good, taking it as a way to keep fit" (S11).

Meanwhile, those who kept "operating by making some adjustments in product categories, sales methods and operating strategies" could not ignore profit as the lifestyle entrepreneurs did because they needed the income to sustain their family livelihoods. They took a variety of coping actions to boost their sales and prevent decreases in profit. However, the challenges they faced included:

- they stuck to their existing businesses, lacking the basic abilities to enter a new business (human capital);
- they found that capital for new investment was limited; and
- they were by the location of existing business and could not sufficiently benefit from the Dujiangyan Irrigation Project (natural capital).

As a result, the coping responses of the "survival" businesses were limited to the existing markets and the effect was not satisfactory. Though they tried to prevent it, decreases in profit were inevitable for the "survival" businesses, and so there were also some changes in their attitudes towards profit.

Different from lifestyle entrepreneurs, economic livelihood was the reason some entrepreneurs used Strategy 2 (i.e., they persisted in operations with a relatively low profit). First, the business was the main income for the owner's family, so they could not quit it easily and, second, the owner was used to having the existing business for their livelihood: "it's not love but custom, I'm accustomed to this business and can't think of any better ideas," (S12) and "in a word, I'll sell the tea no matter what happened, it's my life career" (S1). They could bear the hard work while accepting the low profits: "actually it's really tiring to operate this business, preparing the sources late at night and getting up early in the morning … but it's stable and guarantees a basic living" (S21). In consideration of the "family income" and the "habit" formed over decades, the entrepreneurs would not quit their business as long as it earned enough to sustain their basic living, and some entrepreneurs even expressed a determination to persist in their existing business no matter what the consequences (S1).

For lifestyle entrepreneurs, operating the business was their life and a way to spend time. Their resilience is based on love and enjoying life, and profit is not the main concern. But for livelihood entrepreneurs, operating the business is the way to sustain their life, and they care about the profit. Although their main considerations are different, the acceptable profit range is quite large for both groups, which contributes to their resilience.

The earthquake also caused some changes to the business attitudes of the entrepreneurs (see Table 11.3). The loss of life in the earthquake changed their values towards life and money, and life and living were very much cherished over profit: "At that time (after the earthquake), money became unimportant" (S7); "Eating and dressing well came first for everyone after the earthquake, money is nothing if a disaster strikes" (S1). This attitude prevented them from experiencing psychological depression during and after the earthquake, and they could face the challenges and find a way to rebound. Also, when the reconstruction impacts caused a decline in their business, they accepted it with the belief that "happiness consists in contentment" and so, again, not making them quit easily: "It's good to make a lot of money, but acceptable if there's not that much ... there's no need to pursue too much, happiness consists in contentment" (Y20).

Adapt or grow

There were eight participants who adapted to the crisis and were flexible in their business by temporarily switching to other sectors, going back to former jobs, or changing their main businesses entirely to cope with the market changes caused by the reconstruction. Also, they were innovative in their methods, facing the competition with a positive attitude. Sample 2 is an example of an adaptive business, operated mainly by a mother and her son. The mother had rich business experience, and the son joined the business a few months before the earthquake. When the earthquake struck, they were selling souvenirs, and the business was booming. After the earthquake, the mother persisted in her business by using stalls, selling souvenirs left from before, and selling local snacks, while the son went back

Table 11.3 Coping strategies of the survival businesses, and influential factors in resilience

Coping Strategies	Positive Factors	Negative Factors
The Earthquake		
Restarting the business as quickly as possible by employing alternative sites and materials	Flexibility	A lack of financial, human and social capital
Accessing government assistance, friends' support and personal savings to sustain livelihood, and waiting for proper time to restart the business	Relatively good economic condition; attitudes changed by earthquake	Limited financial, human and social capital
Post-Eathquake Reconstruction		
Adjusting business *attitudes* as some became lifestyle entrepreneurs	Lifestyle factors	Limited financial, human, social and natural capital
Maintaining operations in the existing market by making adjustments in product categories, sales methods and operating strategies	Livelihood factors; attitudes changed by earthquake	Limited financial, human, social and natural capital

Source: Authors

to his former job. In 2011, when reconstruction of West Street was completed, the mother invested in a store there (the prospects were uncertain, and many people were hesitant). In 2013, when business on West Street began to flourish, the son quit the job and rejoined his mother. Nowadays, they have two stores on West Street (one selling normal souvenirs, the other selling expensive ornaments) and an online store (selling souvenirs, ornaments, and home-made beef jerky).

In addition to disturbances, the uncertainty following a disaster and its aftermath can also generate innovation and growth (Holling, 2004; Lew, 2014). In the case of Guanxian Ancient Town, two businesses grew in the crises. The owner of Sample 25 had stayed in Guangzhou for 12 years and was comfortable staying home and taking care of their child before the earthquake. After the earthquake, large numbers of construction workers came to the town, and the owner, sensing a business opportunity, returned to selling fast food. She imitated a form of a buffet, charging by person, not for specific food, and the business flourished because the reconstruction workers liked it. Then, the reconstruction finished and the workers left, so she quickly seized on the needs of tourists and replaced fast foods with more refined ones. Afterwards, the restaurant doubled in scale, making the owner very satisfied with the profits.

The adaptive or growing businesses adopted Strategy 2, "accessing government assistance, friends' support and personal savings to sustain their livelihoods, waiting for a proper time to restart the business," and Strategy 3, "switching to other places or sectors temporarily" after the earthquake. To cope with changes caused by the reconstruction, they took Strategy 2, "keep operating in the existing markets by making some adjustments in product categories, sales methods and operating strategies," and Strategy 3, "seizing new opportunities, switching to promising sectors or expanding business scale." Compared with the survival businesses, they were more flexible in choosing products, business sectors, and operating forms, and were thus more resilient.

What contributes to the resilience of adaptive or growing businesses?

First, adaptive or growing businesses have an advantage in human capital over survival businesses. Some of the adaptive or growing entrepreneurs had worked or operated businesses in different sectors in the past, and so they could find supplementary livelihoods or business sectors more easily when a crisis struck, and were also more experienced in evaluating market conditions and investments. Taking Sample 2 as an example, the mother was experienced in choosing proper operating and investment forms, and the son could go back to his former job when the business was not prosperous. In addition to business experience, mastering key skills in a certain sector is also of great importance. In most cases, the owners of adaptive or growing businesses have a command of the essential business skills or techniques: for Samples 4, 9, 25, and 31, the owner or a family member of the owner was the chef of a restaurant; while for sample 17 the owner had advanced selling skills.

Financial capital to a large extent decided the actions the owners could actually implement; entrepreneurs with plenty of financial capital had the capacity to be fully prepared instead of restarting the business in a rush after the earthquake, and had more choices for new investments. In the case of the small entrepreneurs in Guanxian Ancient Town, they seldom sought external financial support, and profits from the businesses and the owners' family savings were the main financial capital. Adaptive or growing businesses generally had a certain level of savings, which were adequate for sustaining their livelihoods during the shutdown and reopening processes. Also, the profits of the adaptive or growing businesses were higher than those of the survival businesses, and thus they had the capital to improve their facilities, do some remodeling and decorating, invent new products, or switch to more promising markets.

For businesses in Guanxian Ancient Town, natural capital is reflected in their geographic location near the famous and historic Dujiangyan Irrigation Project, which allowed them to gain the benefit brought by this world cultural heritage site. More businesses in tourism-concentrated areas such as South and West Streets adapted or grew in the crisis than businesses in other areas.

Compared with survival businesses, adaptive or growing businesses showed some advantages in social capital as well (see Table 11.4). However, generally speaking, small businesses in Guanxian Ancient Town were so concerned with the business itself, and the owners' families, that they ignored external networks such as relationships with other businesses, the community, and the government. Thus, the social capital of small businesses in Guanxian Ancient Town was relatively limited. Forms of existing social capital included:

- relationships with suppliers;
- trust and support of regular customers; and
- financial or business information support from friends and relatives.

It is with regard to friends and relatives that the adaptive or growing businesses and survival businesses differed: adaptive or growing entrepreneurs have the potential to get key information or advice to help them with their business decisions, while financial support was the main support for survival businesses.

It can be seen that these three types of small business (lack of resilience, surviving, and thriving) are different in their coping strategies, their business attitudes, and their human capital, financial capital, natural capital, and social capital (Figure 11.3). Among the two samples that lacked resilience, one lost most of its financial and social capital because of inappropriate strategies, while the other was crowded out of the market because of financial constraints. Most businesses that survived the crises used "maintaining" as the main strategy, persisting in existing markets. However, those businesses successfully survived the crisis because:

- some businesses shifted from a livelihood to a lifestyle mode, lowering the importance of profits;
- some were influenced by livelihood constraints and business habits, and so were reluctant to quit the business, bearing more work while accepting lower profits; and
- due to the loss of life in the earthquake some changed their values towards life and money.

Table 11.4 Capital of survival versus adapting or growing businesses: respondent comments

Capital	Survival Businesses	Adapting or Growing Businesses
Human Capital		
Business Experience	At that time, many restaurants near the bridge were for sale, but we didn't want to take the risk. Now there's no chance. (S10)	Many people hesitated about opening a store here at that time, but we wanted to have a try. (S2)
	Our knowledge is limited ... Travel notes and forums online is very important nowadays but we have no knowledge about it. (S5)	We also have an online store, selling souvenirs and self-made beef jerky. (S2)
Key Skills	Most of the customers are visitors, the flavor is not that important. (S10)	The flavor is guaranteed because my husband is the chef. (S9)
Financial Capital		
Savings	We had just a few hundred yuan in hand when the earthquake struck ... So we needed to restart the noodles stall, even though the income was low; we needed it for basic living. (S5)	I went back to my former job and returned to the business two years later, when the business condition was good. (S2)
	I don't have too much knowledge, and the capital is limited. Maintaining it is the only choice... actually, I want to open a tea shop, but the investment is too much. (S5)	I came here in 2013 to take over the original store, selling the things I liked; and my mom opened a new store nearby... [investment] is not a big problem. (S2)
Profits	Decoration? Never mention it. What had been invested has not returned, but another one is needed. (S12)	This restaurant had expanded once ... the profit is enough for new investment. (S4)
Natural Capital		
Location	There are barely any visitors here. They are led by the guides and seldom have lunch here. (S7)	There are more visitors now, we just need to make some changes to the dishes. (S31)
Social Capital		
Suppliers	We have known each other for several years and can owe the money for a period of time, paying it the next time. (S1)	We went to the wholesaler where we used to purchase goods and got the goods without having to pay in full immediately. (S3)
Regular Customers	Compared with the newcomers, we have some regular customers. (S1)	We have plenty of regular customers ... your customers will always be yours. (S9)
Friends and Relatives	Friends and relatives would send some money to show their concerns, but business should rely on yourself. (S24)	One of my friends was doing this business. He recommended it to us. (S3)

Source: Authors

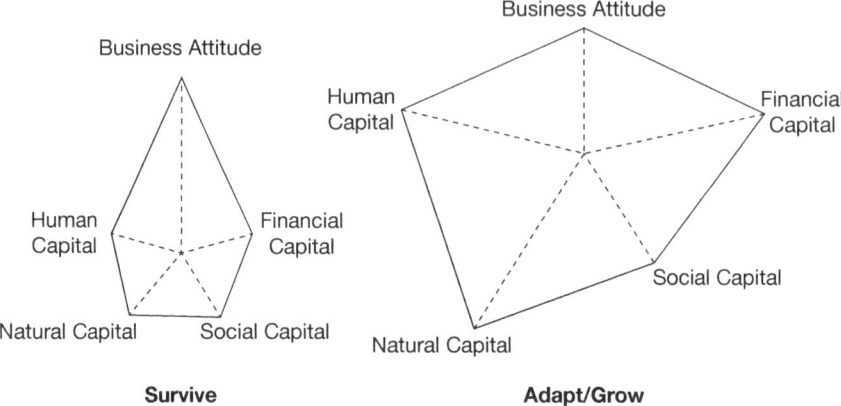

Figure 11.3 Influential factors of small business resilience in Guanxian Ancient Town (longer lines indicate stronger capacities)
Source: Authors

Meanwhile, eight of the samples adapted to external changes or even grew in the crisis. Compared with other businesses, the adaptive or growing businesses had advantages in human, financial, social, and natural capital.

Discussion

In line with Biggs et al. (2015), this study demonstrates the influence of lifestyle values and human, financial, natural, and social capital on the resilience of small businesses. However, resilience is indeed context-dependent (Biggs et al., 2015), and factors influencing resilience differ among social, economic and cultural contexts.

Biggs (2011) conducted his study in Australia, and the lifestyle factors of the small entrepreneurs there included sense of place, personal identity, values, and quality of life (Biggs et al., 2012). For entrepreneurs in lower-income areas, economic livelihood may be their first concern. From the investigation of small businesses in Guanxian Ancient Town, the study suggested that:

- lifestyle values benefited resilience by influencing the business attitudes of owners; and
- besides lifestyle values, livelihood considerations and the shocks brought by the earthquake also had an effect on their business attitudes, and thus on business resilience.

The lifestyle entrepreneurs in this study were different from the typical ones in developed country tourism destinations; they started the business for their livelihood and then shifted to a lifestyle mode when the family's economic conditions improved, resulting in profits no longer being a main concern. However, for livelihood entrepreneurs, profits are still the main concern. The livelihood entrepreneurs persisted in their businesses even though profits decreased because it was their family income and they had gotten used to operating the existing business. In evolutionary economic geography terms, they became path dependent, rather than path creative (Neumayer et al., 2014). Also, the earthquake caused some shocks to their values, and so life and living, instead of profit, became more cherished.

It is worth mentioning that there is no distinct division between lifestyle and livelihood entrepreneurs. For one thing, they are both capable of transformation: livelihood entrepreneurs may turn into lifestyle entrepreneurs when their personal or family conditions change, and vice-versa. For another, some entrepreneurs may have the characteristics of livelihood and lifestyle entrepreneurs at the same time; for example, Sample 21 mentioned that "doing this business is very tiring" but "I take it as an enjoyment."

Biggs et al., (2012) considered the relationships between enterprises, different levels of government, family members and friends, local community groups, and other enterprises when assessing social capital. However, in the case of Guanxian Ancient Town, social capital mainly existed between the small businesses and their suppliers, customers, friends, and relatives; while connections with local community groups, government and other enterprises were weak.

Conclusions

This chapter has explored the crises that natural hazard disasters and major human-caused disturbances can impose on small businesses in the Chinese context. Small businesses in Guanxian Ancient Town showed relatively strong resilience in the Wenchuan Earthquake and the reconstruction that followed. Results from the enquiry suggested that:

- for small businesses in Guanxian Ancient Town, lifestyle was not the only factor that influenced entrepreneur attitudes toward profit, thus influencing their resilience; livelihood constraints and shocks to values caused by the earthquake were other factors that were taken into consideration;
- the businesses that survived the crisis had limited human, financial, natural, and social capital, but a flexible business attitude contributed to their resilience; and
- adaptive or growing businesses had an advantage in business capital, and thus could cope with market changes more effectively.

Two main types of small business emerged from this study, and their policy implications differed. Lifestyle entrepreneurs chose to stay in their existing businesses because of their love and identification with the business, which they wanted to encourage and strengthen. For livelihood entrepreneurs, constrained by financial, human, and social capital, government sponsored economic and training assistance was necessary. The cost of innovating and switching to more promising markets should be reduced. The study also found that, although most small businesses in Guanxian Ancient Town survived, adapted, or grew after the earthquake and reconstruction, their social, human, and financial capital were constrained to internal relationships. To maintain resilience and build it into a community-level achievement, external relationships with local community groups, government, and other small enterprises also need to be strengthened.

Acknowledgement

This research was funded by the China Natural Science Foundation (41371156).

References

Adger, W. N. (2000). Social and ecological resilience: are they related? *Progress in Human Geography*, 24(3), 347–64.

Amann, B. & J. Jaussaud (2012). Family and non-family business resilience in an economic downturn. *Asia Pacific Business Review* 18(2), 203–23.

Baker, K. & A. Coulter (2007). Terrorism and tourism: the vulnerability of beach vendors' livelihoods in Bali. *Journal of Sustainable Tourism* 15(3), 249–66.

Becken, S. (2013). Developing a framework for assessing resilience of tourism sub-systems to climatic factors. *Annals of Tourism Research* 43, 506–28.

Biggs, D. (2011). Understanding resilience in a vulnerable industry: the case of reef tourism in Australia. *Ecology and Society*, 16(1), 30.

Biggs, D., N. C. Ban & C. M. Hall (2012). Lifestyle values, resilience, and nature-based tourism's contribution to conservation on Australia's Great Barrier Reef. *Environmental Conservation* 39(04), 370–79.

Biggs, D., C. M. Hall & N. Stoeckl (2012). The resilience of formal and informal tourism enterprises to disasters: reef tourism in Phuket, Thailand. *Journal of Sustainable Tourism* 20(5), 645–65.

Biggs, D., C. C. Hicks, J. E. Cinner & C. M. Hall (2015). Marine tourism in the face of global change: The resilience of enterprises to crises in Thailand and Australia. *Ocean & Coastal Management* 105, 65–74.

Bureau of Statistics of Dujiangyan (2007–14). National economy and social developed statistical bulletin of Dujiangyan. Dujiangyan, Chengdu: Bureau of Statistics of Dujiangyan. Retrieved from http://gk.chengdu.gov.cn/govInfoPub/dept.action?classId=07130228.

Calgaro, E., D. Dominey-Howes & K. Lloyd (2013). Application of the Destination Sustainability Framework to explore the drivers of vulnerability and resilience in Thailand following the 2004 Indian Ocean tsunami. *Journal of Sustainable Tourism* 22(3), 361–83.

Cioccio, L. & E. J. Michael (2007). Hazard or disaster: tourism management for the inevitable in northeast Victoria. *Tourism Management* 28(1), 1–11.

Dahles, H. & T. P. Susilowati (2015). Business resilience in times of growth and crisis. *Annals of Tourism Research* 51, 34–50.

Danes, S. M., J. Lee & S. Amarapurkar (2009). Determinants of family business resilience after a natural disaster by gender of business owner. *Journal of Developmental Entrepreneurship*, 14(4), 333–54.

Hillmer-Pegram, K. C. (2013). Understanding the resilience of dive tourism to complex change. *Tourism Geographies* 16(4), 598–614.

Holling, C. S. (1973). Resilience and stability of ecological systems. *Annual Review of Ecology and Systematics* 4, 1–23.

Holling, C. S. (2001). Understanding the complexity of economic, *Ecological, and Social Systems. Ecosystems* 4(5), 390–405.

Holling, C. S. (2004). From complex regions to complex worlds. *Ecology and Society* 9(1), 11.

Kizos, T., V. Detsis, T. Iosifides & M. Metaxakis (2014). Surmountable chasms: networks and social innovation for resilient systems. *Ecology and Society* 19(1), 40.

Ledogar, R. J. & J. Fleming (2008). Social capital and resilience: a review of concepts and selected literature relevant to aboriginal youth resilience research. *Pimatisiwin: A Journal of Aboriginal and Indigenous Community Health*, 6(2), 25–46.

Lew, A. A. (2014). Scale, change and resilience in community tourism planning. *Tourism Geographies*, 16(1), 14–22.

Liang, Wei, Xu, Hong-gang & Thomas, Rhodri (2010). Research on the motive and objective of lifestyle tourism entrepreneurs in Dali Ancient Town, *Tourism Tribune* (2), 47–53. (in Chinese)

Nahapiet, J. & S. Ghoshal (1998). Social capital, intellectual capital, and the organizational advantage. *The Academy of Management Review* 23(2), 242–66.

Neumayer, E., T. Plümper & G. Barthel (2014). The political economy of natural disaster damage. *Global Environmental Change* 24, 8–19.

Prasad, S., H. C. Su, N. Altay & J. Tata (2015). Building disaster-resilient micro enterprises in the developing world. *Disasters* 39(3), 447–66.

Ruiz-Ballesteros, E. (2011). Social-ecological resilience and community-based tourism. *Tourism Management* 32(3), 655–66.

Schianetz, K. & L. Kavanagh (2008). Sustainability indicators for tourism destinations: a complex adaptive systems approach using systemic indicator systems. *Journal of Sustainable Tourism*, 16(6), 601–28

Sydnor-Bousso, S. B. (2009). *Assessing the impact of industry resilience as a function of community resilience: The case of natural disasters*. Columbus, OH: Ohio State University.

Williams, N. & T. Vorley (2014). Economic resilience and entrepreneurship: lessons from the Sheffield City Region. *Entrepreneurship & Regional Development* 26(3–4), 257–81.

12 Death and disaster as moments of liminality

Towards collective agency and community resilience in Solukhumbu, Nepal

Maggie C. Miller

Introduction

> Our Camp II was set, and we were about to take the equipment to the upper camp.
> I woke up at 3:00am and at around 6:00am the ladder on the way was broken …
> The place was congested. All of a sudden an avalanche came. If the avalanche
> came fifteen minutes or twenty minutes before then it would have killed around
> sixty or seventy people. We were in the middle and at that time I heard the noise
> and saw it coming from above. After that I was buried by the snow. I could not
> remember what else happened there … I had so much pain on my head, so I was
> taken to Lukla from Basecamp via Helicopter. From Lukla, I was taken to
> Kathmandu by airplane to the B & B hospital … Lots of Sherpas died on the same
> day *so everyone decided not to continue* … Since that route was not safe, *we were
> not ready to risk our life.* Many Sherpas died there.
>
> (Mingma Sherpa, 2015 interview)

Mountaineering is the cornerstone of Nepal's US$370 million-a-year adventure
tourism industry (CBC, 2012; Payne & Shrestha, 2014; Schaffer, 2013). Sherpa[1] and
Nepali climbers commit themselves to securing and saving the lives of their clients,
often endangering their own (Davis, 2014; NPR, 2013; Peedom, 2015). Their
willingness to proceed up the mountain as high-altitude climbing guides and porters,
amongst dangerous and precarious environmental conditions, is perhaps reflective
of socio-economic pressures faced off the mountain (Bott, 2009; Ortner, 1999;
Peedom, 2015). However, on April 18, 2014, these industry complexities were
suddenly exposed. As Jennifer Peedom accounted in her documentary, *Sherpa:
Trouble on Everest*, the avalanche that unexpectedly surged through the slopes of
Mt. Everest that day incited others to challenge the status quo of the "Everest
Industry." Sixteen Sherpa and Nepali climbers were killed and ten more injured; all
were reported to have been fixing rope and carrying loads for commercial
mountaineering teams (Krakauer, 2014; Prettyman, 2014). Mingma Sherpa's
excerpt (epigraph at top of chapter) alludes to, for the first time in the mountain's
history, the Sherpa and Nepali climbing community finding a moment for pause to
band together and successfully enforce a mountain closure for the remainder of the

2014 climbing season, despite the inevitable economic sacrifice in forfeiting their annual income. The ensuing contentious meetings among mountaineering stakeholder groups, accompanied by Sherpa-led strikes, enabled Sherpas to put forth demands for increased consideration of their own safety as workers who are integral to the success of the mountaineering industry, and as human beings worthy of such consideration. Social resilience was demonstrated in these adaptive enactments—Sherpas' demonstration of agency to say "no" to industrial interests in mountaineering (and tourism) helped to leverage their ability to cope with the adverse and dynamic outcomes of these industries, which at times compromise their lives.

Resilience, rooted in physical sciences, was a term that originally denoted the stability of materials and described the capacity of these materials to return to equilibrium after a displacement (Bodin & Wiman, 2004; Norris et al., 2008). The application of the term shifted from physics to ecology (cf. Holling, 1973) with a primary focus on the ability to persist and adapt, versus "bouncing back," to a single stable state (Adger, 2000; Davoudi, 2011; Holling 1996; Norris et al., 2008). Since this fundamental shift, the concept of resilience has been influential in an array of social science disciplines, further conceptualizing the adaptive capacities of individuals, communities, and larger societies (Bonanno, 2004; Davoudi, 2012; Norris et al., 2008). More recently, socio-ecological discourse challenged the ideas of equilibrium and fixity, advocating more emergent and contingent understandings of resilience. This view, sometimes understood as evolutionary resilience, "suggests that faced with adversities, we hardly ever return to where we were" (Davoudi, 2012, p. 302). Moments of death, crisis, and rupture—the corollaries of disaster—provide opportunities for reflection on how life is being lived *and could be lived differently*. Such opportunities can create moments of *liminality*, "suspensions of quotidian reality, occupying privileged spaces where people are allowed to think about how they think, about the terms in which they conduct their thinking, or to feel about how they feel in daily life" (Turner, 1987, p. 102). Howard-Grenville, Golden-Biddle, Irwin and Mao, (2011) suggest that liminality allows individuals to actively consider the possibilities for constructing new cultural resources and altering strategies of action. Individuals—in this case Sherpas—are propelled into a rethinking (and perhaps one day a restructuring) of their community as they adapt to slow and sudden changes encountered on the mountainside.

Adaptability and transformability are important characteristics of resilient systems that begin to shift away from the simplistic "bounce back" or "bounce forward" equilibristic views (Davoudi, 2011, p. 301; Mulligan et al., 2016). Hiller (2015, p.11) suggests that transformability is an aspect of resilience in which communities "steer away from undesirable trajectories by creatively transforming structure." Applied in planning studies, she suggests that this capacity of resilience is a performative process that can be seen as the most progressive in conditions of uncertainty and change. Therefore, this chapter proposes that disaster and existential concerns such as death, as liminal moments, can be seen as central to the adaptive capacities that contribute to community resilience as a process of "becoming" (Mulligan, Steele, Rickards & Funfgeld, 2016, p. 5; Rogers, 2012). Situated within non-equilibrium thinking and recognizing resilience processes as emergent, fluid, reflexive, and complex (cf. Davoudi, 2012; Davoudi & Strange, 2009; Massey, 2005), I use this chapter to illuminate liminality as a part of communities' adaptive capacities and a potential site for "transformability" (Hiller, 2015, p. 11). Drawing on fieldwork observations and narrative interviews conducted in Nepal between February and April 2015, I examine disasters and negotiations of unexpected death to glean insights around how Sherpa climbers and their communities adapt to the sudden changes (death, accidents, disasters, etc.) and the slow and continuous (mountaineering, tourism) development within the Solukhumbu, commonly known as the Mt. Everest region.

Community complexities

Mulligan et al. (2016, p. 11) illuminate the dynamic and multilayered understandings of "community", cautioning "it is important to be aware of the tensions that need to be consciously negotiated in order to develop a more emergent and inclusive sense of community resilience." Therefore, prior to unpacking moments of liminality, I provide brief insight into some of the complex dynamics of Nepal, which are attributed to ways in which slow and sudden changes are experienced and further implicate how community resilience is enacted by Sherpa climbers.

Slow change: restructuring economies

Nepal, land-locked between the Chinese region of Xizang (Tibet) and India along the Himalayan range, has a population of over 28 million (World Bank, 2014). Historically, this nation is understood as having a low-income economy. In 1993, the Untied Nations Human Development Index (HDI) placed Nepal in the 151st position of 173 nations for overall level of socio-economic development, whereby 40–60 percent of Nepal's population was surviving on incomes below that of absolute poverty.[2] Nepal shifted to the 145th HDI ranking position in 2014 and, although there has been an increase in socio-economic development, this nation is still burdened with high levels of hunger and poverty (UNDP, 2015). These historically impoverished conditions, along with Nepal's land-locked geographical position, poor resource base, increasing population density, and low industrial and services output, all contributed to the nation's desire and need to identify a development solution (Bhattarai et al., 2005; Nepal, 2003).

According to Raj Panday (1999), development is based on a premise that with the right social, political, economic, and administrative processes a nation will find opportunities for advancement with whatever resources it has at its command. In Nepal 8 of the 14 world-famous mountain peaks over 8,000 m stand tall as part of the larger Himalayan range. Shrestha and Shrestha (2012, p. 59) discuss the importance of the Nepali Himalaya and indicate "tourism in Nepal began with mountain tourism." As tourists recognized Nepal's geographic splendour, entrepreneurs and the government began to capitalize on the demands for the natural beauty, and the potential economic gains from it (Bhattarai et al., 2005; Rogers & Aitchison, 1998). Furthermore, the successful assault of the summit of Mt. Everest in 1953, by Tenzing Norgay Sherpa, a Nepali native, and Edmund Hillary, a visiting New Zealander, and subsequent summit pursuits, contributed to the remarkable growth of adventure tourism in Nepal (D'Aliesio, 2012; Rogers & Aitchison, 1998; Schaffer, 2013; Shackley, 1999). The allure of the Everest summit continues to attract visitors, so much so that Sherpa climbers admittedly take on their first few pursuits as a personal challenge or goal. Phurba, a fairly young Sherpa climbing guide recalled during his interview, "I have the big aim, like a goal. That [first] time I don't look the money, just the top. That time I'm thinking just one time, one day I climb the Everest."

Prior to the advent of tourism, the economies of the mountain regions relied heavily on agriculture, animal husbandry, and trade (Nepal, 2005; Ortner, 1999). This was illuminated in Dawa's narrative as he enthusiastically explained:

> In my father's time there were no expeditions from Nepal ... We had no tourists so it was all trading systems, exchanging goods. We used to nurture baby yaks and use to trade them for salts. We used to go to Tibet.

Yet, as political and economic shifts in neighbouring nations led to the decline of Trans-Himalayan trading and the government struggled to improve economic and social conditions for Nepali people, tourism entered the political consciousness of the Himalayan regions as a solution to regional disparities (Nepal, 2005; Ortner, 1999; Singh, 1980). Socio-economic situations of Sherpas and their adaptability to high-altitudes positioned them as ideal employees to porter and guide foreign mountaineering and trekking expeditions (Bhattarai et al., 2005; Ortner, 1999). For instance, Dorchi, a retired Sherpa climber, recalled the sizeable loads he and other Sherpas were expected to carry up and down the dangerous terrains for paying clients of commercial mountaineering expeditions:

> Thirty kilograms while climbing up, and some carried forty, fifty kilos while descending down ... due to money, some [Sherpas] would often carry one hundred kilograms. They [Clients] use to pay in American dollars, and if they pay $20, $50, or $100 then people would carry double the weight they usually could.

Therein the introduction of tourism to the Solukhumbu restructured the traditional Sherpa economy, playing a critical role in reshaping employment and income opportunities (Adams, 1992; Bhattarai et al., 2005; Nepal, 2005, Ortner, 1999). Repurposing livestock transhumance as climbing gear transport, and crop production for the growing demands of Mt. Everest climbers and trekkers, are additional examples of how these communities transformed their economy in attempts to advance their resources (Raj Panday, 1999). The Nepali Himalaya became a primary "resource" for tourism development in Nepal, which contributed to the slow changes that began to occur in the Sherpa communities.

Sudden change: endangering livelihoods

As the socio-economic contexts are considered, it is important to briefly acknowledge ecological circumstances of these Sherpa communities. Nepal, situated along the Himalaya, the boundary of two massive tectonic plates, is a country fraught with geophysical and geomorphological concerns making it extremely vulnerable to natural disasters (Bjønness, 1986). Mountain hazards can be categorized as natural events including landslides, avalanches, floods, and earthquakes (Bjønness, 1986). When societies, their economy, and their infrastructure are affected, these mountain hazards can quickly turn into disasters and have the potential to lead to an incredibly disruptive state (Biran, et al., 2014; Sauerborn & Ebi, 2012). Disasters can be understood as traumatic events with a sudden onset that are collectively experienced (McFarlane & Norris, 2006). Tashi, a high-altitude guide, described his tireless rescue attempt after an avalanche suddenly plunged through the slopes of Manaslu, a Nepali mountain standing 8,156 m above sea level.

> I got to Camp III to survey the avalanche. There were thirty-two peoples. Immediately out of there I get nineteen alive, but most of them are hurt … I find nineteen are alive, and ten are dead body, and three are missing, but I didn't found them there. I searched for the whole day.

Avalanches in the Himalayas are known for their vicious frequency and their destructive potential, but have not been considered a major natural threat (Ganju & Dimri, 2004). Ganju and Dimri (2004) argue that in part this is due to the limited interactions humans have with these high-mountain environments. However, as the adventure tourism industry of Nepal continues to draw mountaineers and trekkers (and thus their Sherpa guides and porters) to high peaks, interactions with these environments increase, ultimately contributing to the disastrous potential of Himalayan avalanches. For instance, the April 18, 2014, Mt. Everest avalanche was the largest one-day death toll historically recorded during that time. These smaller-scale disasters, may be perceived as less severe than larger natural disasters (i.e., earthquakes, tsunamis, floods); however, the physical and economic losses incurred can still be seen as quite devastating and disastrous to the individuals and communities affected. Moreover, small-scale events are often repetitive or chronic, creating greater concern among local communities, further necessitating socio-ecological resilience (Aryal, 2012; Gaillard et al., 2010).

   Da Gelje also recounted the unfortunate loss suffered during the 2014 avalanche, and went on to describe the hazardous nature of the disaster site, an unavoidable section of the commercial climbing route. While drawing an imaginary map with his hands, Da Gelje explained:

> This is icefall, this is Nuptse, this is Lola. So last year, the Icefall Doctor opened the route like that. And when the avalanche came here, people died here. All the Sherpas climbing with loads go through this place [Khumbu Icefall]. From here cracks ice and then people died. Sixteen people died.

Icefall Doctors are an elite team of local guides and climbers charged with securing and maintaining the routes up to the Everest summit, particularly through the notorious Khumbu Icefall. Da Gelje pointed out that the way up is indeed through this icefall, located at the head of the Khumbu glacier. Due to treacherous crevasses, ever-shifting ice blocks, and the hanging seracs the Khumbu Icefall is known as one of the most dangerous sections of the well-traveled commercial route on the south side of Mt. Everest. Da Gelje pointed out that "*all* the Sherpas climbing with loads go through this place." Certainly international mountaineers (who climb as part of commercial expeditions) pass by the Icefall as well, but those whose lives are in greatest jeopardy are the mountain workers who diligently fix ropes and repair routes there, and the Sherpa climbers who repeatedly carry loads of gear and equipment across this precarious section some 18–30 times per expedition (cf. Arnette, 2016; Bisharat, 2016; Peedom, 2015).

Excerpts from Sherpa climbers' narratives depict the fragile and precarious contexts that Sherpas individually engage within their heavily tourism-based livelihoods of guiding and working as porters for foreign mountaineers. Their understandings of the inherent risk assumed in their positions on the mountain are increasing, with awareness that, as the tourism industry (and subsequently the mountaineering industry) further develops within the Solukhumbu, disasters, accidents, and death tolls rise.

Moments of liminality

Liminality, an anthropological concept introduced by Arnold van Gennep (1960) and further conceptualized by Victor Turner (1969), may be regarded as a cultural apparatus, characterized by heightened reflexivity in which individuals are able to reflect on and critique the normative social structure, as well as explore new possibilities (Turner, 1987). The notion of liminality appeared in van Gennep's (1960) *rites de passage*, tribal initiation rituals that provoke "transition" through three distinct phases: separation, margin, and aggregation (or reincorporation). Turner (1969, p. 351) further developed understandings of the second phase, which he called a "liminal period," recognizing that the characteristics of the individual undergoing the ritual become ambiguous as he or she "passes through a cultural realm that has few or none of the attributes of past or coming state." In these moments it is as though individuals are reduced to a universal or uniform condition to be transformed, emerging from their symbolic ceremony with "additional powers to enable them to cope with their new station in life" (Turner 1969, p. 351).

More recently ideas of liminality have been adopted from cultural studies and applied within disciplines such as consumer sciences (cf. Cheung & McColl-Kennedy, 2015; Elliot, Harris & Baron, 2005; Kennett-Hensel et al., 2012), organizational studies (cf. Howard-Grenville et al., 2011), and most relevant to this chapter disaster studies (cf. Jencson, 2001; Oliver-Smith & Hoffman, 1999) to explore the impacts of social and structural disruption. According to Oliver-Smith and Hoffman (1999, p. 1), "Disasters take a people back to fundamentals." In these devastating moments "victims" expand their sense of self, community, and purpose-revealing experiences of transformation and survival (Jencson, 2001). Throughout the remainder of this section I draw on Sherpa climbers' confrontations with death and negotiations of disaster to explore the transformational potency of "liminality" from which adaptive behaviours and resilient decisions emerged.

Confronting death

A glimpse of death or a taste of risk illuminates the mutability of human beings (Lewis, 2000; Varley, 2006). These confrontations with mortality are propelled by "urgent" experiences or what American psychotherapist Irvin Yalom (1980, p. 8) identifies as "boundary situations," which include, but are not limited to, "one's own death, some major irreversible decision, or the collapse of some fundamental meaning-providing schema." Consequently from this rupture come contemplations of one's own existential position within the material–physical world (Berger 1967; Yalom, 1980). This is highlighted in Dawa's unexpected encounter with an avalanche while working for a commercial expedition team on Mt. Everest:

> I remember an avalanche. I tried to stop myself from being taken with an ice axe, trying to get it into snow, but it was fresh snow so it took all of us down. I was found at the head of avalanche with my body upside down. When I woke up, it was morning, and I was lying down with oxygen and glucose water on … I was pretty homesick after that accident. I wanted to go home.

Scholars who take up notions of liminality suggest that those affected by disasters, death, and the like find themselves "betwixt and between" their life prior to the event and an uncertain sense of the future (Cheung & McColl-Kennedy, 2015; Turner, 1979, p. 465). According to Jencson (2001) these moments mark the transition of an individual from one status to another, which is often accompanied by considerable stress, doubts, and fear about an uncertain future. For instance, Dawa's near miss triggered what he articulated as "homesickness," a longing for what he had known before the expedition while he contemplated his current position on the mountain. This sense of fear and uncertainty was emphasized in other Sherpa climbers' narratives as well. Lhakpa Dorji described his own "boundary situation" when clients stranded him alone on the south summit of Mt. Everest:

> they [the clients] leave me behind, and they run before, down. Then I am alone … I'm so tired, and I get the ice in my goggle, the sweat. When I almost get to the last camp I couldn't see, then I fell down. I fell down from that. It's about 250 m, about nearly 300 m I fell down, like rolling down. I don't know how long I was dead. I thought [it's] like a dream, like when I woke up I had no goggles. I had no ice axe … Then after that I get to [camp at] South Col at 6:30 pm. Then other members, about three members in the South Col, they don't care me. They didn't know where I was … When I was there I am so cold, I no can walk that time. I tried to make a cup of ice and tried to make the water. I couldn't eat anything, I didn't drink the water. The whole night I couldn't sleep.

According to Yalom (1980, p. 159), "Though the physicality of death destroys an individual, the *idea* of death can save him." As a moment of liminality, a close encounter with death encourages individuals to return to their fundamental priorities to thoughtfully deliberate what is truly important and meaningful (Cheung & McColl-Kennedy, 2015; Oliver-Smith & Hoffman, 1999). Death enacts a vast influence upon existence and our conduct, in which we can understand the way we live and grow (Yalom, 1998). Dawa indicated a change in his own conduct after his brush with death. His homesickness led him to eventually walk away from his position on the mountain, forfeiting his income from that particular expedition. Dawa remembered:

> The next morning, I walked down to my home in Phortse through Pangboche, where I had a cousin. I got to my cousin's home and at the same moment they were talking about me being taken by the avalanche. One of my cousins was crying thinking I was gone forever. They were happy to see me back ... After that I didn't return up.

Dawa explained that this Everest expedition marked his final mountaineering expedition, in part due to much family pressure. Here Dawa's individual experience of knowing death propelled family and other community members into questioning his employment on the mountain. In some sense, liminal moments allow communities to pause, to turn the lens on their condition, prompting new possibilities with hopes to ultimately alter cultural resources (Howard-Grenville et al., 2011; Oliver-Smith & Hoffman, 1999).

Death becomes a catalyst that shifts perceptions around realities. Boundary situations, what Berger (1967) calls "marginal situations," radically challenge social constructions and "objectivated" definitions of reality and how one understands the world, others, and self (Berger, 1967, p. 44). As Phuri Sherpa described his boundary situation, he admitted that it was not until that very moment that he decided to quit working on mountaineering expeditions. Phuri explained:

> While climbing Cho Oyu, I fell into a crevasse. There was heavy snow above and I didn't see the crevasse, and then I fell into it ... I felt very scared. One American doctor and I, two of us fell down. The [other] climbers were very quick. As soon as it occurred they were ready with rope to pull us up. I was lucky. They had really good hearts. We returned back from there ... After that I didn't go.

Climbing Cho Oyu, the world's sixth tallest mountain, was Phuri's last commercial mountaineering expedition. Similar to Dawa, his accident provoked a shift. When asked why he retired, Phuri responded, "I felt scared. After that I had to go to Makalu. There was a phone call and message from the office, but I didn't go, I was too scared." Phuri's time in the crevasse can be understood as an unexpected rupture, a shocking break or happening in his usual routine of his work on the mountain (Watkins & Shulman, 2008).

In those frightening moments of facing possible death, lies the potency of liminality. Phuri was challenged to make sense of life, which led to a disruption in how he engaged his responsibilities on the mountain. Betwixt-and-between, Phuri displayed a capacity to step back and consider consciously what regulates his behaviour (Howard-Grenville et al., 2011). Confronting one's mortality, or the anticipated risk of dying contributed to the ways in which Sherpas enacted freedoms on the mountainside. Existential thinking suggests it is the condition of death that "makes it possible for us to live life in an authentic fashion" (Yalom, 1998, p.187). Here, rather than thinking of authentic as an undisputed origin, we might think of authentic as a purposive and responsible mode of human life.

Transforming communities

Individual Sherpa climbers were propelled into liminal moments, a period of time in which they could rethink pre-existing practices, assumptions, and beliefs (Cheung & McColl-Kennedy, 2015). Mälksoo (2012, p. 489) recognizes, "the strength of liminality as the phase of pure possibility underscores the potential power of agency in the liminal process," whereby existing realities are restructured to create new ones. Such narratives of agency emerged within Sherpa climbers' negotiations of death. For example, Lhakpa Dorji, who was left by his clients and fell nearly 300 m while descending Mt. Everest, explained his actions the following mountaineering season.

> I thought that time, no more … I was in Camp II, and they pressured me [to join] another team for another summit. I said "NO!" Then they bring me to Basecamp and they talk a lot of questions. "Please go do another summit." And I said, "NO!"

The positive and transformative aspect of death and disaster is illuminated in the power and agency enacted by Lhakpa Dorji. As he navigated away from the undesirable trajectory of the mountaineering industry, despite the potential economic loss he might have incurred, Lhakpa Dorji created a new reality for himself.

The prevailing agency of Sherpa climbers can also be seen on a larger community scale. For instance, in 2014 the mountaineering industry was effectively blamed for the deaths of the Sherpa and Nepali climbers who were laying ropes and safety lines in preparation for the hundreds of foreign climbers who would inevitably attempt the Everest summit in the coming season. That year, the Nepali and Sherpa climbing community moved from the initial shock of the disaster through a liminal moment which constituted a formative experience for the collective. Tashi Sherpa recalled, "That time I was on my way to basecamp, the other, my friends, all of them did meet for a meeting, then after all of the expeditions quit." Lakpa Sherpa also reflected on how the climbing community transitioned from this tragic accident, recognizing the reasons why climbers chose to stop climbing after the 2014 Mt. Everest disaster:

That year, you know it's so many accidents, so all the climbers they decide to close. It was black year, bad luck, so many friends lost ... You lose some business, but sixteen, seventeen people died, and after if you continue you know people psychologically affected not good feeling you know.

Cheung and McColl-Kennedy (2015) suggest that during periods of disaster and displacement, a strong collective bond is formed. Those affected by a difficult event may be stripped of their familiar institutions, routines, and resources, a grim situation that entails individuals to come together to find new ways to deal with the challenges of the circumstances of their new emerging worlds (Cheung & McColl-Kennedy, 2015; Oliver-Smith & Hoffman, 1999). A social resilience is activated as Sherpas disregard the pressures of commercial climbing teams. Rinchen Sherpa acknowledged that working on Everest is not the only employment opportunity in the Solukhumbu region as he scorned the single Chinese climber who continued to pursue the summit and tried hiring a Sherpa guide just days after the 2014 mountain closure:

You take a lot of money, pay my porter, like money is nothing. Money is pay, okay, but ... safety is life. Life is life. Life is important. Money we will make next year. Next day. Another job, many jobs, not only mountaineering Everest. We climb another mountain. Or by jeep, by driver, another business. I think that money is important, but must take care the life more.

In moments of liminality, communities find a source of renewal, as they begin to acknowledge the power of their collective agency to create a new setting or "structure," one that can be regarded as better than the old (Mälksoo, 2012).

Concluding thoughts

Liminality, analogous to the subjunctive, is "full of potency and potentiality" (Turner, 1979, p. 466). As illuminated in the excerpts from the Sherpa climbers' narratives, moments of liminality become adaptive and transformative as individuals and communities use these moments of uncertainty to take stock of the purpose and significance of their lives. Evoked by the very act of imagining new boundaries, economic concerns that were previously understood as one of the forces contributing to Sherpas working in the mountaineering industry are suspended. Rather than being seen as static and always vulnerable, Sherpa communities harness the inherent resilience of systems deemed vital to life—in this case their individual and collective agency—to help themselves self-organize in a bottom-up fashion in their constant process of becoming something else (Rogers, 2012; Zebrowski, 2015).

These stories of social-ecological resilience are part of what contributes to the complexities of these mountain communities, where at times death and disaster provide a pause for Sherpas to consider the status quo—their continued involvement in the mountaineering industry, an industry that so often places their lives at risks. Nevertheless, I importantly reemphasize Mulligan et al.'s (2016) concern: as resilience discourses embrace a shift from equilibrium bounce-back to performative transformation in a world fraught with socio-economic and environmental uncertainties, it is necessary to continuously negotiate a range of countervailing possibilities. Moreover, disaster and death, as moments of liminality, which can be conceptualized as an integral part of a community's adaptive capacity, may not always generate new and emerging possibilities for communities.

On April 25, 2015, the 7.8 magnitude earthquake that hit the interior of Nepal killed over 8,800 people, injured more than 22,000, and displaced hundreds of thousands from their damaged homes (NEOC, 2015). In this disastrous moment, the reflexive pause that was brought about by previous confrontations with death was quickly replaced with appeals for increased tourism. As an integral contributing factor to Nepal's economy, the tourism industry was not pardoned from the devastation, which prompted a rapid decrease in arriving tourist and mountaineering numbers. Ironically, this decrease in demand was met with increasing pleas by global organizations and local community members alike, who encouraged continued tourism to Nepal, despite its broken state, due to its dependency on tourist dollars for survival (Tshering Sherpa, 2015; Tourism Concern, 2015). A resilience approach focuses on how to build capacity to deal with unexpected change, yet the aftermath of the 2015 Nepali earthquake illuminates the tensions and contradictions of tourism development within Nepal, raising urgent questions of resilience, sustainability, and transformability.

Acknowledgement

To all of the Sherpa climbers who graciously shared their stories with me, thank you. I would also like to thank Dr. Susan Arai and my colleagues in the REC 640 Community Development graduate course who provided instrumental insight and feedback that assisted in the writing of this manuscript.

Notes

1   It should be briefly noted that the word Sherpa has multiple meanings within a mountain context. Originally, and more often than not, this word signifies a member belonging to a specific ethnic group in the Himalayas of Nepal (Ortner, 1999). However, the category of "Sherpa" has undergone changes as these Sherpa natives have been recognized to be well suited for supporting commercialized climbing expeditions. Thus, "Sherpa" as an identifier was adopted to indicate individuals who assume a role and status as a specialized high-altitude porter with at least some (and sometimes a lot of) climbing expertise (Ortner, 1999). I use Sherpa throughout this document to signify ethnicity of the mountain populations.

2   Although being contested more recently, since the 1990 World Development Report (WDR) on Poverty, the World Bank has anchored its international poverty line as "$1 a day," a typical line amongst low-income countries in the data available at the time of the 1990 WDR (Ravallion, 2010).

References

Adams, V. (1992). Tourism and Sherpas. *Annals of Tourism Research,* 19(3), 534–54.

Adger, W.N. (2000). Social and ecological resilience: Are they related? *Progress in Human Geography,* 24, 347–74.

Arnette, A. (2016). Everest/Lhotse 2016: Arrival at basecamp, April 10. Retrieved from www.alanarnette.com.

Aryal, K.R. (2012). The history of disaster incidents and impacts in Nepal: 1900–2005. *International Journal of Disaster Risk Sciences,* 3(3), 147–54.

Berger, P.L. (1967). *The Sacred Canopy: Elements of a Sociological Theory of Religion.* New York: Doubleday.

Bhattarai, K., Conway, D. & Shrestha, N. (2005). Tourism, terrorism, and turmoil in Nepal. *Annals of Tourism Research,* 32(3), 669–88.

Biran, A., Liu, W., Li, G. & Eichhorn, V. (2014). Consuming post-disaster destinations: The case of Sichuan, China. *Annals of Tourism Research,* 47, 1–17.

Bisharat, A. (2016). After two years of deadly Everest avalanches, Nepal introduces new safety measures. *National Geographic,* April. Retrieved from http://adventureblog.nationalgeographic.com/

Bjønness, I. M. (1986). Mountain hazard perception and risk-avoiding strategies among the Sherpas of Khumbu Himal, Nepal. *Mountain Research and Development,* 277–92.

Bodin, P. & Wiman, B. (2004). Resilience and other stability concepts in ecology: Notes on their origin, validity, and usefulness. *ESS Bulletin,* 2, 33–43.

Bonanno, G.A. (2004). Loss, trauma, and human resilience: Have we underestimated the human capacity to thrive after extremely aversive events? *American Psychologist,* 59(1), 20–28.

Bott, E. (2009). Big mountain, big name: Globalised relations of risk in Himalayan mountaineering. *Journal of Tourism and Cultural Change,* 7(4), 287–301.CBC (2012). Canadian Everest victim used inexperienced company, lacked oxygen: Senior Sherpa expressed fears about climber dying in summit attempt. Canadian Broadcasting Corporation. *CBC News,* September 13. Retrieved from www.cbcnews.com.

Cheung, L. & McColl-Kennedy, J.R. (2015). Resource integration in liminal periods: Transitioning to transformative service. *Journal of Services Marketing,* 29(6/7), 485–97.

D'Aliesio, R. (2012). Grieving man rebukes guides for failing to halt wife's Everest trek. *The Globe and Mail,* September 13. Retrieved from www.theglobeandmail.com.

Davis, W. (2014). As equals on the mountain, the Sherpas deserve better. *The Globe and Mail,* April 26. Retrieved from www.theglobeandmail.com

Davoudi, S. (2011). The legacy of positivism and the emergence of interpretive tradition in spatial planning. *Regional Studies,* 46(4), 429–41.

Davoudi, S. (2012b). Resilience: a bridging concept or a dead end? "Reframing" resilience: challenges for planning theory and practice interacting traps: resilience assessment of a pasture management system in northern afghanistan urban resilience: what does it mean in planning practice? Resilience as a useful concept for climate change adaptation? The politics of resilience for planning: a cautionary note. *Planning Theory & Practice,* 13(2), 299–333.

Davoudi, S. & Strange, I. (2009). Space and place in twentieth century planning. In S. Davoudi and I. Strange (Eds.), *Conceptions of Space and Place in Strategic Spatial Planning* (pp.1–6). London: Routledge.

Elliot, D., Harris, K. & Baron, S. (2005) Crisis management and services marketing. *Journal of Services Marketing,* 19(5), 336–45.

Ganju, A. & Dimri, A.P. (2004). Prevention and mitigation of avalanche disasters in Western Himalayan region. *Natural Hazards,* 31, 357–71.

Gaillard, J.C., Wisener, B., Benouar, D., Cannon, T., Lawrence, C., Dekens, J., ... Vallette, C. (2010). Alternatives for sustained disaster risk reduction. *Human Geography*, 3(1), 66–88.

van Gennep, A. (1960). The Rites of Passage. Chicago: University of Chicago Press.

Hiller, J. (2015). Performances and performativities of resilience. In R. Neunen, K. van Assche & M. Duinevald (Eds.), *Evolutionary Governance Theory: Theory and Applications* (pp. 167–84). London: Springer.

Holling, C.S. (1973). Resilience and stability of ecological systems. *Annual Review of Ecology and Systematics*, 4, 1–23.

Holling, C.S. (1996). Engineering resilience versus ecologica resilience. In P.C. Schulze (Ed.) *Engineering Within Ecological Constraints* (pp. 25–62). Washington, DC: National Academy Press.

Howard-Grenville, J., Golden-Biddle, K., Irwin, J. & Mao, J. (2011). Liminality as cultural process for cultural change. *Organization Science*, 22(2), 522–39.

Jencson, L. (2001). *Anthropology and Humanism*, 26(1), 46–58.

Kennett-Hensel, P., Sneath, J.Z. & Lacey, R. (2012). Liminality and consumption in the aftermath of a natural disaster. *Journal of Consumer Marketing*, 29(1), 52–63.

Krakauer, J. (2014). Death and anger on Everest. *The New Yorker*, April 21. Retrieved from www.newyorker.com/news/news-desk/death-and-anger-on-everest.

Lewis, N. (2000). The climbing body, nature, and the experience of modernity. *Body & Society*, 6(3–4), 58–80.

McFarlane, A.C. & Norris, F. (2006). Definitions and concepts in disaster research. In F. Norris, S. Galea, M. Friedman & P. Watson (Eds.), *Methods for Disaster Mental Health Research* (pp. 3–19). New York: Guildford Press.

Mälksoo, M. (2012). The challenge of liminality for international relations theory. *Review of International Studies*, 38, 481–94.

Massey, D. (2005). *For Space*. London: Sage.

Mulligan, M., Steele, W., Rickards, L. & Funfgeld, H. (2016). Keywords in planning: What do we mean by "community resilience"? *International Planning Studies*, DOI:10.1080/13563475.2016.1155974

NEOC (2015). Death toll. Injured toll. National Emergency Operation Centre (NEoCOffical) #NepalEarthquake [Tweet], 16 July. Retrieved from https://twitter.com/NEoCOfficial.

Nepal, S.K. (2003). *Tourism and the Environment: Perspectives from the Nepal Himalaya*. Nepal: Himal Books.

Nepal, S.K. (2005). Tourism and remote mountain settlements: Spatial and temporal development of tourist infrastructure in the Mt. Everest Region, Nepal. *Tourism Geographies*, 7(2), 205–27.

Norris, F. H., Stevens, S. P., Pfefferbaum, B., Wyche, K.F. & Pfefferbaum, R.L. (2008). Community resilience as a metaphor, theory, set of capacities, and strategy for disaster readiness. *American Journal of Community Psychology*, 41, 127–50.

NPR (2013). On Mount Everest, Sherpa guides bear the brunt of the danger. National Public Radio. *NPR Fresh Air*, August 14. Retrieved from www.npr.org/2013/08/14/206704533/on-mounteverest-sherpa-guides-bear-the-brunt-of-the-danger.

Oliver-Smith, A. & Hoffman, S.M. (1999), *The Angry Earth: Disaster in Anthropological Perspective*. New York: Routledge.

Ortner, S.B. (1999). *Life and Death on Mt. Everest: Sherpas and Himalayan Mountaineering*. Princeton, NJ: Princeton University Press.

Payne, E. & Shrestha, M. (2014). Avalanche kills 12 in single deadliest accidents on Mount Everest. *CNN World*, April 20. Retrieved from www.cnn.com/2014/04/18/world/asia/nepal-everest-avalanche/.

Prettyman, B. (2014). Apa Sherpa: After the deadly avalanche, "leave Everest alone." *The Salt Lake Tribune*, April 29. Retrieved from http://archive.sltrib.com/story.php?ref=/sltrib/news/57877542-78/apa-sherpa-everest-mountain.html.csp.

Raj Panday, D. (1999). *Nepal's Failed Development: Reflections on the Mission and the Maladies*. Kathmandu: Nepal South Asia Centre.

Ravallion, M. (2010). World Bank's $1.25/day poverty measure-countering the latest criticisms. *Research at the World Bank*. Retrieved from http://econ.worldbank.org.

Rogers, P. (2012). *Resilience and the City: Change, (Dis)Order and Disaster*. Farnham: Ashgate.

Rogers, P. & Aitchison, J. (1998). *Towards Sustainable Tourism in the Everest Region of Nepal*. Kathmandu: IUCN Nepal.

Sauerborn, R. & Ebi, K. (2012). Climate change and natural disasters: integrating science and practice to protect health. *Global Health Action* 5. DOI: 10.3402/gha.v5i0.19295

Schaffer, G. (2013). The disposable man: A western history of Sherpas on Everest. *Outside Magazine*, July 10. Retrieved from www.outsideonline.com/outdoor-adventure/climbing/mountaineering/Disposable-Man-History-of-the-Sherpa-on-Everest.html.

Shackley, M. (1999). Managing the cultural impacts of religious tourism in the Himalayas, Tibet, and Nepal. In M. Robinson and P. Boniface (Eds.), *Tourism and cultural conflicts*. Wallingford: CABI Publishing.

Shrestha, H.P. & Shrestha, P. (2012). Tourism in Nepal: A historical perspective and present trend of development. *Himalayan Journal of Sociology & Anthropology*, 5, 54–75.

Singh, T. (1980). *Studies in Himalayan Ecology and Development Strategies*. New Delhi: The English Book Store.

Smithson, J., Ikin, B. (Producers) & Peedom, J. (Director) (2015). *Sherpa: Trouble on Everest* [Documentary]. Australia: Arrow Media.

Tourism Concern (2015). Tourism concern: Action for ethical tourism. *Tourism Concern Newsletter*, May. Retrieved from http://us4.campaign-archive2.com/?u=e44b83e52db7ad342ba8d66ed&id=61716cffc4

Tshering Sherpa, A. (2015). Earthquake update from SoluKhumbu (Mount Everest Region). *Climate Alliance of Himalayan Communities*. Retrieved from www.alanarnette.com/blog/2015/05/13/update-on-latest-nepal-earthquakes/#jp-carousel-21131.

Turner, V. (1969). *The Ritual Process: Structure and Anti-Structure*. Chicago: Aldine Press.

Turner, V. (1979). Frame, flow and reflection: Ritual and drama as public liminality. *Japanese Journal of Religious Studies*, 6(4), 465–99.

Turner, V. (1987). *The Anthropology of Performance*. New York: Performance Arts Publications.

UNDP (2015). *Human Development Report*. United Nations Development Programme. Retrieved from http://hdr.undp.org/sites/all/themes/hdr_theme/country-notes/NPL.pdf.

Varley, P. (2006). Confecting adventure and playing with meaning: The adventure commodification continuum. *Journal of Sport & Tourism*, 11(2), 173–94.

Watkins, M. & Shulman, H. (2008). *Toward Psychologies of Liberation*. New York: Palgrave Macmillan.

World Bank (2014). Data: Nepal population. *World Bank Organization*. Retrieved from http://data.worldbank.org/country/nepal.

Yalom, I.D. (1980). *Existential Psychotherapy*. New York: Basic Books.

Yalom, I.D. (1998). *The Yalom Reader: Selections from the Work of a Master Therapist and Storyteller*. New York: Basic Books.

Zebrowski, C. (2015). *The Value of Resilience: Securing Life in the Twenty-First Century*. London: Routledge.

13 Tourism and the psychologically resilient city

Christchurch after the earthquake

Irina Herrschner and Phoebe Honey

Introduction: a city's capacity to adapt

This chapter takes an industry-focused approach, analyzing the response of the tourism industry to the earthquakes in Christchurch in 2010 and 2011. Findings presented in this chapter are based on a two-week fieldwork period in Christchurch and over 200 hours of semi-structured interviews with key-stakeholders of the city's tourism industry. By doing so, the chapter adds to the discussion surrounding crises management in the tourism sector and city resilience following a crisis. In particular we examine the factors that have influenced Christchurch's adaptive capacity and adaptive cycle to respond to the earthquakes of 2010 and 2011 and analyze the risks and opportunities that arise from urban crises. Thus, we are posing the question, if recovery is the right term to use for cities "bouncing back from disaster", as its suggested return to its original form has not been observed for this case study, but more importantly also limits the scope of urban responses to crises. In this chapter, crises are analyzed for their challenges, but also their transformative capacity that changes the fabric of a city and its society. This stands in contrast to the metaphorical use of the term "resilience," which suggests the capacity of a material or system to return to equilibrium after a displacement (Norris, Stevens, Pfefferbaum, Wyche & Pfefferbaum, 2008). A resilient material, for example, bends and bounces back, rather than breaks, when stressed (Bodin & Wiman, 2004; Gordon, 1978).

Cities are the nodal points of global cultures, information and innovation (Castells, 1994, Sassen, 2001). Their fabric is a close network of social and economic connections and routines. Change is a necessary part of this network, as innovation stands at the heart of urbanism. Recent changes, in particular the increasing globalization of economy and culture have transformed many cities into post-industrial knowledge-based economies, where the gap between production (food and energy) and city life has significantly widened (Clifford et al., 2016). Whilst a city is in many ways an independent network, it also tends to depend heavily on the resources from the non-urban regions, providing food in particular. This dependence necessitates not only organized trade processes with the surrounding country, but also the collaboration and sharing of resources by the residents of a city. The dependency on one another for food, medicine and social care created a network of interdependencies that transformed a loose collection of cohabitants into parts of a community. Especially in the global cities of the West, this disjuncture has led to highly specialized knowledge in cities, but a lack of basic skills of survival (Bontje & Musterd, 2016). Supported by a post-industrial system, exchanging knowledge for products, cities are kept alive by not only the surrounding country, but also a global network of production, thereby emphasizing the interconnectedness of cities, regions and their inhabitants.

In understanding cities in this way, we relate the response of Christchurch to the earthquakes in 2010 and 2011 to the theory of panarchy, thereby highlighting the complexity of its response to change in ecological and human systems. Panarchy is a conceptual framework that transcends boundaries of scale and discipline and organizes our understanding of economic, ecological and institutional systems (Gunderson, 2001, p. 5). It describes the ways in which complex systems of people and nature are dynamically organized and structured across scales of space and time (Gunderson, 2001; Allen, Angeler, Garmestani, Gunderson & Holling, 2014). Under normal circumstances, this intertwined system is beneficial for city and country, functioning as an economic and social symbiosis. Due to unforeseen circumstances affecting this symbiosis, however, the city can be quickly pushed to its limits in terms of providing for basic human needs. This is what is referred to as a traumatic experience, when a person, or in this case a city, relies on itself to survive a catastrophe (van der Kolk, 2003; Fraisl & Stromberger, 2004)

After such a trauma, the modern post-industrial city quickly needs to learn, adapt and transform itself in order to survive, for which the aforementioned panarchy of social networks are crucial, where citizens act independently of their usual status and roles and collaborate under new circumstances and within a new system. This is also illustrated in the four success factors of the resilience of urban, human-dominated ecosystems:

- economic development;
- social capital;
- information and communication; and
- community (Norris et al., 2008).

Economic development in this case refers to the economic safety that a developed, affluent city offers its inhabitants, as well as the economic potential to react to the crisis afforded by urban wealth. Social capital describes the high educational standard of the city as well as the capacity of many to think creatively in reacting to the earthquakes. Information and communication is the ability to share relevant information with the right people. It lies at the heart of Christchurch's post-2011 tourism campaign, but it is particularly crucial in contemporary mediated societies, where information can be shared instantaneously with unlimited numbers of people. The speed of information is thus bi-fold, where its potential is significant in causing damage and panic, but also in sharing vital information with society, industry and politics. The fourth success factor, community, refers to the social network of the community and the sense of belonging. In the case of Christchurch, the community-driven response to the earthquake significantly facilitated its recovery process.

For this reason, we focus in particular on the role of social capital and the community of Christchurch. For this, we apply psychological resilience theory and use it in metaphorical terms for urban resilience and the case of Christchurch. We follow Paton and Johnston's (2016) positivistic approach to resilience as the capacity to react positively to a catalyst to change, rather than simply bouncing back in the term's more literal meaning. For this, we impose the same principles of personal resilience theory to urban resilience, in superimposing a body onto cityscapes (Reich, 2006). This is useful, as personal resilience theory mainly focuses on the building of resilience in individuals (Cohn, Fredrickson, Brown, Mikels & Conway, 2009), whilst literature on urban resilience retrospectively analyzes resilient reactions to disaster (Leichenko, 2011; Campanella, 2006). When adopting a key finding from personal resilience, that is, that pre-traumatic experiences are significant in shaping a persons' resilience, we argue that it is also the social capital of a city that defines its response to crises. In particular a close network of social connections and a sense of belonging are positively linked to a resilient and creative response to crises (Aldrich, 2012; Berkes & Ross, 2013). An example for a fast, urban transformation and adaptive capacity of a community was Christchurch's response to the series of earthquakes that shook the city in 2010 and 2011.

The earthquakes (2010 and 2011) in Christchurch, New Zealand

Christchurch is located in the Canterbury region on South Island of New Zealand and refers to itself as the "Garden City." The city has a long history, being first inhabited by moa-hunting[1] tribes, followed by the Waitaha who migrated from the east coast of North Island in the sixteenth century. This migration was joined by the Ngati Mamoe and Ngai Tahu[2] and continued until about 1830. The first Europeans landed in Canterbury in 1815 and in 1840 settled on the plains. By 1850 whaling ships operated out of Lyttelton.[3] During 1850–51 the first organized groups of English settlers, the founders of Christchurch, arrived on the "first four ships" into Lyttelton Harbour. Christchurch became a city by Royal Charter on July 31, 1856, making it officially the oldest established city in New Zealand. Today, Christchurch has a resident population of 341,469 (Statistics New Zealand, 2013a), making it New Zealand's second largest city.

Canterbury has long been recognized as living "off the sheep's back" (Pawson & Perkins, 2013) and originally Christchurch was built on primary products. Contemporary Christchurch has a diversified service-based economy, with most citizens employed in retail trade and tourism forming a major part of the city's economy, although construction has become the largest industry in Christchurch after the earthquakes (Statistics New Zealand, 2013b). Christchurch is the major gateway to South Island, promoted by Tourism New Zealand as a premier road trip destination (Tourism New Zealand, 2012). Tourism is one of New Zealand's largest export industries, second only to the dairy industry in terms of foreign exchange earnings. It directly employs 4.7 percent of the New Zealand workforce and indirectly employs a further 3.1 percent. Tourism directly employs 168,012 people (6.9 percent of the total employment in New Zealand). In the year ending

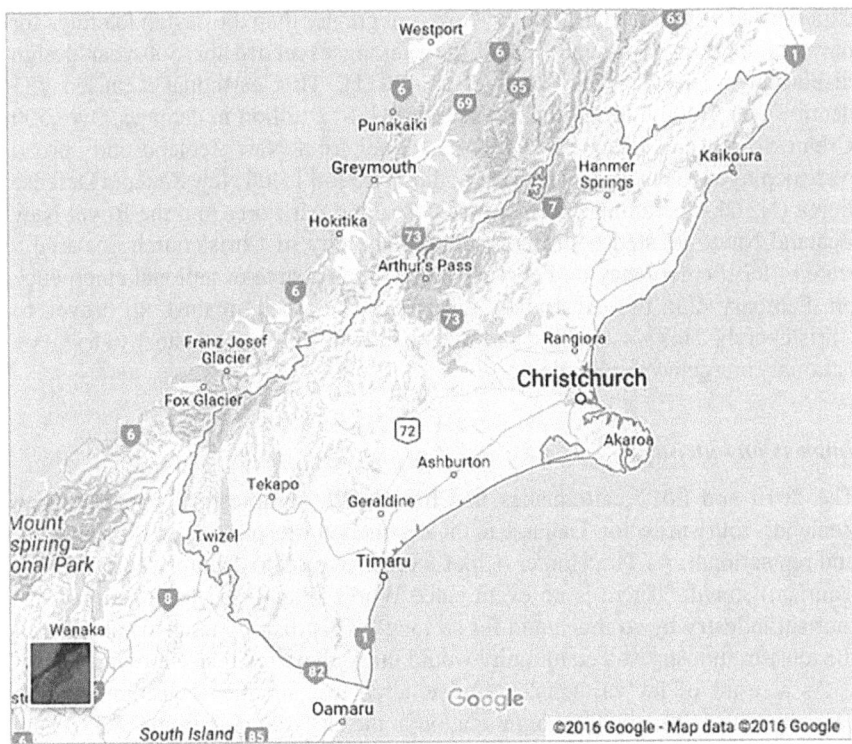

Figure 13.1 Christchurch and Canterbury Region, South Island, New Zealand
Source: Google Maps data

September 2016, International tourist expenditure accounted for NZ$11.8 billion or 17.4 percent of New Zealand total export earnings. Tourism directly contributes NZ$10.6 billion (or 4.9 percent) to New Zealand total GDP; further NZ$7.9 billion (or 3.6 percent) are indirectly contributed (Statistics New Zealand, 2015).

Despite Christchurch being located on the so-called Ring of Fire,[4] the city experienced relatively low seismicity during the previous 100 years (Stevenson et al., 2011). This meant that the earthquake in September 2010 struck the city by surprise as it was ill-prepared (Orchiston, 2013). The human loss sustained from the earthquake was relatively small, as it occurred during the night; property losses, however, exceeded US$4 billion. Only six months later, on February 22, 2011, when many businesses were reopening after repairs from the September event, a second earthquake – an aftershock of the first quake – struck Christchurch.

The epicenter of this second earthquake was much closer to central Christchurch than the first one, and at a depth of 5 km, making it the biggest earthquake in New Zealand's history. The earthquake produced peak ground accelerations in Christchurch's central business district (CBD) that were 2.5 times greater than the accelerations felt during the September earthquake. Peak ground accelerations experienced within the CBD were 50 percent greater than the design loadings for new buildings in Christchurch and the shaking exceeded the 500-year design displacement spectra (Stevenson et al., 2011). This earthquake caused 185 deaths, over 7,000 injuries and in excess of US$12 billion in damage. Over 500 Urban Search and Rescue (USAR) personnel from New Zealand and abroad were deployed to rescue and recover bodies. Around 1,400 New Zealand Defense Force (NZDF) including the Royal New Zealand Air Force and the Royal New Zealand Navy assisted with the response. The city of Christchurch declared a state of local emergency on February 22 and then a state of national emergency on February 23; the Australian government warned against all travel to Christchurch. This was the first earthquake in New Zealand's history to trigger a national emergency declaration.

Impacts on tourism

The 2010 and 2011 earthquakes in Christchurch significantly reshaped New Zealand's tourism sector. Damage to the destination was physical, developmental and reputational. As Tim Hunter (Chief Executive of Christchurch & Canterbury Tourism) stated: "There is no event since World War II that has disrupted our tourism industry by so much and for so long". This disruption changed the way the tourism industry and community would think about resilience forever.

As a result of the earthquake, Christchurch lost significant accommodation capacity with commercial accommodation in the city dropping from 7,809 rooms to 4,432 rooms (57 percent of pre-earthquake capacity). Many visitor experiences were impacted due to loss of infrastructure with sports stadiums, swimming pools, tourism infrastructure and conference centers all damaged; key attractions of the city, such as the Cathedral, City Square and the Arts Centre also had to close. Christchurch became a city with an almost empty CBD, with only 10 restaurants remaining out of the previously 100. Guest nights dropped by 105,000, accommodation capacity dropped by 57 percent, and 40 percent of CBD restaurants in the CBD were destroyed.

One year after the disaster, Australian arrivals in Christchurch had reduced by 34 percent and airline capacity between Australia and "the Garden City" dropped by 25 percent. (Tim Hunter, Christchurch and Canterbury Tourism, personal communication, 2015). In 2015, four years after the earthquake, 80 percent of buildings within the CBD had been demolished, anchor projects remained on-hold and both government and private enterprise were uncertain about investment. Although the recovery of the industry has been slow, the adaptive capacity of New Zealand's tourism operators and the resilience of Christchurch illustrate the often-unforeseeable opportunities and risks that come with a crisis. Distinct in the recovery of Christchurch are three time-periods: the immediate responses to the crisis, temporary urban renewals and businesses, and the permanent rebuild of the city. In this we follow Hollings (1986) adaptive resilience thinking and apply the model to the social, economic and political responses to the earthquakes sequence in Christchurch (2010 and 2011).

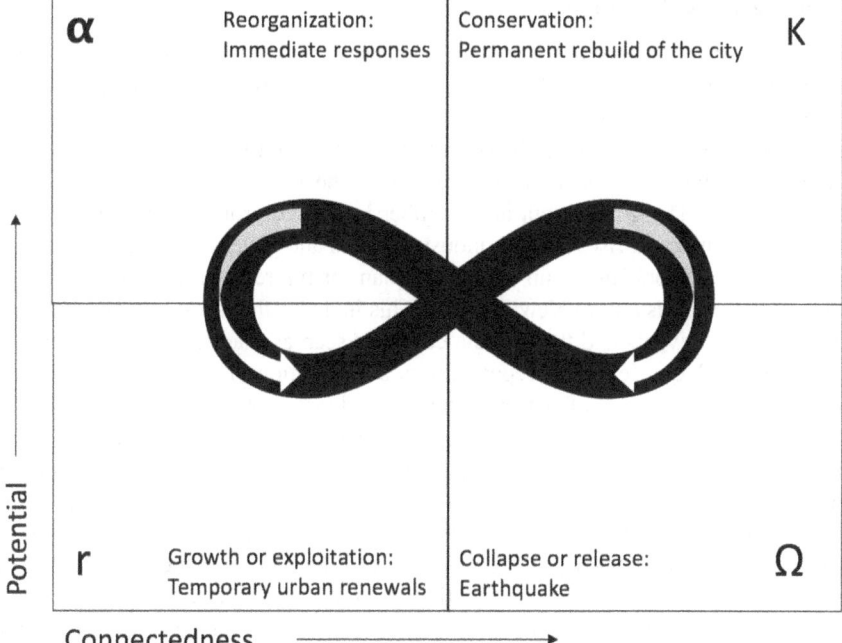

Figure 13.2 Adaptive Cycle applied to Christchurch
Source: Authors

Emergency and immediate responses

Top-down government responses

In line with Maslow's hierarchy of needs, response to a crisis follows a similar trajectory as an individual, where safety takes first priority, followed by shelter. In Christchurch, safety was established by creating a "tri-color placard tagging" around the CBD that regulated and restricted access to areas deemed dangerous to enter (Stevenson, Seville & Vargo, 2012). The area became known as the Red Zone and an access pass system was set up by the Ministry of Civil Defence and Emergency Management to enable people to travel in and out of the CBD. Business owners, hoteliers and people who had belongings within this area were not permitted access and thus could not retrieve their belongings. This was one of the greatest challenges in the initial recovery phase especially for the tourists wanting to recover their luggage and passports. Although tourism "wasn't a major part of the planning" (Charlie Ives, Regional Tourism Organizations New Zealand), the tourism industry played an important role in the recovery process, with many people performing roles outside of their job, sharing resources and using the tourism information infrastructure as general points of information for residents and visitors alike. This has helped make tourism an important element in this disaster/post-disaster environment and essential to the resilience of a city.

Creating new collaborations

The collaborative nature of the tourism industry facilitated the emergence of post-disaster networks such as the Visitor Sector Emergency Advisory Group (VSEAG). The group was set up after the earthquake in September, as a national organization that collectively plans to be ready for and able to respond to emergencies of national significance that occur locally or in a region that hosts large numbers of visitors to New Zealand. The government and membership-based group is headquartered in Wellington and comprises tourism industry bodies and government agencies active in the visitor sector. The group's aim is to plan for future emergencies that impact on international visitors to New Zealand. This includes threats to the visitor sector by unforeseen events and their media coverage seen as inaccurate and potentially damaging to New Zealand's reputation as a destination. The group calls on the international networks of its members to "disseminate accurate and timely information offshore" and "to assist with national and local responses during emergencies" (OECD, 2014). Calling on the group's close-knit social network, a member of the group stated: "It is likely the relationship-based nature of the tourism industry played a major role in the rapid and effective response" (Visitor Sector Emergency Advisory Group). The role of the community and the pre-established networks is highlighted here, facilitating a quick and effective response. These diverse groups of actors with different roles are critical in the resilience of social-ecological systems, as they provide overlapping functions with different strengths.

The unofficial and unplanned response

Following on from this relationship-based understanding of responding to crisis, the hotel sector set up an unofficial human-resource network, sharing staff and resources between functioning and non-functioning hotels:

> The HR initiative is a classic example of a community that got together and came up with an initiative to help everybody. You could easily say "my hotel hasn't been affected and good luck to the rest of you" but we all got together and decided how we could help one another ... Particularly to see those who had been through trauma in the city to come out and help, was an absolute credit
>
> (Bronwyn Knutson, Peppers Clearwater Resort)

This strategy not only allowed for a large number of businesses to continue functioning as well as for staff to remain in employed; it also addressed the psychological side of the disaster by continuing routine communication and providing a sense of purpose to staff (Tim Dearsley, the All Seasons Hotel – now Ibis Christchurch). As part of the especially established Canterbury Earthquake Recovery Authority (CERA) acting as a connection between these unofficial responses and the official government initiative, the tourism sector worked closely with government and Christchurch City Council. This government and private partnership allowed for an effective response to the crisis, e.g., the return of over 1,000 pieces of luggage to their owners in 37 countries.

Illustrated here is the particular importance of flexible partnerships and creativity. The ability to cope in an effective manner depends on the strength of partnerships and sense of belonging to a place. Similar to personal resilience depending on inner strength and a strong physical constitution, the resilience of cities depends on their communal strength and strong physical infrastructure (Newman et al., 2009, p. 1). For Christchurch this meant that, despite its physical damage, the city was able to function effectively by calling on social networks and personal relationships. The ability and preparedness of the government to form new partnerships and to allow for bottom-up initiatives further illustrates the stable, democratic foundation of the city. This is in line with what Norris et al. (2008) describe as the adaptive capacity of a city, comprising the four primary sets of networked resources mentioned above (economic development, social capital, information and communication, and community competence).

Handling the media and social media

Whilst the ability to quickly share information via the Internet was a crucial part of Christchurch's efficient longer-term response to the earthquake (dissemination of news and sharing of resources), the challenges of the media and social media in particular have been acknowledged by many scholars and practitioners alike (Tim Hunter, CEO Christchurch & Canterbury Tourism; Capozzi, 2013; Coombs, 2014).

Tourism is especially susceptible to reputational damage by inaccurate and/or sensationalist press reports, as the industry mainly depends on its intangible attractiveness (Gonzalez, 2008). Most disasters affecting a tourism destination, however, will attract unwelcome and often unfavorable attention with the potential to impact negatively on the perception of a destination. In the context of instant and global communication networks, messages are spread quickly and uncontrollably. In the case of New Zealand, largely dependent on its nature-based tourism industry, an effective and controlled media response was crucial. In the immediate response to the earthquake, however, the global image of Christchurch was that of a warzone: "The city's image of natural beauty and traditional architecture had been replaced with imagery of a devastated city." But it was the coverage of the continuing aftershocks that "cemented the broken city image" (Christchurch & Canterbury Tourism).

As part of the country's medium- to long-term strategy managing the country's international brand, Tourism New Zealand implemented a Search Engine Optimization campaign to suppress negative coverage. Whilst it is nearly impossible to remain in control over the global media, VSEAG agreed on one key message that was to be communicated by all members and associates: "Christchurch had suffered a terrible tragedy, but you can still come to New Zealand." By doing so, the organization reflected attention away from the crisis, putting into perspective the disaster that had devastated one city on one of the two islands. At the core of this message is damage minimization and the honesty to only market the destination as it is, or, as Christchurch and Canterbury Tourism states, "no attempt should be made to re-market a destination until the destination is convinced that the visitor experience is acceptable." One year after the earthquake Christchurch was again marketed to the Australian market with an NZ$1.6 million marketing campaign. Complementing the official marketing campaign, a quirky YouTube campaign led by Sir Robert Parker, a former Christchurch Mayor, aimed to attract online "talkability" and conversation. The series "Bob Thinks Big: Christchurch reimaged" called on Australia to lend Christchurch one of its iconic "Big Things."[5]

Creativity and adaptive urbanization in the three years following the earthquake

The "Bob Thinks Big" campaign illustrates well the spirit driving urban renewal in Christchurch. Following the idea of the popular proverb "Necessity is the mother of invention," Christchurch discovered its creativity through the crisis it had no choice but to experience. This confirms and expands on what Young and Stevenson (2013) describe as the "innovative milieu" of a city that allows cities to be creative in accelerated change. The period of "accelerated change" here refers to the period following on from the initial emergency response, where basic human needs had to be resolved in the most effective and efficient manner. Referring back to Maslow's hierarchy of needs, this period (the three years following the earthquake) is concerned with "belonging" and "esteem," rather than "survival" and "shelter." Urban renewal shaped this period, when multiple pop-up businesses and temporary structures changed the face of the city. As well as providing a space for the city to function, it has also created a space for the community to find their sense of belonging and rediscover their home city.

The large urban spaces that were vacant due to the destruction of original buildings meant that the earthquake also presented new opportunities for creative business development. An example of this is the "Smash Palace," a bar inside a bus on High Street as a placeholder for the original bar that was destroyed. The bar has become an icon for the "new Christchurch" and illustrates the spirit of creativity and temporality defining Christchurch during this period. Continuing from private–public partnerships during the initial phases of crisis management, many of these pop-up businesses were enabled by policy changes in regard to business start-ups implemented. Former Mayor Sir Robert Parker refers to these partnerships and the streamlining of business regulations when stating that "We [the Christchurch City Council] broadened the things that people could do and decreased the amount of difficulty they would have to go through to do it."

Re-establishing a routine

Another icon of the "new Christchurch" is the temporary shopping mall Re:START, an outdoor retail space consisting of temporary buildings made from shipping containers painted in a bright and colorful palette (Figure 13.3). Many interviewees noted that it has become the new heart of the city and expressed their gratitude that the development was permitted. It has been the foundation for the tourist industry in Christchurch and helped launch the city to number six in the Lonely Planet guide to the Top 10 cities to travel in 2013 (Atkinson et al., 2012). The title of the recommendation given to Christchurch "rising from the rubble with flair" exemplifies the renewal of the city, literally "bouncing back." Harnessing the creativity of the time, CERA and the Christchurch City Council made the space a priority area, so that people could come into the CBD, despite the red zone still in place. Whilst providing visitor attractions to the city in order to entice tourists back into the city, the shopping mall, as well as a temporary cathedral made from cardboard (Figure 13.4), facilitated an important recovery process for residents: the return to routine. Widely acknowledged as a vital part of long-term recovery from personal trauma (Partridge et al., 2012), re-establishing routine also helped the city to heal and recover.

Filling the gaps left and opened up by the earthquake

Broadening participation through engagement of all stakeholders is fundamental to building resilience and has been the key to Christchurch's recovery. The tourism industry was a great example of an industry that was involved in all levels of the response and recovery ensuring a more resilient future for Christchurch. A creative bottom-up approach drove much of the innovation in Christchurch, but despite the City Council's willingness to support small-scale businesses, a gap remained between innovators and policy makers. This gap did not, however, remain vacant for a long time, but was filled by so-called "enablers." Describing themselves as "enabler, do-er and champion of and for awesome," the Ministry of Awesome is a group of highly qualified and well-connected people helping to connect ideas with the resources needed to make them reality.

Figure 13.3 The Re:START Shopping Mall
Photography: Phoebe Honey

Figure 13.4 Rebuilding of ChristChurch Cathedral
Photography: Phoebe Honey

The principle of broadening participation is highlighted by case studies such as an independent trust "Life in Vacant Spaces" (LIVS), who acted as "site brokers for creative Christchurch." LIVS manages privately owned property for landowners and finds short- and medium-term uses for the many vacant sites and buildings of Christchurch. The objective of the trust is to make vacant space available to creative projects until the owner finds a permanent tenant or is ready to proceed with redevelopment. The project follows many of the principles of Renew Newcastle (Australia), where the CBD has been left empty by stalled redevelopments during the global financial crisis and the rise of suburban shopping centers (Stevenson, 2013, p. 69). "Greening the Rubble" and "Gap Fillers" are other projects using temporary vacant spaces to bring life back to the CBD, by transforming a "landscape of trauma" into a "landscape of creativity." "Greening the Rubble" creates and maintains temporary public parks on cleared sites in Christchurch, whilst "Gap Fillers" transformed a specific site in the CBD into a creative space that hosted amongst others a temporary garden café, live music, poetry readings and an outdoor cinema (powered by a bicycle). Building diversity and redundancy into Governance systems is essential to the urban resilience of a city. In Christchurch is also allowed for smaller scale development to occur during the planning processes of larger anchor projects. The mural in Figure 13.5 illustrates the communal aspect of the rebuild of Christchurch and draws attention to the intangible and community aspects of urban resilience.

Figure 13.5 Mural motivating a communal rebuilding of Christchurch
Photography: Phoebe Honey

Whilst literature on "creative cities" is shaped by a top-down approach and best-practice examples to "create creativity", initiatives such as the "Ministry of Awesome" illustrate a bottom-up approach that has creativity at the heart of its objectives as opposed to the means for productivity. This honest and earnest character shapes the recovery, where the city was forced to refocus on community and creativity. These projects also clearly demonstrate the creative potential of the city, but also the temporality at the core of every initiative. The projects are defined by their temporality and in themselves fill the gap between the earthquake and long-term rebuilding of the city. Drawing on the "Production of Social Space" proposed by French sociologist Henri Lefebvre (1974), the temporary urban landscape of Christchurch mirrors the society that created it. The small, low-risk and often only half-official buildings allow citizens to experiment with their new environment, as such refamiliarizing themselves with their city. The focus on transience makes also clear the desire for the permanent and the challenges that are still to come for rebuilding and reconnecting the social and physical fabric of the city. The current temporary structures of the city represent as such also a conversation of the citizens of Christchurch on their demands and wishes for a new city. A crucial part of rebuilding the city and community of Christchurch as a resilient city are also the ways the "old city" should be commemorated in the "new." In a similar stance to the Erinnerungskirche in Berlin, a physical memory of war in the heart of Berlin, the debate surrounding the rebuild of the ChristChurch Cathedral reflects the desire for a permanent manifestation of crisis integrated into the new city.[6] This debate in many ways crystallizes the debates of physical memory, heritage conservation and rebuilding the city anew (Fuge, Hering & Schmidt, 2014, p. 74). Five years after the disastrous earthquakes, these dimensions interplay in the long-term city planning.

The long-term plan for the city and tourism

It has long been recognized that recovery from trauma is an ongoing process, and, again applying the metaphorical person to Christchurch, complete recovery and return to an original pre-earthquake city is neither desired nor possible. The transitional city movement of Christchurch has emerged as one of the best in the world and has become a constant in the city, a point of interest for the community and visitors alike. The many creative projects have helped the city to become more inclusive, collaborative and confident in its creativity. Christchurch has as such become a place of possibility and a place of discovery for a curious visitor. This poses the question: what will be left when permanent buildings have replaced the temporary structures? How will the visitor experience develop, and what position will Christchurch permanently take in New Zealand's tourism industry?

In Christchurch the recovery has been slow, particularly in the city center, though managing slow variables and feedbacks has been crucial. By introducing systems to enable businesses to operate in a new environment has helped build community resilience, economic resilience and social resilience which has become a loop of self-reinforcing feedback. Christchurch's architecture has been modified and strengthened to withstand damage from future earthquakes. Further, the architecture focusing mainly on glass and steel not only provides a safe alternative for future earthquakes, but also acts as a constant reassurance for the city's traumatized residents.

Despite the generally slow recovery, there are signs of a recovering tourism industry. Visitation has increased by 3.6 percent (2015) on the previous year and the Australian market is recovering. Infrastructure and accommodation capacity is also recovering. Continuing capacity issues are slowing the sector's recovery, but new hotels are opening regularly. Michael Esposito points towards these positive developments and the significant increase in capacity over the past three years, mentioning that "Before the quake we had over 500,000 visitors per year. A year after we were down to 15,000. Now we are back up to around 350,000" (Michael Esposito, Welcome Aboard). Welcome Aboard[7] has also reopened all their attractions, visitation numbers are increasing and signs of economic recovery have become visible.

On July 30, 2012, the Christchurch Central Recovery Plan was released. The plan, developed by the Christchurch Central Development Unit (CCDU), includes input from residents, community groups and various government authorities charged with the task of developing the vision for the new Christchurch. The plan aims to make central Christchurch:

> the thriving heart of an international city. Drawing on its rich natural and cultural heritage, and the skills and passion of our communities, to embrace

opportunities for innovation and growth. Clearly stating the creative capital of the city, the plan is to build the area better than it was before the earthquake by increasing its social and cultural value.

(CCC, 2012)

By 2015, 80 percent of the CBD was demolished, likening the rebuild of the city to a "blank canvas." Tim Dearsley (Ibis Christchurch) sees an opportunity in this, announcing "Out of disaster comes good. It gives us the opportunity to do things better." For him, Christchurch now has a unique opportunity to build a stronger, more resilient and creative city; a chance to reimagine the city. Certainly, Christchurch will probably become a reimagined city with the most up to-date accommodation inventory in New Zealand. However, in times of a "global village" and "traveling cultures," is this enough for a city to be successful in a global competition for tourism and innovation?

A new convention center is planned on a prime site in the heart of the city and will cater for up to 2,000 delegates and include two hotels and retail. A new covered stadium is going to replace the old damaged AMI stadium and will be able to seat 35,000 people. A metro sports hub will allow further participation in sports with a competition-sized pool, leisure pools and eight indoor courts; the performing arts and music will also be catered for in new venues. Additional precincts of the new Christchurch are planned to include business, innovation and health, and an earthquake memorial will commemorate those who lost their lives in the tragedy in 2011. The positioning of business and sports at the heart of the new Christchurch is interesting, reflecting on the nature of the many temporary projects, with arts and culture at their center. However, major criticism of the development plan mentions the priority of profit over sustainability and a lack of grassroots involvement into the planning process (Anderson, 2014). Coralie Winn (the founder of Gap Filler) is criticizing the redevelopment plan for neglecting the Arts and ignoring the importance of the transitional movement for the rebuilding of the city.

With old heritage buildings destroyed and temporary structures disassembled, the new, permanent Christchurch will be a modern, citizen-friendly, but also generic global city. The character provided to the city by old or temporary buildings is being replaced by the generic demands of a postmodern city. Lacking the complexity of history or innovation, Christchurch has the potential to develop into a planned city, similar to Canberra, Brasilia or Nay Phi Taw. For the citizens, these cities might provide the most practical and efficient urban life; for a successful tourism industry, however, they lack character and a unique quality. For Christchurch, being already situated with a mainly nature-based destination, the risk lies in the development towards a sole gateway destination.

Conclusions: connectivity, creativity and community

Connectivity has been the core to the recovery of the people of Christchurch and is displayed through social inclusion, art projects and events. It has become clear that natural disasters such as the Canterbury earthquakes have a major impact on people's mental health, and psychosocial recovery after a disaster can take five to ten years. Analyzing urban resilience through a psychoanalytical lens derived from personal trauma, therefore, allows for an understanding of urban resilience and recovery in its physical and mental dimensions. Applying trauma theory to the urban resilience of Christchurch has further highlighted the role of community, creativity and connectivity for the ongoing recovery process. This trauma metaphor and relation between city and individual has currently been acknowledged in a campaign of the mental health department of Canterbury. The campaign reminds citizens to "check in" with family and community, and recognizes the role of each individual's trauma in the recovery process of the city.

A Maori proverb has become a much-cited illustration for new Christchurch and the transformational nature of the 2010 and 2011 earthquakes and illustrates the approach of this chapter as well as the urban resilience of Christchurch:

Hutia te rito	If you remove the heart
Hutia te rito o te harakeke	Of the flax bush
Kei whea te ko(ri)mako e ko	From where will the Bellbird sing?
Ki mai ki ahau	If you say to me
He aha te mea nui	What is the most important thing
He aha te mea nui o te ao	In this world?
Maku e kii atu	I will reply to you
He tangata, he tangata,	It is people, it is people,
He tangata, hei!	It is people!

In this chapter we have followed this idea as the basis for exploring the human dimension of urban resilience, a dimension often overlooked in a postmodern world, where the tangible is often attributed more importance than the emotional. Resilience is here understood not to be the return to what was, but the capacity to adapt and evolve with time and through disaster. For Christchurch the recovery process is an ongoing one and there are still many questions to be asked and answered, but over the five years since 2010 the importance of human capital and close community has become clear. The question of individual and physical memory within the cityscape remains to be discussed and will ensure an ongoing debate of urban resilience in the city. In combining the urban and personal aspects of resilience for analyzing the response to the Christchurch earthquakes we add a practical case study to the theoretical concept of panarchy and the connected adaptive response cycle.

Notes

1   Moas are a species of large, flightless birds native to New Zealand that became extinct in the late thirteenth century.
2   Migrating from North Island's east coast, Ngāi Tahu thrived in South Island. They intermarried with local tribes, and adopted their beliefs. Their lands cover much of Te Wai Pounamu (South Island) and are New Zealand's largest single tribal territory. As Ngāi Tahu moved down the island they fought several battles with two tribes already living there: Ngāti Māmoe and Waitaha.

3 Lyttelton is a port town on the north shore of Lyttelton Harbour, at the north-western end of Banks Peninsula and close to Christchurch, on the eastern coast of South Island of New Zealand.

4 The Ring of Fire is a string of volcanoes and sites of seismic activity, or earthquakes, around the edges of the Pacific Ocean. Globally, roughly 90 percent of all earthquakes occur along the Ring of Fire, and the ring is dotted with 75 percent of Earth's active volcanoes. It is shaped like a 40,000 km (25,000 mile) horseshoe – a string of 452 volcanoes stretching from the southern tip of South America, along the coast of North America, across the Bering Strait, down through Japan and into New Zealand.

5 The "Big Things" of Australia are a loosely related set of large road-side sculptures showcasing an iconic feature of the region/city. Big things include multiple "Big Fruits" (mango, apple, banana), as well as "Big Cultural Icons" (miners' lamp, soldier, Ugg boots). The first "big thing" was the "big Scotsmen" in Adelaide.

6 ChristChurch Cathedral, a deconsecrated Anglican cathedral in the city square was damaged during the earthquakes. The community is divided as to whether it should be rebuilt or reconstructed

7 Welcome aboard runs the six most iconic tourism activities in the city of Christchurch: the gondola, the historic tram, the tram restaurant, river punting, tours in the botanic gardens and a one-day fully guided city bus tour.

References

Aldrich, D. P. (2012). *Building Resilience: Social Capital in Post-Disaster Recovery.* Chicago: University of Chicago Press.

Allen, Craig R., Angeler, David G., Garmestani, Ahjond S., Gunderson, Lance H. & Holling, C.S. (2014) Panarchy: theory and application. *Ecosystems* 17(4), 578–89.

Anderson, C. (2014). Christchurch: after the earthquake, a city rebuilt in whose image? *Guardian*, 27 January. Retrieved from www.theguardian.com/cities/2014/jan/27/.

Atkinson, B., Baxter, S., Ver Berkmoes, R. & Bindloss, J. (2012) *Lonely Planet's Best in Travel 2013: The Best Trends, Destinations, Journey and Experiences for the upcoming Year.* Melbourne: Lonely Planet

Berkes, Fikret & Helen Ross (2013). Community resilience: toward an integrated approach. *Society & Natural Resources* 26(1), 5–20.

Bodin, P. & Wiman, B. (2004). Resilience and other stability concepts in ecology: Notes on their origin, validity, and usefulness. *ESS bulletin, 2*(2), 33–43.

Bontje, Marco & Musterd, Sako (2016) *Inventive City-Regions: Path Dependence and Creative Knowledge Strategies.* London: Routledge.

Campanella, Thomas J. (2006). Urban resilience and the recovery of New Orleans. *Journal of the American Planning Association* 72(2), 141–6.

Capozzi, L. (2013) *Crisis Management in the Age of Social Media.* New York: Business Expert Press.

Castells, Manuel (1994). European cities, the informational society, and the global economy. *New Left Review* 204, 18.

CCC (2011). *Central City Plan: Draft Central City Recovery Plan for Ministerial Approval.* Christchurch: Christchurch City Council.

Clifford, Nicholas, Cope, Meghan, Gillespie, Thomas & French, Shaun (eds.) (2016). *Key Methods in Geography.* London: Sage.

Cohn, Michael A., Fredrickson, Barbara, L., Brown, Stephanie L., Mikels, Joseph A. & Conway, Anne M. (2009). Happiness unpacked: positive emotions increase life satisfaction by building resilience. *Emotion* 9(3), 361.

Coombs, T. (2014) *Ongoing Crisis Communication: Planning, Managing, and Responding.* Washington: SAGE Publications.

Fraisl, Bettina & Stromberger, Monica (2004) *City and Trauma.* Königshausen & Neumann.

Fuge, J., Hering, R. & Schmidt, H. (2014). *Gedächtnisräume: Geschichtsbilder und Erinnerungskulturen in Norddeutschland.* Göttingen: Vandenhoeck & Ruprecht.

Gonzalez, M. V. (2008). Intangible heritage tourism and identity. *Tourism Management,* 29(4), 807–10.

Gordon, J. (1978). *Structures.* Harmondsworth: Penguin Books.

Gunderson, Lance H. (2001). *Panarchy: Understanding Transformations in Human and Natural Systems.* Washington, DC: Island Press.

van der Kolk, Bessel A. (2003) *Psychological Trauma.* Arlington, VA: American Psychiatric Press Inc.

Leichenko, Robin (2011). Climate change and urban resilience. *Current Opinion in Environmental Sustainability* 3(3), 164–8.

Lefebvre, H. (1974). La production de l'espace. *L'Homme et la société,* 31(1), 15–32.

Murray, Jennex (2012). *Managing Crises and Disasters with Emerging Technologies: Advancements.* Hershey: IGI Global

Newman, P., Beatley, T. & Boyer, H. (2009) *Resilient Cities: Responding to Peak Oil and Climate Change.* Washington: Island Press.

Norris, F. H., Stevens, S. P., Pfefferbaum, B., Wyche, K. F. & Pfefferbaum, R. L. (2008). Community resilience as a metaphor, theory, set of capacities, and strategy for disaster readiness. *American Journal of Community Psychology, 41*(1–2), 127–50.

OECD (2014). New Zealand, in *OECD Tourism Trends and Policies 2014,* OECD Publishing. http://dx.doi.org/10.1787/tour-2014-30-en.

Orchiston, C. (2013). Tourism business preparedness, resilience and disaster planning in a region of high seismic risk: the case of the Southern Alps, New Zealand. *Current Issues in Tourism,* 16(5), 477–94.

Partridge, R., Proano, L., Marcozzi, D. Garza, A. G. & Weinstein, E.S. (2012) *Oxford American Handbook of Disaster Medicine.* New York: Oxford University Press

Paton, Douglas & Johnston, David (2016). *Disaster Resilience: An Integrated Approach.* Springfield, IL: Charles C. Thomas.

Pawson, E. & Perkins, H. (2013). Worlds of wool: Recreating value off the sheep's back. *New Zealand Geographer,* 69(3), 208–20.

Reich, John W. (2006). Three psychological principles of resilience in natural disasters. *Disaster Prevention and Management: An International Journal* 15(5), 793–8.

Sassen, Saskia (2001) *The Global City: New York, London, Tokyo.* Princeton, NJ: Princeton University Press.

Statistics New Zealand (2013a). 2013 Census population and dwelling. December 3. Retrieved from www.stats.govt.nz/Census/2013-census/data-tables/population-dwelling-tables.aspx.

Statistics New Zealand (2013b). Census 2013. December 3. Retrieved from www.stats.govt.nz/Census/2013-census.aspx?gclid=Cj0KEQjw2-bHBRDEh6qk5b6yqKIBEiQA FUz29pxxthB3kqTNg-lBYqE7kK97L9dbIQEY7p0UAQX8jscaAsKj8P8HAQ.

Statistics New Zealand (2015). *Tourism Satellite Account: 2015.* Wellington: Statistics New Zealand. Retrieved from www.stats.govt.nz.

Stevenson, D. (2013) *Cities of Culture: A Global Perspective.* London: Routledge.

Stevenson, J. R., Kachali, H., Whitman, Z., Seville, E., Vargo, J. & Wilson, T. (2011). Preliminary Observations of the Impacts the 22 February Christchurch Earthquake had on Organisations and the Economy: A Report from the Field (22 February–22 March). *Bulletin of the New Zealand Society for Earthquake Engineering,* 44(2), 65.

Stevenson, J. R., Seville, E. & Vargo, J. (2012). *The Canterbury Earthquakes: Challenges and Opportunities for Central Business District Organisations*. Resilient Organisations. Retrieved from www.resorgs.org.nz/images/stories/pdfs/apec_report2_cbdeffects_final.pdf

Tourism New Zealand (2012). Christchurch. Retrieved from http://www.newzealand.com/au/christchurch/.

Young, G., Stevenson, D. (2013) *The Ashgate Research Companion to Planning and Culture*. Burlington: Ashgate Publishing.

14 Restoring spiritual resilience in post-disaster recovery in Fukushima

Kumi Kato

Introduction

In Japan, a *Kanji* (Chinese character) of the Year is selected at the end of each year by public nomination (JPES, 2011). The selection is one representation of the public sentiments, thoughts and opinions for the year's significant events, and for 2011 *Kizuna* (絆) – a connection or bond between people – was selected. The word is often used for family bond, comradeship, team and community unity to express a sense of trust, care, empathy, compassion, affinity and love. The 2011 nomination reflected what was most strongly recognized after the East Japan Earthquake and Tsunami, which devastated over 500 km of the north-east coast of Japan and caused the nuclear meltdown at Fukushima in March of that year. *Kizuna* was what people lost, as over 18,000 died or remain missing. It is also what saved the people, helped them survive and kept them going through the recovery phase. *Kizuna* was nurtured and developed among the community, as well as with those 1.4 million people who came as volunteers for relief and reconstruction from all parts of Japan and internationally (NSWC, 2016).[1] Initially, for those living outside Tohoku region, there was little that could be done except to follow what was reported in the media. Organisations started to accept goods and donations of clothes, blankets and other daily essentials. Celebrations, sporting events and entertainment were reduced in scale, suspended or cancelled all together, and people did whatever they could: sending goods, making donations, fund-raising, cancelling their holidays to send money, or going there as relief volunteers.

Evacuated communities in Fukushima and their land

The Magnitude 9 earthquake that hit the north-east coast of the Tohoku region in Japan at 14:46 on 11 March 2011 is now widely referred to as '3/11' in Japan. Six years later, over 109,000 people are still in a state of evacuation, nearly 32,500 of which are removed from their homes in Fukushima due to radiation contamination caused by the meltdown at the Fukushima Daiichi Nuclear Power Plant (Reconstruction Agency, 2017; Fukushima Prefecture, 2017). Immediately after the accident, the evacuation order and warning was issued to the 10 km and 20 km radius zones, and in the months following, wider areas of the region were defined into three zones in July 2012, according to the possible accumulated radiation dose per year. The three categories used are (Iitate-mura, 2012):

● Difficulty in returning (*Kitaku-konnnan*). The annual radiation dose is expected to exceed 50mSV and remain to be over 20mSV for more than 5 years.
● Restricted residential area (*Kyoju-kise*). These are areas with expected radiation levels of 20–50mSV/year.
● Preparation to remove restriction (*Hinan-kaijo-junbi*). This consists of areas with radiation levels under 20mSV per year (3.8µSv/Hour).

The natural radiation level in Japan is set as 0.04µSv/Hour, with monitoring required for areas with 0.23µSv/Hour (= 1mSV/year)(0.19µSv/Hour). Construction and other work permits are only issued for the areas below 5mSV/year (2.5µSV/Hour = 5mSV/year) (Valentine, 2007).

Iitate Village

Iitate is located over 50 km north-west of the Fukushima Daiichi nuclear plant. It is on a plateau about 450 m above sea level between Mt Ryozen on the north and Mt Hiyama on the south. Located on high land, its productive landscape was famous for quality beef and dairy products, sold under the Iitate brand. Rice fields, vegetable farms, flower gardens and orchards (apples, peaches, persimmons, pears) filled the rolling green hills of the region. However, because of its topography, as well as the wind and snowfall on the days following the accident, the village received a high level of radiation, forcing its evacuation. Initially, the villagers were busy accommodating people evacuated from the 20 km zone, as well as the police, self-defence forces and other relief workers, only to be told a week later that the radiation level in Iitate was also at the dangerous level and they themselves had to evacuate. Many had to abandon their farm animals, although some of them managed to relocate them to other farms. Iitate's zoning did not change over the years until April 2017, when the restriction was removed except for the southernmost district, which remains restricted (METI, 2017).

From 2014 to 2015, the districts in Iitate were 'decontaminated', which involved collecting and removing the surface soil (about 5cm) in black plastic bags. The government has not identified a final storage area for these bags, and (as of mid-2016) they have been left at roadsides and in some cases on farmland.[2] Despite this, the Government contends that 'Japan overcame the Great East Japan Earthquake and a nuclear accident.'[3] In this statement, Japan's Prime Minister Abe refers to climate change threats to 'beautiful islands, such as Tuvalu'. Iitate, in fact, was nominated to be part of the 'Most Beautiful Village Alliance in Japan' only six months prior to the accident, and they continue to face long-term threats. The evacuation order to the village was lifted on 31 March 2017, but the prospect of their land recovering to its pre-disaster productivity remains highly uncertain.

For this farming community, their life has revolved around the land for many generations, and thus the evacuation meant a separation, not only from their land, but from their family and community. A village elder, 90-year-old Uncle Tsugio, has lived in Iitate all his life, caring for his rice fields, vegetables, flower farms and his animals. He and many other villagers are still living in the 30 or so sites of temporary housing and rented houses in surrounding areas. A man representing one housing group said:

> The stress of living in small houses, having no gardens to work on, eating supermarket food, and of course not knowing when we can go home, if ever, all build up. People are stressed and tired.
>
> Honestly [another man said], being separated from the family is the hardest thing. Three or four generation households were not unusual in our village, but the temporary houses or apartments are too small for everyone to live together; and the younger ones had to start a new life somewhere else.

Another village elder explained how it is difficult to keep contact with villagers, as:

> people are living in all different places, and there are no occasions to gather – taking your produce to the market, helping to repair tools and equipment, doing big jobs (e.g. roof repairs) as a community, planning festivals – we did things together, talked to each other, got together, always.

Clearly, for the community of Iitate, their *kizuna* (bonding) is land-based. The land-based community, its identity, and its sense of place are the underpinnings that support their 'spiritual resilience'. Spiritual resilience is a reliance on traditional faith and belief systems to strengthen one's resolve in a post-disaster scenario (Béné et al., 2016; Masten & Narayan, 2012; O'Grady et al., 2016; Walsh, 2007). For Japanese, the traditional rural agricultural landscape has a strong spiritual component, both in specific religious beliefs (land gods), and in the balance among its components, including plants, animals, water, air and humans, which is reflected in the concept of *Satoyama* (Duraipappah et al., 2012; Saito & Ichikawa, 2014). As such, much of their traditional Japanese religious beliefs are deeply embedded in the land and landscape in which they live. Maintaining or restoring the spirituality of the rural landscape should be fundamental in post-disaster recovery, even though it may not be itemised in Japan's ¥19 trillion reconstruction plan for the region.

Spiritual resilience is mostly a psychological concept applied to individuals. In disaster resilience, few have dealt with this area, and most of the studies related to tourism are concerned with economic recovery, focusing on the recovery of industry and arrival numbers following disasters and crises (Lew, 2014). Further, social capital and connection between people are identified as fundamental in community resilience, but the role of 'place' and 'spirituality' has not had a strong presence (Cox & Perry, 2011; Hanna, Dale & Ling, 2009; Harner, 2001; Harvey, 2002). Spiritual resilience, defined as the religious connection between people and their land, is the basis of the sense and identity for many rural communities throughout the world. This is illustrated in Iitate, where a project to restore a set of wolf paintings installed in the village shrine ceiling, which symbolizes a local nature worship, has come to express the community's enduring connection with their land. Before detailing the project, a discussion of spiritual resilience and place connection will be revisited at the end of the chapter.

Resilience, community and spirituality

Community resilience refers to the capacity of a group of individuals, or a community, to cope with stress, overcome adversity and adapt positively to change. The ability to bounce back from negative experiences may reflect the innate qualities of individuals or be the result of learning and experience. Regardless of the origin of resilience, there is evidence to suggest that it can be developed and enhanced to promote greater well-being. Cox and Perry (2011) present a useful definition of community resilience, in the specific context of disaster recovery and describe it as:

> the capability of a community face a threat, survive and bounce back or, perhaps more accurately, bounce forward into a normalcy newly defined by the disaster losses and changes. ... [It is] a reflection of people's shared and unique capacities to manage and adaptively respond to the extraordinary demands on resources and the losses associated with disasters. ... Adaptive capacities should include the capacity to develop and maintain social capital as it is expressed through a sense of belonging, a sense of community, place attachment and participation in civil society
>
> (Cox & Perry, 2011, pp. 395–6).

For farming communities, life evolves around their land, and their daily interaction with place forms not only their sense of place, but also the connection with their community (social capital). Studies have identified a critical role that social capital plays in disaster recovery and reconstruction as well as mitigation and preparedness. Further, social capital is strengthened by the sense of reciprocity as a mutual obligation, responsibility and affinity in surviving the hardship. Delaney (2015) also identifies social capital and sense of reciprocity as critical ingredients for social sustainability. Cox and Perry (2011) emphasize the salience of 'place, identity and social capital' in building community resilience. Studies, however, have not sufficiently addressed the importance of 'place' as a vital linkage of social capital. This is particularly true for resource-dependent rural communities, whose identity is deeply rooted in their place, which in return maintains community coherence and linkage. As Diaz and Dayal (2008, p. 1174) suggest, 'the most catastrophic impact of natural disasters is an individual feeling of loss of place'. This is the case with the evacuated communities in Fukushima. It is, as Cox and Perry (2011, p. 396) suggest, 'a collective dislocation of community and belongings'.

In rural communities, a sense of reciprocity is acknowledged and expressed not only between people and community, but also between people and their place (land). It is demonstrated that culture and traditional values strengthen livelihood resilience, and adaptation comes from within, through dynamics, which are specific according to values of the people (Daskon, 2010). Daskon further argues for 'a multi-purpose adaptive system that not only encourages individual economic achievements in a community, but also safeguards the embedded values of their societies' (Daskon, 2010, p. 1097). This is 'an evolving complex system that co-adapts to the specifics of the particular place, and especially to the aspirations and values of local people' (Farrell & Twining-Ward, 2005, p. 110). Sense of connection with land and the community is the core of community resilience that may be expressed in various forms, including local spiritual beliefs and religious worship, that define a place-based identity.

Spiritual resilience also relates to the traditional ecological knowledge (TEK), which includes expert skills, knowledge and wisdom gained from living in a particular land and sea. They may be found in life practice, cultural expressions, artefacts, oral history, monuments, folktales, literature and art (Kato, 2007a; 2007b), also closely relating to the earlier mentioned concept of *satoyama* (Indrawan et al., 2014; Ichikawa et al., 2006; Knight, 2010). Traditional knowledge often contains ethics, respect, gratefulness, awe or fear towards the natural world, which are expressed in various forms. TEK also forms a 'collective social memory', which allows reflection on the past and redirection of the future, evaluating the essential values of life. The community resilience that contains a quality of place-based identity (i.e. connection with land in the community) and a capacity for sustainability is particularly important in disaster recovery. For example, Fukao (2015) found the reconstruction of a Hindu temple an essential part of the reconstruction of the Uyyalikuppam community in India after the 2004 Indian Ocean Tsunami. Ishii (2016) identifies the vital role that the 'rain ritual' played in reconstructing community resilience in a northern Chinese village following a series of major disturbances (social unrest) caused by the Cultural Revolution and other social restructuring in Maoist China. Spiritual resilience has a capacity for collective social and cultural memory that not only supports community resilience, but also cultural sustainability.

Lew suggests that resilience planning has emerged as 'a potentially more effective approach to community planning and development than the sustainability paradigm' (Lew, 2014, 14), pointing out that resilience has the capacity to adapt to change by attempting to build capacity to return to a desired state following both anticipated and unanticipated disruptions, whereas sustainability mitigates or prevents change by maintaining resources above a normative safe level. Capacity-building, based on inner strengths, seems to be a key element in resilience. Further, resilience is an ongoing process of development, rather than a fixed quality or a state held by an individual or a group (Luther, 2003). Tourism may be part of resilience planning as a significant driving force in reconstruction after various disturbances. In disaster recovery, support for building resilience also comes from outsiders (volunteers, aid and other kinds of support mechanism), who can help the affected communities reconnect with their land. This may be further articulated as volunteer tourism, which is discussed at the end of this chapter.

Spiritual resilience is part of the social-ecological resilience of a place (Folke, Colding & Berkes, 2003; Walker & Salt, 2006), which tends to embody more of an ecosystem perspective. Nature and human activities should be considered as integrated as complex and adaptive social-ecological systems (SES). From this perspective, various components of an evolving complex system co-adapt dynamically in response to various feedback in a particular place, and especially to feedback that reflect the aspirations and values of local people (Farrell & Twining-Ward, 2005; SRC, 2015). Spirituality is, of course, a major component of local community values, though it is often overlooked in social-ecological system models that come out of Western science. Walker et al (2004) define resilience as a system's capacity 'to absorb disturbance and reorganize while undergoing change so as to still retain essentially the same function, structure, identity, and feedbacks' (Walker et al., 2004). Spirituality, in terms of traditional belief systems, could be considered an indicator of whether or not the essential nature of a community is maintained through a disaster.

Restoring community belief in Iitate

This section is concerned with the specific project conducted in Iitate: an art restoration project for a local village shrine. The project, started in mid-2013, was to restore the local belief system associated with nature worship, namely wolf worship. This was a very personal project, so I will be using a first person narrative approach to describe it.

Wolf and mountain god

The project started with an exploration into the meaning of contemporary extinction, based on my interest in the Japanese wolf extinctions that occurred in the early twentieth century. The last record of a wolf hunt in Honshu mainland was in 1905 in Nara in the western part of Japan, while Hokkaido wolves had been hunted out in 1896 (Hishikawa, 2009; Knight, 2006; Walker, 2008). Andersen, an American researcher, purchased its hide, which is now housed in the British Museum today as one of the five remaining specimens globally. I learned that the Yamatsuni Shrine in Iitate worships the wolf, with a famous set of ceiling paintings completed in 1905. I was intrigued as this was the same year as the last wolf hunt.

At shrines in Japan, a pair of animals are placed at the gateway as guardians or as a holy messengers. The most common animals are a lion-like beast *komainu*, and an *inari* fox. More unusual ones may be a rhino, an elephant or a tortoise (Takuki, 2013). Commonly, one has an open mouth (*ah*) and the other a closed mouth (*un*), symbolizing a Buddhism concept of the beginning and the end of the Universe.[4] Yamatsumi is one of some 50 shrines with the wolf guardians in Japan, most of which are located in the mountain regions of central Japan with the Mitsumine Shrine being most significant. Amongst the wolf shrines, Yamatsumi stands out for its 237 wooden panels with wolf paintings in the ceilings of the prayer hall (*Haiden*). According to the shrine priest, the paintings were commissioned to a local painter when the new prayer hall was completed in 1905 at a substantial cost. The prayer hall is the most visited area of this temple, often containing ceiling paintings, a donation box and bells,[5] lanterns, various kinds of certificates (e.g. acknowledgements of donations) and strips with prayers hanging in the room. The main shrine, where the deity is enshrined, is actually at the top of the mountain, Mt Tora-tori, just behind the shrine, and is less accessible than the prayer hall: though it is only a 30-minute hike, the priest stopped me from attempting the walk saying 'the snow-covered trail is dangerous'.

Wolves appear in the mythological history of Japan through literature, such as *Nihon shoki* (Chronicles of Japan, *c*. 720), *Manyo-shu* (Collection of Ten Thousand leaves, *c*. 759) and *Kofudoki Itsubun* (Lost writing on ancient matters): there, wolves are referred to as *Oguchi-no-magami* or a large-mouthed true god.

Wolves at Yamatsumi, as with other mythological messenger animals, are called *Gokenzoku-sama* – honourable holy messengers – sent by the mountain god to provide protection to farming villages. They chase away pest animals (e.g. deer, monkeys, wild boars) which feast on the crops and damage the land. It is believed that *gokenzoku* comes down to the villages during the farming season and returns deep into the wildness of the forests when their duty is fulfilled. The wolf worship was also believed to prevent various problems, illness and disasters, and it became customary for worshippers to 'borrow' *gokenzoku* for one year in the form of talisman or *gofu*. The borrowing is described as *Gokenzoku-sama haishaku* – 'humbly receive the honour of borrowing honourable gokenzoku'. Gofu are returned at the end of the year to be replaced by a new one for the coming year. This borrowing, returning and receiving a new *gofu* follows the yearly cycle of farming – planting, growing, harvesting and storing. In some villages, a small temporary 'house' is built in the village for the *gokenzoku* to stay during the farming season.

Wolf worship flourished in late Edo period when this 250-year seclusion was coming to an end (1868), and especially in the era called Ansei (1854–60). Literally *peaceful governing*, Ansei was in fact a chaotic and unsettling time, as there was a series of disasters (earthquake, tsunami, flood, draught, fire), cholera and flu epidemics. It was also when foreign expeditions, Russia and America (in 1853), started to arrive on Japan's shores and demanding an opening of the country to trade. The era also ended with the emperor's sudden passing. It was during this time that over 10,000 wolf talisman were issued from a shrine in 1859 for people seeking the protective power of wolves.

The Yamatsumi Shrine in Iitate had 237 wolf paintings installed in 1905. In my initial phone contact, a women, who turned out to be the priest's wife, said they commute to the shrine every day, because the evacuated villagers

> are allowed to visit their homes, but not to stay overnight. People often stop by to pray at the shrine, have a chat over a cup of tea with other villagers. They are all scattered in neighbouring areas, and don't get to see each other regularly. It is our job to keep the shrine open to the people.

Our first visit was at the end of the 2013, as people were preparing for the new year celebration. Villagers were taking wolf talismans home, even though they were living in the temporary housing away from their home and farms. Here I saw a parallel between wolf worship in the unsettled Edo period, wolf extinction at the end of the nineteenth century, and the current state of Fukushima – although it is an irony that the evacuated village overcome by radiation is now watched over by the extinct wolf as a guardian. The shrine remained the only place that connected people and their land, and Yamatsumi (山津見, literally 'mountain, ocean and to watch') means 'that who watches over the mountain and the ocean'. The shrine is, therefore, worshipped by both farmers and fishers.

At Yamatsumi the roofing was originally wooden, which often leaked, resulting in its being replaced in 1975 with bronze tiles. Some parts of the ceiling had been water-damaged and ink smudged, but most of the paintings retained surprisingly distinct colours, especially the blues. No details of the artist or the original paintings of the ceiling have been documented, although the artist was believed to have worked for the former local lord of the Soma-Nakamura Clan, who ruled the area from the early twelfth century to the abolishing of clan system in Meiji 4 (1871). Ten panels showed a personal name and the amount of money donated. Apart from this, there is no documentation or records about the art in the region's museums, art galleries or history archives, which suggests that these paintings are by no means 'high art', but belong to the farming community and their daily prayers.

In response to the priest's family's concerns, in December 2012 and February 2013 I visited Yamatsumi with a photographer to document the paintings. The prayer hall is divided into three sections with pillars, but not partitioned. The wood, apparently cedar panels, had quite a strong sheen, which made photographing from directly underneath a challenge (see Figure 14.1). On both visits sunny winter days occurred, and the reflection from the ground snow outside provided good natural light. From the two visits, each panel was photographed, in the order they appear on the ceiling, grouped and edited into three photographic books. It was our intention to provide a documented archive to the shrine priest and his wife, who were concerned about the conservation status of the paintings. This, however, took a tragic turn when the shrine burnt down just three weeks after we photographed the paintings, and the priest's wife died in the fire. The cause of fire was unknown.

Restoring spiritual connection with the land

Our digital record became the only remaining evidence to this community asset. Fortunately, I was able to secure funding from a foundation to restore the paintings.[6] With the help of 26 artists (professors, graduate students) from Tokyo University of Arts, we started recreating paintings using the digital record. The artists visited the village, walked to the top of the mountain to see the entire village, and spoke to the community members before they started the restoration process. The first 100 paintings were completed in September 2015 and a further 137 in March, 2016. The works are re-creations, rather than restorations, as the original works and wood panels no longer existed. However, I use the term 'restoration' here, as

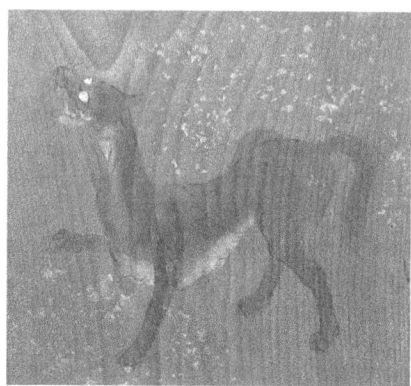

Figure 14.1 Wolf paintings at Yamatsumi Shrine before the fire
Source: Simon Wearne

I consider the process of re-creating the paintings as a 'restoration' of the connection between a people and their land, a local belief that pays respect to nature, and the practice of people making offerings to shrines for their wishes. In the work, it was important to use techniques and materials used in the period (around 1905), but not necessarily to reproduce exactly the same work. Paintings are 'living art', reflecting people's wishes and prayer, and the work done in 2015–16 should reflect the wishes, respect and prayers for the people in Fukushima.

The first 100 paintings were shown to the villagers at the village community hall in November 2015, when the shrine held its annual festival for the first time after the fire (Figure 14.2). The first people came to see the paintings were a group of fishers from a coastal village devastated by the tsunami. One fisher said he came to receive the new wolf *gofu*. Although his house was taken by the tsunami, their garden shrine with *gofu* survived. He had been wanting to come to thank the god, as his family were all safe. 'You have come home,' another villager whispered to the wolves.

The complete set of paintings were exhibited at the Fukushima Prefectural Art Gallery before being handed over to the Shrine. It was important to hold the exhibition in Fukushima City, as many of the villagers were living there and in the vicinity. The exhibition was intended to create a place where villagers can gather again, and pray to their mountain god, if only symbolically. For this reason, a space representing the central section of old shrine was installed using the prints of photographs of the original paintings[7] and the large shrine photograph at the exhibition entrance. The six-week exhibition attracted 5,600 visitors and 204 feedback sheets were returned, of which 92 per cent were rated

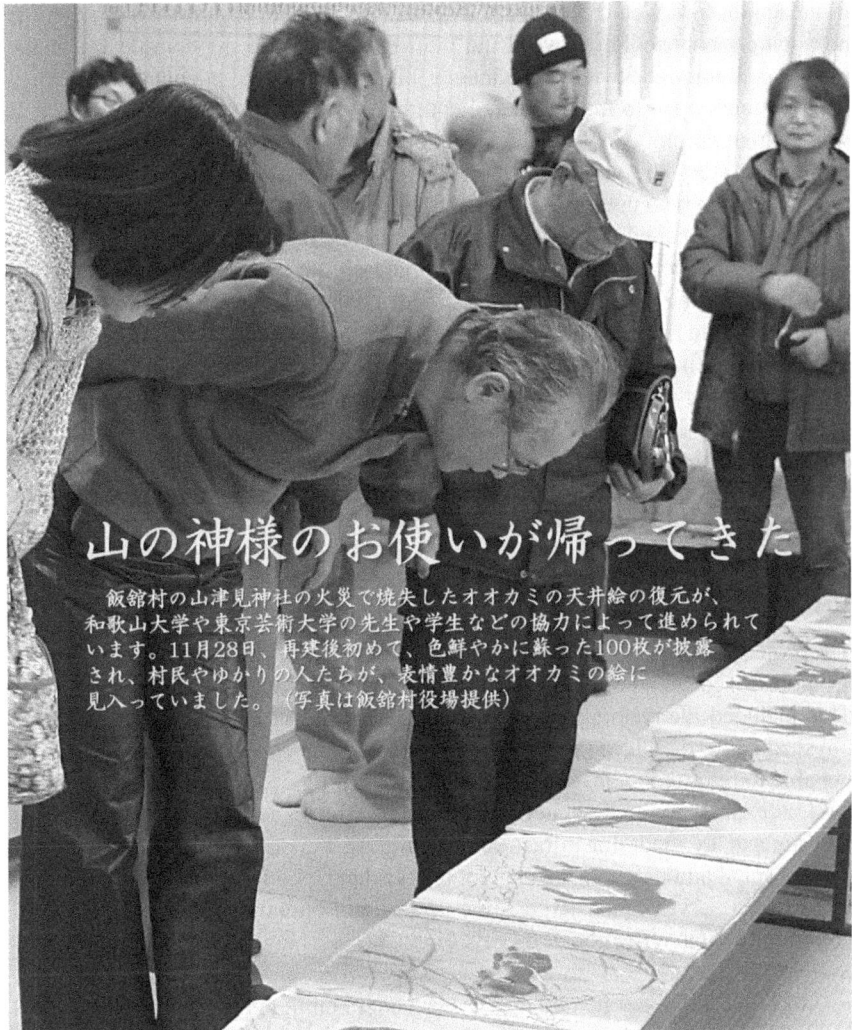

山の神様のお使いが帰ってきた

飯舘村の山津見神社の火災で焼失したオオカミの天井絵の復元が、
和歌山大学や東京芸術大学の先生や学生などの協力によって進められて
います。11月28日、再建後初めて、色鮮やかに蘇った100枚が披露
され、村民やゆかりの人たちが、表情豊かなオオカミの絵に
見入っていました。（写真は飯舘村役場提供）

Figure 14.2 Public viewing of paintings at Iitate Community Hall
Source: Newsletter No. 58 *Mountain God's messenger has come home* (Support Center, 2015)

good or excellent. Most importantly, many villagers visited the exhibition, and many of them prayed in front of the shrine photograph at the entrance and talked about the village life at the exhibition, some saying 'They have come home.'

Personally, this is my tribute to the priest's wife. I remember our last farewell – I was wishing her that the whole community would return to the village soon. As we were leaving, she came running to our car with a box of apples. She said 'Here, please take them. They are the SAFE Fukushima apples.' I feel responsible for carrying part of her legacy, supporting the community in some way, as we promised that we would record the wolf paintings. With regard to 1905, there is no apparent connection between the wolf extinction and the completion of the paintings. It might be a pure coincidence, but the stories about Yamatsumi, which kept its doors open to the evacuated community of Iitate, needs to be told, as it was their community's spiritual connection with their land.

The last story the priest's wife told me was about her early memory of an elderly woman who used to visit the shrine from the next village. She was a keen worshiper and visited the shrine regularly, but her legs were weak, and the local bus only came to the other side of the mountain. She would walk the mountain with her husband's help, and always said, 'Every time we come towards the shrine, the wind rises and I hear the sound – the movement through the dense long grasses. From then on, I am able to walk by myself.' The woman said that 'It was the wolf coming to greet and walk with me, though we never see it.' Wolves were believed by the Japanese to accompany people to their home, and the Latin name of the Japanese wolf is *Canis lupus hodo-philax: hodo* (path), *philax* (accompanying) – *a path* guide. So my hope is that these new wolves will walk the villagers home and watch over their life for the hundreds of years to come.

Discussion

The project was to restore the cultural assets of a community shrine that represented a nature-based belief symbolic to the community's connection with their land. This included the restoration of the shrine's symbolism, as well as its role as a physical gathering place (a place of prayer and connection) for a community not yet able to return to their home. Through this, it provided a connection with their land as expressed in the beliefs. In disaster recovery and reconstruction, large investments are made for infrastructure building, but spiritual resilience, linking social capital and people's attachment to their land, is often overlooked. Spiritual resilience deserves much larger investments and attention in building resilient communities. On reflection, the goals of this project may be summarized as:

- to bring hope and justice to the evacuated communities, using the shrine's wolf as a symbol that reconnects the community's stories, livelihoods and their land, and thereby strengthens their individual and community resilience;
- to invite the participation of volunteers in the community resilience building efforts; these included art students, researchers and supporters with various skills, interests and motivations; and
- to conceptualize and actually conduct this project as an academic and social contribution, building a moral, ethical and hopeful tourism research agenda (Mostafanezhad, 2013; Pritchard, Morgan & Ateljevic, 2011; Butcher, 2003).

In disaster recovery, it is reasonable to assume that volunteers are motivated by compassion, care and a sense of justice, especially in the case of Fukushima. Those involved in the project also expressed their wish to make professional and personal contributions to the current situation of Fukushima. Those participants, together with the local communities, form a 'Community of Compassion', adopting the concept of a 'Geography of Compassion' by Mary Mostafanezhad (2013), and extending it to define the positive force that supports recovery and reconstruction of the areas and people affected by disasters, conflicts and other kinds of destruction. For the current Fukushima residents, especially in the evacuated area, the volunteers help keep their place names on the map, as well as address concerns over issues related to nuclear power.[8]

Strengthening a *community of compassion*, formed with volunteer contributors and local people, is another important dimension to support spiritual resilience. This happens when there is a joint effort to regain not just a lost homeland, but more importantly, their dignity, identity, sense of justice and hope. The creative capacity of this collective power to enhance spiritual resilience is fundamental in reconstruction, providing a transformative long-term vision. This role of spiritual resilience deserves further investigation at other disaster sites to fully appreciate its function and impacts.

Notes

1 Assistance was provided by 156 countries and 41 organisations; 28 countries sent rescue teams and 53 countries donated goods (Ministry of Foreign Affairs of Japan, December 28, 2012, available at www.mofa.go.jp/mofaj/saigai/pdfs/bussisien.pdf).
2 Compensation is paid to farm owners, who may consider a cash income for the unused farm space is worthwhile.
3 Press conference by Prime Minister Shinzo Abe following G7 Summit in Schloss Elmau (June 8, 2015, http://japan.kantei.go.jp/97{_}abe/statement/201506/0608speech.html).
4 Ah is the first sound voiced when opening the mouth, and Un is the last sound emitted when closing the mouth.
5 When praying at a shrine, it is customary to ring a bell and throw an offering of coins into the saisen box before praying.
6 Mitsui Environmental Foundation (October 2013 to September 2016).
7 Designed by the photographer Simon Wearne.
8 The current government's pro-nuclear position remains firm, and by May, 2017, four reactors have been restarted in operation, including the Takahama Station in Fukui and Sendai Station in Kagoshima in southern Kyushu. Kyushu was devastated by a series of major quakes in April 2016, but the government insists on the safety of the reactors that 'have passed the most strict (revised in 2013) regulation of the world' (comment by PM Abe, July 10, 2014).

References

Béné, C., Al-Hassan, R. M., Ramatsu, M., Amarasinghe, O., Fong, P., Ocran, J., Onumah, E., Ratuniata, R., Tuyen, T., Truong, V., Mcgregor, J. & Mills, D. (2016). Is resilience socially constructed? Empirical evidence from Fiji, Ghana, Sri Lanka, and Vietnam. *Global Environmental Change, 38*, 153–70.

Butcher, J. (2003). *Moralisation of Tourism: Sun, Sand. ... and Saving the World?* London: Routledge.

Cox, R. & Perry, K. (2011). Like a fish out of water: reconsidering disaster recovery and the role of place and social capital in community disaster resilience, *American Journal of Community Psychology, 48*, 395–411.

Daskon, C. (2010). Cultural resilience – the role of cultural traditions in sustaining rural livelihood: a case from rural Kandyan villages in Central Sri Lanka. *Sustainability*, 2, 1080–1100.

Delaney, A. (2015). Social sustainability in post-3.11 coastal Japan: the significance of social capital. In Assmann, S. (Ed). *Sustainability in Contemporary Rural Japan: Challenges and Opportunities* (pp. 3–17). London: Routledge.

Diaz, J. & Dayal, A. (2008). Sense of place: a model for community based psychosocial support programs. *The Australian Journal of Disaster and Trauma Studies*, 1.

Duraipappah, K., Takeuchi, K., Watanabe, M., Nishi, M. & Nakamura, K. (Eds). (2012). *Satoyama–Satoumi Ecosystems and Human Well-being: Socio-Ecological Production Landscapes of Japan*. New York: United Nations University Press.

Farrell, B. & Twining-Ward, L. (2005). Seven steps towards sustainability: tourism in the context of new knowledge, *Journal of Sustainable Tourism*, 13(2), 109–22.

Folke, C., Colding, J. & Berkes, F. (2003). Synthesis: building resilience and adaptive capacity in social-ecological systems. In Berkes, F., Colding, J. & Folke, C. (Eds.), *Navigating Social-Ecological Systems: Building Resilience for Complexity and Change* (pp. 352–87). Cambridge: Cambridge University Press.

Fukao, J. (2015). Community relocation and recreation of cultural environment: a case of Southern Indian region affected by the Indian Ocean Tsunami. In Hayashi, I. (ed). *Disaster Reconstruction in the Asia Pacific Nations: Humanitarian Support, Community Relocation, Disaster Prevention and Culture*. Tokyo: Akashi Publishing.

Fukushima Prefecture (2017). Information related to the East Japan Earthquake (in Japanese). Retrieved from http://www.pref.fukushima.lg.jp/site/portal/list271-840.html.

Hanna, K., Dale, A. & Ling, C. (2009). Social capital and quality of place: reflection on growth and change in a small town. *Local Environment*, 14 (1), 31–44.

Harner, J (2001). Place identity and cooper mining in Sonora, Mexico. *Annals of the Association of American Geographers*, 91 (4), 660–80.

Harvey, D. (2002). Memories and desires. In Could, P. and Pitts, F. (eds) *Geographical Voices: Fourteen Autobiographical Essays*. Syracuse University Press.

Hishikawa, A. (2009). *Folklore of Japanese Wolves*. Tokyo: Tokyo University Press.

Ichikawa, K., Okubo, N., Okubo, S. & Takeuchi, K. (2006). Transition of the Satoyama landscape in the urban fringe of the Tokyo Metropolitan area from 1880 to 2001. *Landscape and Urban Planning*, 78, 398–410.

Iitate-mura. (2012). *Evacuation Zones*. Available at www.vill.iitate.fukushima.jp/saigai/?p=3801.

Indrawan, M., Yabe, M., Nomura, H. & Harrison, R. (2014). Deconstructing satoyama: the socio-ecological landscape in Japan. *Ecological Engineering*, 64, 77–84.

Ishii, Y. (2016). Resilience seen in a village, northern China: revitalisation of rain ritual. In Kawakita, A. & Nishi, Y. *Resilience as History: War, Independence, Disaster* (in Japanese) (pp. 97–127). Kyoto: Kyoto University Press.

JPES (2011). *Kanji of the Year*. Japan Kanji Proficiency Examination Society. Available at www.kanken.or.jp/project/edification/years_kanji/2011.html.

Kato, K. (2007a). Prayer for the whales: spirituality and ethics of former whaling community. Intangible Cultural Heritage for Sustainability. *International Journal of Cultural Property*, 14, 283–313.

Kato, K. (2007b). Community, connection and conservation: intangible cultural values in natural heritage – the case of Shirakami-sahcni World Heritage Area. *International Journal of Heritage Studies*, 12.5, 458–73.

Knight, C. (2010). The discourse of 'Encultured nature' in Japan: the concept of Satoyama and its role in 21st century nature conservation, *Asian Studies Review*, 34 (4), 421–41.

Knight, J. (2006). *Waiting for Wolves in Japan: An Anthropological Study of People–Wildlife Relations*. Honolulu: University of Hawaii Press.

Lew, A. A. (2014). Scale, change and resilience in community tourism planning, *Tourism Geographies: An International Journal of Tourism Space, Place and Environment*, 16(1), 14–22.

Luther, S. (2003). *Resilience and Vulnerability*. Cambridge: Cambridge University Press.

Masten, A. & Narayan, A. (2012). Child development in the context of disaster, war and terrorism: pathways of risk and resilience. *Annual Review of Psychology*, 63, 227–57.

METI (2017). Evacuation order. Current situation. Retrieved from www.meti.go.jp/earthquake/nuclear/kinkyu.html, accessed 10 April 2017.

Mostafanezhad, M. (2013). The geography of compassion in volunteer tourism. *Tourism Geographies*, 15(2), 318–37.

NSWC (2016). *Changes in volunteer numbers*. National Social Welfare Council. Retrieved from www.saigaivc.com/ボランティア活動者数の推移/

O'Grady, K. A., Orton, J. D., White, K. & Snyder, N. (2016). A way forward for spirituality, resilience, and international social science. *Journal of Psychology & Theology*, 44(2), 66–72.

Pritchard, A., Morgan, N. & Ateljevic, I. (2011). Hopeful tourism: a new transformative perspective. *Annals of Tourism Research*, 38(3): 941–63.

Reconstruction Agency (2017). Number of evacuees (28 April). Retrieved from http://reconstruction.go.jp/topics/main-cat2/sub-cat2-1/20170428_hinansha.pdf.

Ruiz-Ballesteros, E. (2011). Social-ecological resilience and community-based tourism: an approach from Agua Blanca, Ecuador, *Tourism Management*, 32, 655–66.

Saito, O. & Ichikawa, K. (2014). Socio-ecological systems in paddy-dominated landscape in Asian monsoon. In Ushio, N. & Miyashita, T. (eds). *Socio-Ecological Restoration in Paddy-dominated Landscapes* (pp. 17–38). Tokyo: Springer Japan.

SRC (2015). *Applying Resilience Thinking: Seven Principles for Building Resilience to Social-Ecological Systems*. Stockholm: Stockholm Resilience Center. Retrieved from www.stockholmresilience.su.se.

Support Center. (2015). Mountain God's messengers have come home. (For people living in nuclear affected areas.) *Fureai Newsletter*, 58: 1. Retrieved from www.city.iwaki.lg.jp/www/contents/1001000004156/simple/letter_fureai58.pdf.

Takuki, Y. (2013). *A Complete Guide to Komainu*. Tokyo: Banana Books.

Valentine, J. (Ed). (2007). *The 2007 Recommendations of the International Commission on Radiological Protection. ICRP (International Commission on Radiological Protection), Publication 103*, Retrieved from www.icrp.org/docs/ICRP_Publication_103-Annals_of_the_ICRP_37(2–4)-Free_extract.pdf.

Walsh, F. (2007). Traumatic loss and major disasters: strengthening family and community resilience. *Family Process*, 46(2), 207–27.

Walker, B. (2008). *Lost Wolves of Japan*. Seattle, WA: University of Washington Press.

Walker, B. & Salt, D. (2006). *Resilience Thinking: Sustaining Ecosystems and People in a Changing World*. Washington, DC: Island Press.

Walker, B., Holling, C. S., Carpenter, S. R. & Kinzig, A. (2004). Resilience, adaptability and transformability in social-ecological systems. *Ecology and Society* 9(2): 5. Available at www.ecologyandsociety.org/vol9/iss2/art5.

15 Fast and slow resilience in the New Zealand tourism industry

Caroline Orchiston and Stephen Espiner

Introduction

Resilience has generated appeal in the academic tourism literature as a term that captures core aspects of adaptation and change across a range of social, cultural, economic, ecological and physical contexts. In this arena, resilience describes the capacity of communities to adapt to changing conditions and ultimately sustain their tourism enterprises (Lew, Ng, Ni & Wu, 2016). In terms of its origins, this concept borrows heavily from ecology, yet encompasses an inclusive and integrative 'human-environment systems' approach, which gives it a firm interdisciplinary underpinning in its application to the tourism context.

In line with the evolutionary resilience concept, resilience demands adaptability and systems thinking within the wider social-ecological system (Lew, 2014; Becken & Hughey, 2013), by 'attempting to build capacity to rebound to a desired state following both anticipated and unanticipated disruptions' (Lew, 2014, p. 14). Resilience theory can be applied to the full range of crises precipitated by natural events, from frequent yet unpredictable events, to longer-term incremental hazards associated with climate change (McCool, 2015; Becken & Hughey, 2013). From a resilience perspective, tourism cannot be considered in isolation because tourism destinations exist within a wider social, economic, political and environmental context.

In the tourism literature, there is considerable emphasis on resilience to immediate challenges (e.g. natural disasters or financial shocks), yet there is merit in conceptualizing resilience as a dynamic long-term state, which exists across different levels of tourism from individual business owners, to destination and national-level activities. The concept of fast- and slow-onset disasters is relevant here (Lew, 2014), where the rate of change has significant implications for tourism destination management and resilience planning. Two New Zealand case study sites are presented below, to illustrate the interesting contrasts between the outcomes of a fast-onset disaster and an emerging slow-onset natural event from a resilience planning perspective. The first case study is from the West Coast of South Island, a peripheral region with a popular glacier and nature-based tourism industry. This area faces many challenges to its delivery of tourism experiences, from high-consequence (low-frequency) seismic events, to incremental changes in glacier dynamics from the impacts of climate change (Orchiston, 2012; Espiner & Becken, 2014). Tourism businesses have had to build adaptive, resilient planning strategies in order to survive in very challenging social and environmental conditions.

The second case study is the post-earthquake city of Christchurch. Data collected one year after the destructive 2010–11 earthquake sequence are presented which investigate the resilience and preparedness of tourism businesses in the worst affected areas, and the ways their owners' experiences of the earthquakes changed their business practices (Orchiston, 2013; Orchiston & Higham, 2014). These data reveal interesting insights into post-disaster resilience in the tourism industry, as well as sub-sector differences in resilient practices, and the challenges and opportunities presented by major business interruptions.

Taking a case study approach to a conceptual discussion about slow- and fast-onset environmental change has value in two key areas:

- it provides grounded illustrations of the complex outcomes resulting from disruptions to social-ecological systems; and
- this approach tests conceptual understanding by contributing examples of resilience planning across various spatial, temporal and organisational scales.

In doing so, this chapter complements Lew's (2014) scale, change and resilience model for community resilience planning. The West Coast case study describes a social-ecological system almost entirely reliant on nature-based tourism to sustain itself, acknowledging the close connection between tourism activity and community resilience against a background of (typically) incremental (although occasionally rapid) environmental change. In contrast, the Canterbury earthquakes affected a largely urban area and the impact of this event on tourism operators was immediate and sustained over many years post-earthquake. We suggest this fast-onset disaster was followed by slow recovery, and a growing awareness of the importance of adaptability and resilience to future events.

Resilience and tourism in New Zealand

As a relatively small nation, New Zealand faces a number of challenges as a tourism destination, many of which underline the significance of concepts such as vulnerability, adaptation, resilience and sustainability. Among the greatest of these challenges is its geographic distance from international mass markets, and the source of the country's most significant foreign exchange earnings. In 2015, a record 3.1 million international visitors travelled to New Zealand (Statistics New Zealand, 2015a), edging tourism ahead of the dairy industry as the country's single largest export earner (Statistics New Zealand, 2015b). Australia made up 43 per cent of inbound tourism in 2015, followed by China, the United States, the United Kingdom, alongside traditional (Germany, Japan, Canada) and emerging markets (India, Malaysia, South Korea) (Statistics New Zealand, 2016).

The vast majority of these international visitors were motivated by the desire to view scenery and engage with nature (Tourism New Zealand, 2014) in the country's substantial network of national parks and protected areas. The tourism product in New Zealand is largely built on nature-based activities, with the scenic backdrop created by an active tectonic environment, leading to relatively high risk of earthquakes, tsunami and volcanic activity in some parts of the country. Many of the most celebrated natural attractions are peripheral to New Zealand's main centres, requiring additional air and road transport to access them after the visitors arrive, compounding reliance on fossil fuels and adding to the exposure of the destination to shifts in commodity prices and other global economic and political security risks.

Beyond the resource use conundrum faced by any long-haul destination, New Zealand's medium- to long-term prospects as an international tourism destination are complicated by other vulnerabilities and risks, including the susceptibility of key attractions to a suite of natural hazards and the effects of changing environmental conditions. Floods, earthquakes and volcanic eruptions can disable a region's tourism system for days, weeks or even months – depending on the scale of the event, a feature compounded by the often difficult terrain and limited capacity of transport, power and telecommunications networks in some regions. More gradual disasters also have the potential to undermine the tourism product, such as the effects of changing climate on glacier tourism, the ski industry or marine attractions. These circumstances emphasize the value and relevance of the resilience concept.

The multiple vulnerabilities and adaptive responses of the New Zealand tourism sector have been made apparent in various circumstances and analyses, two of which are discussed later in this chapter. Visitor management at the destination level is carried out by the Department of Conservation within protected areas, while responsibility for visitors in other areas is poorly defined. Links between tourist safety and civil defence planning is relatively new in the New Zealand context (Becken & Hughey, 2013). At a broad scale, the significance of the tourism sector in emergency response planning is recognized at a national level through New Zealand's Civil Defence and Emergency Management (CDEM) Plan. In recent years the Ministry of Civil Defence and Emergency Management (MCDEM) has addressed the issue of international visitor safety by creating a cluster of tourism and government agencies, represented within the Visitor Sector Emergency Advisory Group (OECD, 2014). This group was established after the Christchurch earthquakes to respond to any emergency or crisis that could affect the welfare and travel plans of international visitors to New Zealand, or to crises that generate negative media coverage which could lead to reputational damage to the industry. The MCDEM cluster approach has been used effectively for other sectors, including transport, health, telecommunications and lifelines. The strength of clusters is in building coordinated planning and collective responses, working together to clarify goals, roles and responsibilities; identify and address gaps in capacity; coordinate response and recovery efforts; and facilitate the sharing of relevant information (CDEM, 2016). Bringing together government (central and regional), tourism agencies and industry bodies active in the visitor sector results in a more resilient tourism response to crises, and improves the ability of New Zealand to protect its tourism industry and minimize the impact of future disruption. This approach promotes polycentric governance and fosters complex systems thinking, which are two key elements of community resilience, as promoted by the Stockholm Resilience Centre (2014).

In practical terms, however, attempts to buffer against both immediate and emergent threats to the industry, and create a more resilient industry, have often been locality and sub-sector specific. This is not surprising given the typically small-scale and fragmented nature of the tourism system (Leiper, 1990), but has resulted in uneven responses to the challenges the industry faces. Dimensions of resilience planning, including disaster preparedness, adaptation to emerging change and community coordination are apparent in components of New Zealand's tourism sector, certainly at the local and regional scale. While not formally documented, there is also evidence of community-led initiatives to facilitate information sharing, cooperation between agencies and business continuity (Orchiston, 2012) in order to create mutually beneficial outcomes and better equip communities and their industries to meet the challenges associated with dynamic social-ecological systems.

Notwithstanding the range of sector, community and inter-agency initiatives outlined above, the research literature documenting or monitoring the evidence and extent of 'resilience' in New Zealand tourism is relatively limited. Among the work published, key themes in the resilience and tourism research to date include:

- studies on disaster preparedness for events such as earthquakes, floods and tsunami (Becken & Hughey, 2013; Becken, Wilson & Hughey, 2011; Gregory, Loveridge & Gough, 1997; Orchiston, 2011; 2012);
- studies about the effects of disasters on tourism recovery, especially business (Orchiston & Higham, 2014; Orchiston, Vargo & Seville, 2014; Prayag & Orchiston, 2016); and
- tourism industry and destination community responses to less immediate threats and changing conditions (Espiner & Becken, 2014; Purdie, Gomez & Espiner, 2015; Stewart et al., 2016).

There is a clear geographic focus to this work in New Zealand, with an emphasis on the coastal communities of Northland, South Island's West Coast and the city of Christchurch. These latter two localities form the basis of the case studies reported below (Figure 15.1).

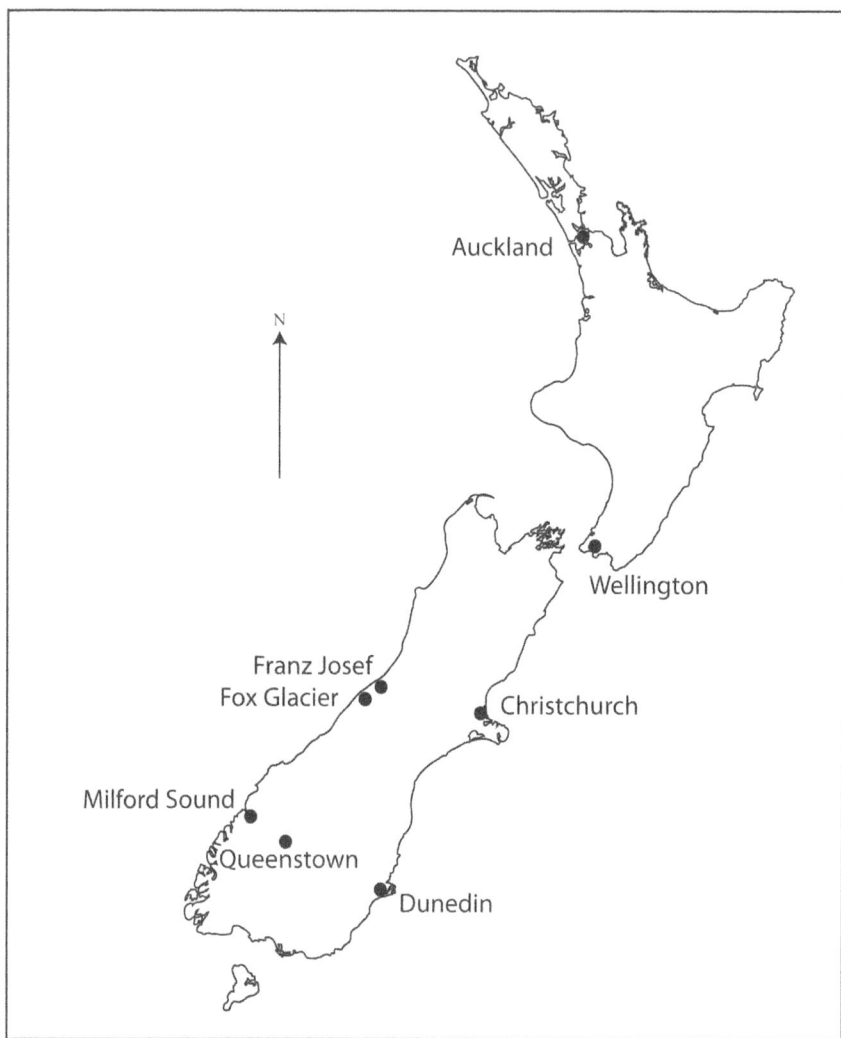

Figure 15.1 Map of New Zealand showing the case study locations of Franz Josef, Fox Glacier and Christchurch
Source: C. Orchiston

Tourist towns on the edge

Case study: Nature-based tourism and resilience on New Zealand's West Coast

The West Coast of New Zealand's South Island is a rugged and remote region where tourism plays a critical role in the economy. With approximately 85 per cent of terrestrial area protected as conservation land (MBIE, 2015), the region's communities – once dependent on timber milling, mining and agriculture – have increasingly looked to nature-based tourism as a source of economic livelihood.

Among the most important of the area's tourism attractions are the Fox and Franz Josef Glaciers (located in the Westland *Tai Poutini* National Park), which have been regarded for decades as the 'engine room' of West Coast tourism. Glacier tourism currently attracts more than 500,000 international visitors to the region annually (Statistics New Zealand, 2015a) and is core component of an industry that delivered NZ$210 million to the Westland District in 2015 (MBIE, 2016). Those living in the adjacent townships of Fox and Franz Josef are heavily reliant on the tourism sector for their economic livelihoods (Statistics New Zealand, 2013).

While currently thriving, these glacier tourism towns face a range of significant challenges, an understanding of which is instructive to the discussion and debate on vulnerability and resilience in tourism settings (Espiner and Becken,2014; Espiner, Orchiston & Higham, 2017; Orchiston, 2011; Stewart et al., 2016). Espiner and Becken (2014) identified key drivers of change influencing the vulnerability of this destination, emphasizing increasing costs of energy, climate change and a range of natural hazards as the most critical among these. Geographic peripherality (in terms of both national and international markets), a resource-intensive visitor activity site (fossil fuel-dependency in the adventure tourism industry and visitor transport), and limited infrastructure were among nationally consistent features increasing the vulnerability of tourism in this setting.

More particularly at these glacier sites, the impact of climate change and rapidly receding glaciers (Purdie et al., 2014) have led to a range of visitor access issues, including cessation of guided walking access onto the ice from the valley floor, and substantially increasing the necessity for, and number of, helicopter flights to compensate for these changes in accessibility (Figure 15.2). The challenges for glacier tourism operators and other recreation providers, are exacerbated by various natural hazards from relatively frequent floods and landslides, to high consequence (but low frequency) seismic events, all of which have the potential to seriously disrupt the tourism industry.

Compounding issues of vulnerability are the size and composition of the two glacier communities and the often seasonal character of the attractions (and visitation in general). The resident populations of Fox and Franz Josef are 375 and 444 respectively (Statistics New Zealand, 2013), although the peak summer season could see each town accommodating several thousand visitors. As a result of the tourism activity, employment in Fox and Franz Josef is also heavily service-oriented, with 66 per cent of residents working in the accommodation and food sector (Statistics New Zealand, 2013). A low wage, young, transient workforce has the potential to undermine the development and maintenance of the enduring social networks that appear to be at the heart of resilient communities planning (Paton & Johnston, 2006).

Faced with these diverse challenges, the tourism industry in these glacier towns has been required to adapt to both immediate and emerging threats to its longevity. The periodic advance and recessionary phases associated with these glacier attractions (Anderson, Lawson & Owens, 2008) has necessitated novel and innovative approaches to visitor management and maintaining the tourist experience. For instance, commercial access to the glaciers has been maintained

Figure 15.2 Scenic helicopter flights onto the glacier at Franz Josef, Westland, New Zealand
Source: S. Espiner

at various points across more than 100 years through the deployment of mechanisms as diverse as complex cantilevered bridge structures attached high on the rock walls leading to Franz Josef Glacier and guides using row boats to ferry tourists across a proglacial lake that formed at the glacier terminal *circa* 1900 (McCormack, 1999). The current practice of shuttling tourists to and from the glacier via helicopter services based in each of the glacier towns is the latest adaptive innovation. Such an historical record of change and adaptive response has embedded a sense of community confidence to deal with future change (Espiner and Becken, 2014), and is a good illustration of the resilient nature of the tourism industry in this region.

In addition to responding to the dynamic nature of the glaciers' form, there is evidence of resilience planning among the tourism stakeholders in their attempts to diversify products. While the region is heavily dependent on the glaciers to attract tourists (Stewart et al., 2016), this vulnerability is addressed in part through the development of a kiwi house (an indoor wildlife centre), hot pools, several bush walks and a cycleway. As secondary visitor attractions, none of these developments has the primary pulling power of the glacier experience, but such product diversification may serve to strengthen the destination's capacity to cope with the effects of a changing climate (Espiner and Becken, 2014), and other threats to the sustainability of tourism in this region. The maintenance of diversity and redundancy are noted as key principles of building resilience in social-ecological systems (Stockholm Resilience Centre, 2014). While the glacier towns demonstrate some evidence of intra-industry diversity, the region's peripheral character and limited infrastructure mean that diversity beyond tourism is less apparent, leaving the community vulnerable to social and economic shifts that might reduce the flow of visitors to the area.

Finally, the region's exposure to a range of natural hazards (floods, landslides and earthquakes in particular), in combination with the community's small, cohesive character, has resulted in significant involvement in emergency management systems and the development of effective relationships between the council, businesses, park management and the local police (Espiner & Becken, 2014). Because of their prominent presence in the towns, tourism industry stakeholders (including commercial tourism operators and national park management staff) in Fox and Franz Josef are at the forefront of these community-led initiatives, including volunteer fire and ambulance services, civil defence planning and emergency management planning. Such 'connectivity' has been identified as a leading characteristic of resilient human social systems, developed through information sharing, reciprocity and building trust (Stockholm Resilience Centre, 2014). These features are hallmarks of the glacier communities. In addition to emergency response efforts at the local level, longer-term hazard mitigation work is a focus of this region, with an Alpine Fault avoidance zone (Langridge & Beban, 2011) under consideration. One scenario of this would require relocating at least parts of the Franz Josef township to an area beyond the Alpine Fault boundary in order to mitigate the physical effects of a future rupture.

In addition to local-level planning, a recent South Island-wide Civil Defence initiative involved a close collaboration with geoscientists to undertake scenario planning for a future Alpine Fault earthquake. This large scale, multi-agency planning event (Exercise *Te Ripahapa*) was informed by a detailed geomorphic scenario from which a range of agencies could plan their responses to the consequences of a future magnitude 8 Alpine Fault earthquake (Robinson et al., 2014). Such exercises provide an opportunity to plan for uncertain events, and build resilience across a network of stakeholders, including transport, health, telecommunications and emergency services, which supports Principle 4 ('foster complex adaptive systems thinking) for building resilience into complex social-ecological systems (Stockholm Resilience Centre, 2014).

The glacier towns of Westland National Park, New Zealand, provide an intriguing case study of resilience in a nature-based tourism context. Situated on the edge of an active tectonic plate boundary and subject to multiple natural hazards, these communities face continual challenges to the maintenance of their core visitor attractions – two sub-alpine glaciers that are currently receding at unprecedented rates as a result of climate change. At present, these communities have demonstrated considerable effectiveness at harnessing available human resources and in applying innovative strategies to counteract the consequences of changing social and environmental conditions in the dynamism of their social-ecological system. As with all systems, however, there is a tipping point, beyond which lies a transformative change. While there is evidence of considerable resilience in these glacier towns, the future of the tourism industry remains uncertain given the range and scale of inherent vulnerabilities.

Case Study: Canterbury tourism after the earthquakes

Christchurch is the largest city in South Island (pop. 340,000), and the logistical hub of the region's tourism sector. Prior to 2010, the city's gateway airport received 85 per cent of all inbound visitors to the South Island, and the destination generated 16 per cent of total national tourism activity. Tourism in Christchurch at this time was focused on cultural and scenic values, alongside thriving business convention and cruise ship markets.

The Canterbury earthquake sequence began in September 2010, when at 4.30 a.m. a strong, shallow earthquake struck 30 km west of Christchurch. Significant damage was done to heritage buildings in the city, and large quantities of silt emerged from the ground as a consequence of liquefaction (Figure 15.3). There was a feeling that the city had 'dodged a bullet', and the fact that there were no fatalities was attributed to the quake's early morning timing (when people were at home in bed) and the country's world-leading building codes. However, five months later, on 22 February 2011, a shallow, destructive aftershock (magnitude 6.3) occurred directly beneath the city. This event, which occurred on a weekday lunchtime, caused massive damage to Christchurch, and resulted in 185 deaths (New Zealand Police, 2012), with two building collapses causing 133 of the fatalities. The New Zealand government declared a national state of emergency, which saw assistance offered from around the world. Two-thirds of hotel and backpacker accommodations were destroyed, reducing available beds from 3,750 before the quakes, to 1,100 afterwards (The Press, 2012). More than 220 heritage buildings were subsequently demolished, which immeasurably changed Christchurch as a destination.

Figure 15.3 ChristChurch Cathedral before (left) and after (right) the 22 February 2011 magnitude 6.3 earthquake
Source: C. Orchiston

Aftershocks continued to be felt for several years after the event, causing significant stress to local residents and visitors. While the effects of the disaster had less impact on the overall New Zealand economy than expected (a total cost equivalent to 10 per cent of GDP), local and regional hospitality, retail and tourism sectors were particularly badly affected (Doyle & Noy, 2015). International visitor numbers dropped significantly as a consequence of the central business district being destroyed, along with most of the major accommodations in the city (Orchiston, 2013). Christchurch is the airport gateway to South Island, and, although the airport itself was only closed for hours, not days, the impact on arrivals and flow-on visitation caused negative outcomes throughout the rest of South Island in the months that followed the powerful February earthquake.

Five years after the earthquakes, inbound tourism to Christchurch has yet to rebound to pre-earthquake levels (Wood et al., 2016), largely as a consequence of reduced accommodation capacity and the absence of a replacement convention centre. The availability of hotels and other types of accommodation continues to be a significant limiting factor in the recovery of inbound tourism in Christchurch. In addition, the perception of Christchurch as a destination was likely to have been affected by the negative media coverage of the event, and the fear associated with visiting an earthquake-prone city. Conversely, however, some perceive the process of rebuilding and regenerating the city as an attraction in itself, with the Lonely Planet placing Christchurch in the Top 10 'must see' cities of 2013 (Lonely Planet, 2013).

The resilience of tourism operators in the response and recovery phases of the earthquake sequence was severely tested. The ongoing nature of the aftershock sequence created a sense of uncertainty amongst tourists about whether to visit the city. It also caused uncertainty amongst investors as decisions were made about whether to rebuild a traditional central business district or to foster the smaller satellite hubs of business activity that had sprung up organically around the city in the aftermath of the quakes. Finally, a blueprint for the rebuild of Christchurch was released in 2012, 18 months after the earthquakes, which confirmed that the central business district (CBD) would indeed be rebuilt in its original location. Its design has been highlighted as a unique opportunity to build a future city, one that is dynamic and attractive for residents and visitors alike.

An impact and recovery survey was designed and implemented one year after the 22 February 2011 quake, which investigated preparedness, resilience and recovery amongst all tourism operators in Canterbury (Orchiston, Vargo & Seville, 2014). The most disruptive factor identified by operators was the major reduction in visitor numbers to their business. Some 70 per cent of operators reported a decline in international visitor arrivals after the quakes (Orchiston et al., 2014). Across Canterbury, 87 per cent of operators reported a change in the types of visitor to their business, with a greater proportion of domestic (local and regional) visitors compared to international visitors. This was particularly noticeable in the activity and attraction sector, which was heavily reliant on international markets. The change in nature of visitation posed a significant challenge to operators, some of whom responded by adapting their marketing and promotional activities to compensate for the lack of international visitors. The regional and national destination marketing organizations (Christchurch Canterbury Tourism and Tourism New Zealand) together took the decision to 'de-market' Christchurch for six months after the February quakes, after which they initiated a marketing campaign called 'South Island Road Trips', designed to attract visitors into the rest of South Island while Christchurch was effectively out of commission as a destination (Orchiston & Higham, 2014).

The results of the post-earthquake tourism business survey also highlighted the sub-sector differences in recovery among the accommodation, activity/attraction and visitor transport sectors. For businesses located within the CBD (mainly hotel and backpacker accommodation businesses), the impacts were severe and ongoing because of the cordon that remained in place for more than two years as damaged buildings were demolished. In contrast, motel and holiday park accommodations outside the cordon and across various parts of the city experienced a boom in demand due to the overall reduction in available beds in Christchurch. The Commercial Accommodation Monitor for Canterbury (year ending May 2012) described a 26 per cent increase in occupancy rates for motels compared to the previous year (MBIE, 2016). In contrast, hotels and 'backpacker' hostels were down 54 per cent and 35.4 per cent respectively. As a consequence, revenue changes were polarized, with sub-sector variations illustrated by the significant losses in the hotel sector.

In terms of business preparedness, one year after the earthquakes the majority of tourism operators reported that they were satisfied with their current level of preparedness, and they felt better prepared to deal with a future disaster after the experience of the quakes. Ironically, however, almost half of operators had not backed up their data. This was a critical issue for businesses located behind the cordon who could not access their offices to retrieve important documents and files in the aftermath of the quakes. In addition, since the earthquakes few businesses had developed crisis and emergency or business continuity plans, with only a third of businesses having any form of plan in place (Orchiston et al., 2014). Larger businesses with higher incomes were more likely to have a range of planning and preparedness tools. In contrast, fewer micro businesses had disaster plans, staff inductions and other disaster-related policies, relying instead on informal plans and an ability to react as situations present themselves. While the lack of formal planning could be considered to lower business resilience, the adaptive and agile management approach described by tourism operators as being part of their modus operandi is also suggestive of a degree of resilience (Stockholm Resilience Centre, 2014).

One component of the Christchurch's post-earthquake tourism business survey incorporated 13 organizational resilience indicators that were widely tested across a number of business sectors (Lee, Vargo & Seville, 2013; Seville et al., 2014). Subsequent factor analysis using these indicators revealed two dimensions of organizational resilience for tourism organisations in a post-disaster context (Orchiston, Prayag & Brown, 2015):

- The first key dimension of resilience is 'planning and culture', which describes a forward-focused work culture, where preparing for and responding to change is critical to business success and longevity.
- The second dimension of resilience is 'collaboration and innovation', which was found to be a distinct factor in the tourism sector compared to other business sectors. It describes the importance of working across silos, making collective decisions and reducing barriers towards working effectively with other organizations (Orchiston et al., 2015), factors which directly relate to Principle 2 ('Manage Connectivity') of the Stockholm Resilience Centre's (2014) resilience approach.

Discussion

The two case studies presented above highlight that the development of resilient practices across the range of scales in the tourism sector (individual operator to destinations) is often a response to the scale of change, i.e. fast- or slow-onset. However, the net result of developing improved resilience may then become largely independent of the rate of change. In other words, once a business, community or destination has started on the pathway to improved resilience, the evidence suggests each is better equipped to deal with the outcomes of either fast- or slow-onset disasters in future. The resilient practices that are developed in response to different types of crises or disasters can be conceptualized as existing on a continuum from fast to slow rates of change, but the practices born of these responses may have utility in confronting change at a variety of scales.

For example, the tourism industry at the glaciers has to deal with both 'fast' (seismic activity) and 'slow' (climate-related) resilience issues, although much of the recent focus for this nature-based tourism community has been on responses to medium- to long-term environmental change (Figure 15.4). In particular, local tourism businesses and the community have attempted to buffer themselves against slow-burning threats to sustainability through investing in social networks, diversifying the destination's product range, and participating in regional-level planning processes for the townships and the adjacent national park. These adaptive responses have been largely governed by an evolving acknowledgement of the likely future impact of climate change on the region. This is in contrast to the case study of Christchurch tourism, where the resilience issues were initially far more immediate because of the on-going nature of the earthquake sequence.

In Christchurch, the tourism sector needed to respond rapidly to the events as the aftershocks continued to affect the city over many months. Evidence of this is illustrated by the adaptive marketing strategies used by the destination marketing organization, and the collaborative and innovative approaches adopted by tourism business operators. However, there is evidence of 'slow' resilience responses even in the context of an earthquake disaster (Figure 15.5). Once the immediate sequence of earthquakes subsided in 2011, the tourism sector began a process of long-term recovery. Examples of slower approaches to resilience include formal continuity planning with tourism enterprises and destination marketing organizations, as well

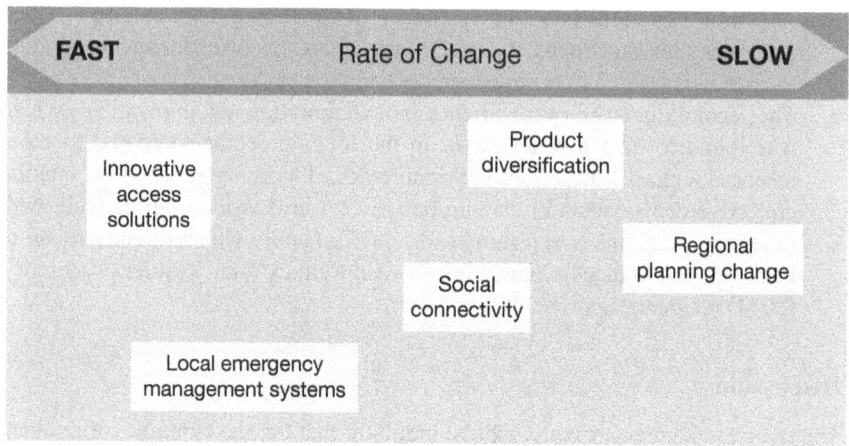

Figure 15.4 Indicators of fast–slow resilience at the glaciers, Westland Taipoutini National
Park, New Zealand
Source: Authors

as the process of contributing to the vision of the future Christchurch through
providing input and advocacy for tourism in the city rebuild plans. There has also
been a realization of the need to capitalize on opportunities that present themselves
following disasters. For example, central Christchurch is being completely rebuilt,
renewed and refreshed, which is an unprecedented opportunity from a tourism
perspective. A proactive approach must continue to be taken by destination
managers to ensure the best outcomes for the tourism sector, including advocating
for a rapid reinstatement of key tourism infrastructure (including a convention
centre and sports stadium). The resilience lessons that were learned in the first
years following the earthquakes are now being embedded and built upon as the
attention shifts to the future city, and to future crisis and disaster management in
the Christchurch tourism sector. Tourism enterprises were severely tested, with
many of the outcomes of the earthquakes lying outside their immediate control
(e.g. inbound tourism recovery). However, those businesses that survived the
disaster are now better prepared to face future disruptions, through the use of
adaptive, agile (both formal and informal) planning approaches.

Hence, there are elements of fast–slow resilience in each of the tourism case
studies, although the current emphasis may differ. This suggests that the most
resilient systems will be capable of responding to both emerging (slow) and rapid
onset (fast) change. These examples support Lew's observation that:

> people perceive and manage slow changes in the environment, culture and
> society in a different manner than they do under sudden major shocks to those
> systems. ... Rates of change can be highly variable over time and at different
> social and geographic scales, which can require different modes of response.
>
> (Lew, 2014, p. 17)

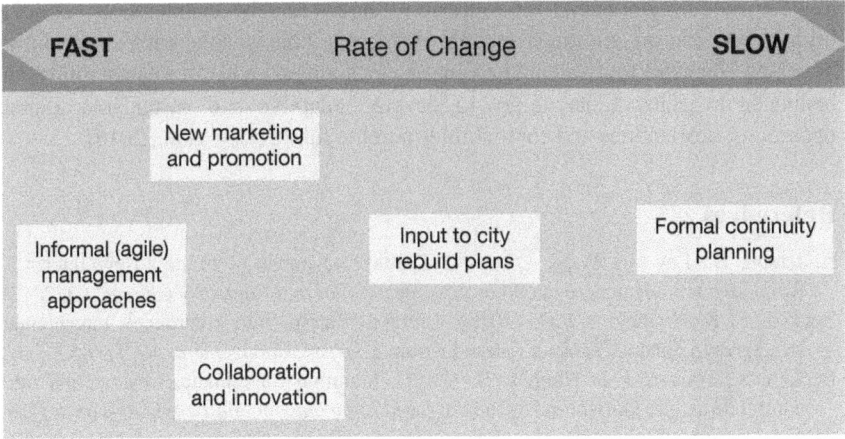

Figure 15.5 Indicators of fast–slow resilience in Christchurch, New Zealand
Source: Authors

Conclusion

This chapter has presented two tourism case studies from New Zealand involving destinations with different geographical, social and historical contexts:

- a post-disaster urban tourism hub recovering from a major earthquake; and
- a peripheral, nature-based destination with a long history of developing highly adaptive approaches to managing inherent vulnerabilities and many slow change variables.

Tourism resilience exists to varying degrees across a range of temporal (e.g. slow-onset versus immediate natural events), spatial (urban versus peripheral), governance (local, regional, national) and business scales (Lew, 2014), and is not easy to measure or quantify. While the empirical measurement of resilience is in its infancy within the academic discourse, there is increasing pressure from both tourism practitioners and funding agencies to improve our understanding of resilience metrics, and the ways in which resilience measures can be applied through adaptive business management practices to improve the social and environmental outcomes of tourism operations. Resilience measures designed to address fast- and slow-onset disasters have been presented in this chapter, and illustrate that developing resilient practices to address the outcomes of a disruption can benefit the long-term resilience of a business enterprise or destination to future disaster.

The challenges presented across multiple scales are most effectively confronted by individuals, enterprises, agencies and sectors that incorporate elements of resilience planning into their business practices. Examples of resilience planning by operators and destination managers in New Zealand include social connectivity and redundancy, fostering complex systems thinking, encouraging learning and promoting polycentric governance systems (Stockholm Resilience Centre, 2014), demonstrating that the diversity and fragmentation of tourism resources can be countered by building bridges between organisations and stakeholders to address inherent vulnerabilities. The New Zealand tourism industry needs to address issues of sustainability and resilience to enhance its ability to develop adaptive and sustainable tourism operations, destinations and communities into the future (Lew et al., 2016).

References

Anderson, B., Lawson, W. & Owens, I. (2008). Response of Franz Josef Glacier Ka Roimata o Hine Hukatere to climate change. *Global and Planetary Change*, 63, 23–30.

Becken, S. & Hughey, K.F.D. (2013). Linking tourism into emergency management structures to enhance disaster risk reduction. *Tourism Management*, 36, 77–85.

Becken, S., Wilson, J. & Hughey, K. (2011). Planning for climate, weather and other natural disasters: tourism in Northland. *Land Environment and People Research Paper No. 1*, Lincoln University, New Zealand.

CDEM (2016). *Civil Defence and Emergency Management Clusters*. Retrieved from www. civildefence.govt.nz/assets/Uploads/publications/the-guide-v1.2-section-11-clusters.pdf

Doyle, L. & Noy, I. (2015). The short-run nationwide macroeconomic effects of the Canterbury earthquakes, *New Zealand Economic Papers*, 49(2), 134–56, DOI: 10.1080/00779954.2014.885379

Espiner, S. & Becken, S. (2014). Tourist towns on the edge: conceptualising vulnerability and resilience in a protected area tourism system. *Journal of Sustainable Tourism*, 22(4), 646–65.

Espiner, S., Orchiston, C. & Higham, J. (2017). Resilience and sustainability in nature-based tourism destinations: a conceptual model. *Journal of Sustainable Tourism*, 1–16. dx.doi.org/10.1080/09669582.2017.1281929.

Gregory, G., Loveridge, A. & Gough, J. (1997). Social and cultural aspects of natural hazards perception and response. *New Zealand Geographer*, 53(1), 47–54.

Langridge, R.M. & Beban, J.G. (2011). *Planning for a Safer Franz Josef-Waiau Community, Westland District: Considering Rupture of the Alpine Fault*. GNS Science Consultancy Report 2011/217.

Lee, A., Vargo, J., and Seville, E. (2013). Developing a tool to measure and compare organizations' resilience. *Natural Hazards Review*, 10.1061/(ASCE)NH.1527–6996.0000075, 29–41

Leiper, N. (1990). *Tourism Systems: An Interdisciplinary Perspective*. Department of Management Systems, Massey University, New Zealand.

Lew, A.A. (2014). Scale change and resilience in community tourism planning. *Tourism Geographies*, 16(1), 14–22.

Lew, A.A., Ng, P.T., Ni, C-C., & Wu, T-C. (2016). Community sustainability and resilience: similarities, differences and indicators. *Tourism Geographies*, 18(1), 18–27.

Lonely Planet (2013). Best in Travel 2013. Retrieved from www.lonelyplanet.com/themes/best-in-travel/top-10-cities.

MBIE (2015). *Regional Economic Activity Report 2015*. Ministry of Business, Innovation and Employment. Wellington: New Zealand Government.

MBIE (2016). Regional tourism summary: West Coast RTO. Ministry of Business, Innovation and Employment. New Zealand Government, Wellington.

McCool, S. (2015). Sustainable tourism: guiding fiction, social trap or path to resilience? In T.V. Singh (Ed.) *Challenges in Tourism Research* (pp. 224–34). Bristol: Channel View Publications.

McCormack, T. (1999). Glacier advance: the development of tourism at Franz Josef Glacier 1865–1965. Unpublished master's thesis, University of Otago, New Zealand.

New Zealand Police (2012). List of deceased. Retrieved from www.police.govt.nz/list-deceased.

OECD (2014). Country profiles: tourism trends and policies – New Zealand, in *Tourism Trends and Policies 2014*. OECD Publishing.

Orchiston, C. (2011). Seismic risk scenario planning and sustainable tourism management: Christchurch and the Alpine fault zone, South Island, New Zealand. *Journal of Sustainable Tourism*, 20(1), 59–79

Orchiston, C. (2012). Tourism business preparedness, resilience and disaster planning in a region of high seismic risk: the case of the Southern Alps, New Zealand, *Current Issues in Tourism*, DOI:10.1080/13683500.2012.741115.

Orchiston, C. (2013). Tourism business preparedness, resilience and disaster planning in a region of high seismic risk: the case of the Southern Alps, New Zealand. *Current Issues in Tourism*, 16(5), 477–94.

Orchiston, C. & Higham, J.E.S. (2014). Knowledge management and tourism recovery (de)marketing: the Christchurch earthquakes 2010–11. *Current Issues in Tourism*, 19(1), 64–84, DOI: 10.1080/13683500.2014.990424

Orchiston, C., Prayag, G. & Brown, C. (2015). Organisational resilience in the tourism sector. *Annals of Tourism Research*, http://dx.doi.org/10.1016/j.annals.2015.11.002

Orchiston, C., Vargo, J & Seville, E. (2014). Regional and sub-sector impacts of the Canterbury earthquake sequence for tourism businesses. *Australian Journal of Emergency Management*. 29(4), 32–37.

Paton, D. & Johnston, D. (2006). *Disaster Resilience: An Integrated Approach*. Springfield, IL: Charles C Thomas Publishing.

Prayag, G. & Orchiston, C. (2016). Earthquake impacts, mitigation, and organisational resilience of business sectors in Canterbury, In Michael Hall, Sanna Malinen, Rob Vosslamber and Russell Wordsworth (eds) *Business and Natural Disasters: Business, Organisational and Consumer Resilience and the Christchurch Earthquakes*. London: Routledge.

Purdie, H., Anderson, B., Chinn, T., Owens, I., Mackintosh, A. & Lawson, W. (2014). Franz Josef and Fox Glaciers, New Zealand: historic length records. *Global and Planetary Change*, 121, 41–52

Purdie, H., Gomez, C. & Espiner, S. (2015). Glacier recession and the changing rockfall hazard: implications for glacier tourism. *New Zealand Geographer*. DOI:10.1111/nzg.12091.

Robinson, T., Davies, T., Wilson, T., Orchiston, C. & Thompson, J. (2014). Design and development of realistic exercise scenarios: a case study of the 2013 Civil Defence Exercise Te Ripahapa. *GNS Science Miscellaneous Series* 69, February 2014.

Seville, E., Stevenson, J., Brown, C., Giovinazzi, S. & Vargo, J. (2014). Disruption and resilience: how organisations coped with the Canterbury earthquakes. *ERI Research Report* 2014/002, December 2014.

Statistics New Zealand (2013). *Census 2013*. Statistics New Zealand, Wellington.

Statistics New Zealand (2015a). *International Visitor Arrivals to New Zealand: December 2015*. Wellington: Statistics New Zealand.

Statistics New Zealand (2015b). *Tourism Satellite Account: The contribution Made by Tourism to the New Zealand Economy*. Wellington: Statistics New Zealand.

Statistics New Zealand (2016). International visitors arrivals to New Zealand. Retrieved from www.stats.govt.nz/browse_for_stats/population/Migration/iva.aspx.

Stewart, E.J., Wilson, J., Espiner, S., Purdie, H., Lemieux, C. & Dawson, J. (2016). Implications of climate change for glacier tourism. *Tourism Geographies*, DOI: 10.1080/14616688.2016.1198416

Stockholm Resilience Centre (2014). *Applying Resilience Thinking: Seven Principles for Building Resilience in Social-Ecological Systems*. Stockholm: Stockholm University.

The Press (2012). Hotel revival taking longer. Retrieved from www.stuff.co.nz/business/rebuilding-christchurch/7992007/Hotel-revival-takinglonger.

Tourism New Zealand (2014). *Visitor Experience Monitor 2014*. Tourism New Zealand, Wellington.

Wood, A., Noy, I. & Parker, M. (2016). The Canterbury rebuild five years on from the Christchurch earthquake. *Reserve Bank of New Zealand Bulletin,* 79(3) February 2016, retrieved from www.rbnz.govt.nz/-/media/ReserveBank/Files/Publications/Bulletins/2016/2016feb79–3.pdf.

Part IV

Indigenous responses to changing environments

16 Within the changing system of Arctic tourism, what should be made resilient to what, and for whom?

Kevin Hillmer-Pegram

Introduction

Change permeates life in the Arctic—materially and discursively—to a degree arguably greater than anywhere on Earth. Climate change and globalization are increasingly impacting terrestrial and marine ecosystems, socio-economic and cultural systems, and coupled social-ecological systems (Arctic Climate Impact Assessment, 2005; Arctic Council, 2013; Larsen et al., 2014; AHDR II, 2015). Tourism in the Arctic is on the rise (Fay & Karlsdottir, 2011; Hall & Saarinen, 2010, Maher et al., 2014) and is a key element of Arctic change, both influencing and being influenced by other drivers. Within this context of complex change, resilience is emerging as a framework for studying and governing the Arctic (Arctic Council, 2013; Chapin, Sommerkorn, Robards & Hillmer-Pegram, 2015), while simultaneously emerging as a novel framework for studying and governing tourism (Hillmer-Pegram, 2014; Lew, 2014). The co-emergence of resilience in Arctic studies and tourism studies makes a resilience-based analysis of Arctic tourism especially timely.

However, resilience has come under scrutiny for its perceived shortcomings, such as naturalizing what are really human-induced processes of change and ignoring inequitable power relations in society (e.g., Kirchhoff, Brand, Hoheisel & Grimm, 2010). Turner (2013, p. 621) suggests that resilience's survival-of-the-fittest mentality—carried over to social analysis from its ecological roots—creates a "disturbing voyeurism—coming too close, for many social scientists, to social Darwinism". Rather than completely rejecting resilience, however, this chapter takes seriously key critiques and attempts to address them, thus supporting the movement to unleash resilience's potential for radical social change (Cretney, 2014; Nelson, 2014).

This chapter's approach to resilient tourism is grounded in critical political economy and concerned with promoting social justice across geographic scales (see Fraser, 2009). It argues that, in the Arctic, resilient tourism work ought to focus on the power of Indigenous communities—which have been historically marginalized from many of the political processes and economic benefits of global change—to control and profit from regional tourism. Other calls for Indigenous self-determination in the Arctic (e.g., Nuttall, 2000) support this claim, as does research from around the globe, demonstrating that Indigenous communities have harnessed tourism development to benefit themselves financially and culturally (e.g., Bunten, 2010; Lemelin, Koster & Youroukos, 2015; Zeppel, 2006).

Steering Arctic change with resilience: promises and problems

Previous studies define the Arctic through the presence of sea ice, particular temperature ranges, the northern tree line, and the Arctic Circle. Other definitions are more flexible and take advantage of existing data collection areas. This chapter uses one such definition, which defines the Arctic as:

> All of Alaska, Canada north of 60°N together with northern Québec and Labrador, all of Greenland, the Faroe Islands, and Iceland and the northernmost counties of Norway, Sweden, and Finland ... [and those parts of Russia that lie closest to the Arctic Circle]
>
> (AHDR, 2004, pp. 17–18).

This definition also includes the Arctic Ocean and its adjacent seas. It is important to acknowledge Arctic environmental and social heterogeneity. There are eight nations with Arctic territory, each possessing unique social-ecological histories: Canada, Denmark (which possess Greenland and the Faroe Islands), Finland, Iceland, Norway, Russia, Sweden, and the United States.

Similarities between countries include seasonal extremes in coldness, light, and dark; unique, highly adapted ecosystems; and recognized Indigenous groups (except in Iceland). There are over 40 Indigenous groups in the Arctic (comprising roughly 10 percent of the population), including: Saami in Finland, Sweden, and Norway; Nenets in Russia; Aleut, Yupik, and Iñupiat Inuit in Alaska; Inuvialuit Inuit in Canada; and Kalaallit Inuit in Greenland (www.Arcticcentre.org).

Arctic change

The most fundamental trait of the Arctic, its coldness, is fading due to anthropogenic climate change (ACIA, 2005; Larsen et al., 2014). Polar amplification, whereby decreasing ice and snow coverage increases absorption of solar radiation, is causing the Arctic to warm at twice the global rate (Walsh, Overland, Groisman & Rudolf, 2011). Arctic sea ice has declined since the late 1970s and is continuing to decline (Stroeve et al., 2012). Reduction in the presence of snow has been reported, as have changes in the condition of permafrost (i.e., permanently frozen soil beneath the surface) (Larsen et al., 2014). Increasing rates of coastal erosion are also attributed to climate change (Larsen et al., 2014). Changes to the physical environment are expected to challenge the adaptive capacities of some species, including polar bears and a wide range of other wildlife (e.g., whales, fish, birds, caribou, reindeer) (Larsen et al., 2014). Increased shrubification and a northward advance of the tree line have been observed (Larsen et al., 2014). Environmental pollutants, habitat fragmentation, industrial development, and unsustainable harvests are also driving environmental changes (Conservation of Arctic Flora and Fauna, 2010). There is sound reason to expect that Arctic ecosystems may soon reach transformational tipping points, beyond which returns to previous states will be impossible (Arctic Council, 2013).

On the human side, the population of around 4 million is experiencing very modest growth (AHDR II, 2015). The level of industrial activity in the Arctic is, however, on the rise, with global demand for oil and gas driving the spread of activities in marine and terrestrial environments (Mikkelsen & Langhelle, 2008)— although oil activity has been temporarily dampened by the recent crash in global oil prices. In addition to energy resources, Arctic landscapes hold significant quantities of rare minerals, which is spurring the expansion of mining in certain regions (AHDR II, 2015). Fishing, shipping, and tourism have also been projected to increase as sea ice shrinks and access to the marine Arctic increases, although limited coastal infrastructure is a major roadblock for marine-based industry (Protection of the Arctic Marine Environment, 2009).

Arctic governance is characterized by several trends:

- devolution of authority;
- Indigenous empowerment; and
- challenges in fiscal and human capacity (AHDR II, 2015).

Indigenous cultures are changing as increasing connectedness to global systems is bringing new livelihood challenges and opportunities (AHDR II, 2015). While Indigenous groups have successfully blended tradition with modernity on many fronts (see Cameron, 2012), concerns remain about human health and safety within many Indigenous communities, which are often small and rural (AHDR II, 2015). The Arctic Council strives to promote scientific collaboration and cooperation among Arctic nations, marking a departure from Cold War geopolitical tensions. Economic development remains a priority in much of the Arctic— including among Indigenous communities—and so the interaction of different industrial sectors in the future will be important (e.g., can landscapes be used for oil production and tourism?)

Resilience's promise

Compelled by the extreme changes taking place in the Arctic and the complexity of interactions between social and ecological variables, decision makers are turning to resilience to guide their choices, largely because the framework embraces change (Arctic Council, 2013). Using mathematical insights from complex system modeling, early resilience theorists rejected notions about ecosystems maintaining equilibrium, and replaced them with models wherein ecosystems undergo regular collapse and renewal via adaptive cycles (Gunderson & Holling, 2002). In resilience thinking, systems can collapse and reorganize around the same structures and functions, or reorganize around profoundly different structures and functions, achieving a new identity and marking a transformation (Folke et al., 2010). Resilience is defined as the amount of perturbation a system can absorb—through adaptation—before transforming.

While ecosystems were the original subjects of resilience, it is now commonly applied to social-ecological systems, which are sets of relationships "in which people depend on ... services provided by ecosystems, and ecosystem dynamics are influenced ... by human activity" (Chapin, Kofinas & Folke, 2009, p. 6). Social-ecological resilience work usually aims to measure resilience in given systems (through proxy indicators), increase resilience in desirable systems, and steer untenable system through transformations. The Stockholm Resilience Center (2014) offers seven principles for applying resilience, while the *Arctic Resilience Report* (http://Arctic-council.org/arr/) presents resilience as the best framework for guiding regional governance.

Resilience can be distinguished from sustainability because resilience is predicated on adaptation (changing in response disturbances), while sustainability is centrally focused on conservation (preventing disturbances) (Lew, Ng, Ni & Wu, 2016). This difference is illustrated by *evolutionary resilience* (Scheffer, 2009), a concept taking the ontological position that change is fundamental to existence and, therefore, ideas about not changing ought to be disregarded. This chapter proceeds from the understanding that change is inevitable, in an evolutionary sense, but that the rate and direction of change must be understood as partially anthropogenic and therefore subject to politics.

Resilience's problems (and a way forward)

Scholars from multiple fields have expressed concern about resilience and the politics it animates due to its incomplete grasp of social power dynamics (e.g., Nadasdy, 2007). One line of criticism shows how resilience thinking fails to recognize and reflect on its own epistemic positionality, inherent normativity, and accordant limitations in revealing Truth (Cote & Nightingale, 2012). A second line of criticism demonstrates that resilience, within the systems it studies, has discounted the role of concepts from critical social theory, such as power, agency, and difference. Anthropologist Alf Hornborg (2013, p. 118) observes, "Resilience discourse generally appears to be ignorant of most of the tenets of modern social science."

One such area of modern social science is critical political economy, which is concerned with the unjust and unsustainable social-ecological relations of capitalism, and the benefits of organizing society differently (see Foster 2002; Harvey, 1982). Scholars hailing from this tradition have argued that resilience's advocates use its ideology to produce neoliberal political and environmental subjects (MacKinnon & Derickson, 2013), where neoliberalism is a set of ideas and policies that expand free markets, invigorate privatization, and minimize government interventions, or, in other words, promote the reproduction and expansion of capitalism that disproportionately benefits an elite global minority (Harvey, 2005).

Capitalists have embraced resilience, the critique goes, because resilience compels local communities—exposed to the volatility of global markets and the environmental insecurities brought about by climate change—to survive on their own and to accept the conditions of capitalism that contribute to their vulnerabilities in the first place (e.g., through industrial greenhouse gas emissions). MacKinnon and Derickson (2013, p. 254) contend that "Resilient spaces are precisely what capitalism needs, spaces that are periodically reinvented to meet the changing demands of capital accumulation in an increasingly globalized society."

MacKinnon and Derickson (2013) reject resilience (and capitalism) *in toto*, but this chapter takes a less radical approach. Here, resilience is improved by increasing the attention paid to power within the system being studied, and by emphasizing the normative imperative for an equitable spatial distribution of capitalism's benefits. In the Arctic, many Indigenous communities have partially embraced capitalism as a means to promote their own well-being, including through tourism development (Hillmer-Pegram, 2016). However, the need for fairer capitalism (in terms of local vs. outside interests) is present in all economic sectors in the Arctic, including tourism, and can be advanced through Indigenous empowerment (see e.g., Hall, 2013).

The Arctic tourism system

This section describes Arctic tourism—defined as voluntary temporary intra- and interregional travel for a range of purposes—in terms of a social-ecological system. Social-ecological systems analysis emphasizes two categories of relationships:

- the ecosystem services that humans derive from natural environments; and
- the influences that humans have on local ecosystems.

An examination of select Arctic tourism literature, including the author's research in Barrow, Alaska (Hillmer-Pegram, 2016), is used to develop an image of the system's identity. Because this chapter's scope of analysis is the entire Arctic region, it describes a single regional-scale system.

Ecosystem services

Attractions and stakeholders

The ecosystems of the Arctic—which range from boreal forests, to Arctic tundra, to coastal and marine environments—provide some of the most important attractions for tourists. Both Arctic physical environments (e.g., ice-laden seascapes) and the biology they support (e.g., polar bears) serve as key draws. Maher (2010) found that Northwest Passage cruise tourists most hoped to experience wildlife and icebergs, followed closely by pristine vistas, national parks, and silence. However, Arctic ecosystems are not the only source of attraction for tourists. The study cited above found that Indigenous cultures and western heritage are also key Arctic tourism attractions.

Davidson (2005, p. 9) describes an overarching *idea of the north* that combines Arctic natural, cultural, and historical elements to draw visitors to the region:

> Everyone carries their own idea of the north within them ... thoughts of a
> harder place, a place of dearth: uplands, adverse weather, remoteness from
> cities ... a willingness to encounter the intractable elements of climate,
> topography and humanity ...

The *idea of the north*, however, may be contingent. Changing Arctic ecosystems
may be serving as tourism attractions as well, at least in the short term, under the
paradigm of last change tourism, which Lemelin, Dawson, and Stewart (2012)
define as increased demand for an attraction based on the perception of the
attraction's vulnerability.

Data from Barrow, Alaska, tell a similar story about Arctic tourism attractions
(Hillmer-Pegram, 2016). Arctic ecosystems play an important role in drawing
tourists—especially for bird watching in that case—but so too do the Indigenous
culture, western heritage, and latitudinal features that are not climate-dependent
(e.g., visiting the 'top of the world', viewing the Aurora Borealis).

The willingness of people to travel to and within the Arctic to experience its
attractions is the foundation of the Arctic tourism system. Tourists, however, are
only one among several stakeholder groups. Enzenbacher (2011) identifies 20
additional Arctic tourism stakeholder groups, including host communities, tourism
businesses, governments, and non-government organizations—all of which
influence regional ecosystems in different ways.

Visitor numbers and economic contributions

Quantifying Arctic tourists and their regional economic contribution is an
important element of assessing the system's ecosystem services. Many tourism
scholars have undertaken such quantifications (Fay & Karlsdottir, 2011; Hall &
Saarinen, 2010; Maher, 2013), demonstrating a trend of increasing Arctic tourism
over the last decades that is "already substantial ... and continuing to grow" (Hall
& Saarinen, 2010, p. 448). However, limited and conflicting tourism accounting
methods in the Arctic nations make exact quantifications elusive.

During 2014–15, visitor spending in Alaska was estimated at US$1.94 billion,
with 47,000 jobs being supported at the peak of tourism season (McDowell Group
Inc., 2016). Numbers for the Canadian Arctic were estimated at 528,000 visitors
per years, spending approximately CA$388 million (Maher et al., 2014). In Iceland,
around 807,000 visitors supported approximately 5 percent of the workforce (7,000
jobs) in 2013, while the Faroe Islands saw approximately 100,000 visitors spend
€55 million in 2012 (Maher et al., 2014). In northern Sweden visitations increased
115 percent between 2000 and 2012, with spending reaching SEK4.5 billion, while
northern Finland hosted 2.4 million visitors in 2013 (Maher et al., 2014). Estimates
put the number of arrivals in Arctic Russia at roughly 500,000 (Tzekina, 2014),
while visitors in northern Norway (including Svalbard) have been reported at
roughly 2.5 million per year (Maher, 2013). Cruise ships are the most common
mode of visitation in the Arctic (Lück, Maher & Stewart, 2010), but terrestrial and
air travel account for significant portions of visitors as well.

Human influence on ecosystems

Tourism's impacts

The body of literature concerned with potential environmental impacts from Arctic tourism is significant, while research documenting its actual impacts on various local-scale ecosystems is smaller. Malcom and Penner (2011) track how whale-watching vessels in Arctic Canada disturb Beluga whale behavior. In addition to Arctic tourism impacting local-scale ecosystems, it also affects the regional environment through the greenhouse gasses emitted by travel to and within the Arctic. Dawson, Stewart, Lemelin and Scott (2010) estimate the carbon cost of polar bear viewing in Arctic Canada, capturing the tragic irony of last chance tourism's self-perpetuating feedback cycle.

While a resilience framework emphasizes the interrelationship between the social and ecological components of a system, it is also imperative to address the manner in which different aspects of the social component influence each other. In the case of Arctic tourism, this includes understanding the impacts of tourism on host communities. Cerveny (2007) provides ethnographic data about the socio-cultural changes that Hoonah, Alaska, experienced after becoming a cruise ship destination. Stewart et al. (2011) compare and contrast the attitudes of residents of two Indigenous villages in Arctic Canada toward increasing cruise tourism, revealing concerns but an overall supportive attitude.

The author's research in Barrow found that tourist numbers are small enough that direct environment and social impacts are fairly benign. However, local stakeholders did express dismay about visitors trampling tundra, invading locals' privacy, and disturbing Indigenous subsistence hunting (Hillmer-Pegram, 2016).

Tourism's governance

Resilience understands governance—meaning the formal and informal rules that control human behavior—to be the mechanism by which humans can steer social-ecological systems in desired directions. It is thus important to examine the literature on Arctic tourism governance to gain a fuller understanding of the system. In general terms, Arctic tourism operates within a global governance framework that first and foremost encourages the private pursuit of profit. Within this capitalist political economy, tourism governance can take the form of mitigating tourism's negative environmental and social impacts, promoting tourism development, or trying to balance the countervailing forces of economic growth and conservation.

On the environmental side, Tuulentie and Rantala (2013) present a case from Arctic Sweden and Finland, where free access to public forests is being challenged by local perceptions of overuse by Asian tourists (namely berry pickers), creating nationalist sentiment for privatization. Meanwhile, Johnston (2011) reviews multiple pan-Arctic sustainable tourism initiatives, such as the Arctic Council's SMART: Sustainable Model for Arctic Region Tourism. SMART offers governance strategies for both environment and social issues related to tourism development. In contrast to seeking governance that mitigates the social impacts of tourism, Pashkevich and Stjernström (2014) identify governance barriers to developing more tourism in Arctic Russia. The role of governance in the Arctic tourism system is complex, operating at multiple scales simultaneously and changing across space, sometimes trying to expand tourism development and other times trying to mitigate its impacts.

Summary

Key dynamics of the Arctic tourism system (Figure 16.1) include:

- Arctic ecosystems attract visitors to the region (although they are not the only source of attractions); and
- Arctic tourists impact local ecosystems through direct contact and the regional environment through the emission of greenhouse gasses (while also affecting destinations socially).

Functionally, this system provides economic benefits for numerous stakeholders, both local and non-local, and benefits for millions of tourists per year. Structurally, the system is governed by a multi-scale web of policies and norms, tourists' demands, and the realities of operating in the Arctic environment.

Overall, the number of Arctic tourists has grown, as has the amount of money they spend, both of which are due to, in part, the draw of Arctic ecosystems. However, important questions remain about the equity of profit sharing between local and non-local stakeholders in the system (Enzenbacher, 2011).

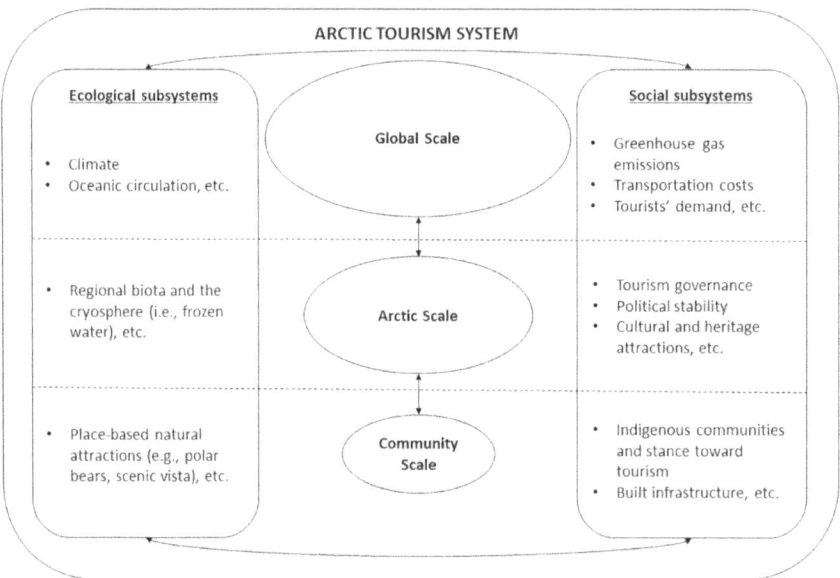

Figure 16.1 Conceptual multi-scale model of the Arctic tourism system

A conventional resilience-based study of this system might examine the impacts of climate change (a key perturbation) and seek adaptive strategies that would help the system continue to function as it changes, i.e., to remain resilient. There are many fascinating questions that could be asked about the impacts of global warming on the *idea of the north*, the role of last chance tourism in current and future visitor numbers, and post-transformation Arctic tourism attractions. However, because this study is guided by critical political economy, it focuses on the resilience of one set of relationships within the Arctic tourism system: the power of Indigenous communities to control and profit from tourism.

Why focus on Indigenous control of tourism?

The reason for focusing on Indigenous power over tourism (or what can be called Indigenous tourism governance) as a sub-system within the larger Arctic tourism system is twofold:

- Indigenous control of tourism would partly rectify the marginalization of Indigenous people from economic and political power, which began with the colonization of their territories and cultures, and continues in less explicit forms today—such as systemic racism—by increasing their powers of self-determination (AHDR II, 2015).
- Scholarly research from around the world shows that, while tourism development can provide an important source of income and pride for Indigenous communities, it can also have extremely negative social-ecological affects if controlled by outside forces or done in a manner that produces inequitable community benefits (e.g., Bianchi, 2011; Johnston, 2006; Strickland-Munro & Moore, 2013; Weaver, 2010). While Indigenous communities being in charge of tourism development does not necessarily assure a fair distribution of benefits within those communities, it is still a better option than having outside forces driving the process.

Ultimately, focusing on the resilience of Indigenous tourism governance is an effective response to MacKinnon and Derickson's (2013) concern that resilience thinking disproportionately benefits political economic elites at the expense of the marginalized.

Future scenarios and implications for resilience

Much of the thinking about the future of Arctic tourism assumes that it will continue to increase as the climate warms and sea ice recedes, due to more access to Arctic destinations. The Arctic Marine Shipping Assessment states that "Arctic marine tourism's most likely future is that larger numbers of tourists, traveling aboard increased numbers of ships of all types, will be spending more time at more locations" (PAME, 2009, p. 100). However, this thinking fails to fully consider the ways that climate change is altering the ecosystems that underpin many regional attractions. While studies have attempted to address this issue (e.g., Kaján, 2013), there is significant uncertainty about future demand for Arctic tourism, as well as opaqueness about interactions between tourism and other industrial sectors and the influence of non-Arctic drivers. Inaccurate projections about Arctic tourism's future could lead Indigenous communities to over- or underinvest in tourism preparedness, thus compromising their power to control and profit from tourism. When planning in uncertain contexts, it is useful to consider multiple future scenarios in order to maintain directional flexibility (Carpenter, Bennett & Peterson, 2006). The following section presents two extreme-case scenarios for the Arctic tourism system for the purpose of contrast (Table 16.1), but the real future may very well lie somewhere in between.

Table 16.1 Future scenarios for the Arctic tourism system and implications for Indigenous sovereignty over tourism

	Maximum tourism: *"Arctic Disneyland"* *(easy to profit, hard to control)*	*Minimum tourism:* *"The Chill is Gone"* *(easy to control, hard to profit)*
State of key drivers	• Arctic warming and associated changes proceed slowly • Access increases and tourists find new Arctic conditions desirable • Non-Arctic variables remain in favorable states • Interaction between tourism and other industrial sectors favors tourism	• Arctic warming and associated changes proceed rapidly • Access increases but tourists find new Arctic conditions undesirable • Non-Arctic variables turn unfavorable • Interaction between tourism and other industrial sectors favors other industrial sectors
Opportunities for Indigenous sovereignty over tourism	• Ample tourists to extract profit from	• Lack of outside competition • Chance to develop small-scale, high-end product
Challenges for Indigenous sovereignty over tourism	• Being overwhelmed and outcompeted by outside tourism companies	• Few tourists to extract profit from
Adaptations to promote the resilience of Indigenous sovereignty over tourism	• Invest in the struggle for self-determination more generally • Use existing power to limit competition from outside companies • Develop local capacity and desire to do tourism work	• Explore effectiveness of alternative Arctic attractions and develop an "idea of the *new*-north" • Develop vertically integrated tourism businesses

Maximum tourism

In the maximum tourism scenario, Arctic warming and associated biophysical changes proceed at a relatively slow pace. Sea ice continues to shrink, but in a way that facilitates greater access for marine tourists while still providing the ice-laden seascapes that many visitors expect. Iconic Arctic species are able to adapt and remain present for tourists to view. In this scenario, tourists tolerate and enjoy the inevitable changes that do take place in the Arctic, such as the emergence of new biomes and species (e.g., hybrid polar and grizzly bears called *pizzly* bears), and new opportunities for ice-free activities. Other drivers of Arctic tourism (e.g., travel costs, political stability, greenhouse gas regulations, desirability of competing destinations) are in favorable states for increased visitor numbers. Tourism in this scenario would increase across the Arctic, including the establishment of a consistent industry in Russia and the continued maturation and expansion of the industry in North America and Europe.

This scenario would produce a specific set of opportunities and challenges for the resilience of Indigenous groups to maintain (or gain) political–economic sovereignty over tourism in their communities. On the positive side, there would be plenty of tourists to extract profit from. However, if the Arctic tourism experience becomes a mass tourism experience, then the tourists that small communities would be dealing with might have low levels of cultural and environmental sensitivity. Residents of Barrow, Alaska, labeled this scenario "Arctic Disneyland" in a workshop conducted by the author's research team in 2015. The main challenge for Indigenous communities in Arctic Disneyland would be to hold back the commoditizing and exploitative forces of global capitalism, so that they could develop at their own desired pace and to their own desired extent. If the profit potential is high enough, non-local tourism operators (e.g., cruise lines, international hotels) could overwhelm politically marginalized Indigenous communities, while leaving few avenues for profit to trickle down.

Political–economic adaptations that Indigenous communities could make in this scenario to enhance their ability to control and profit from tourism include:

- investing in the ongoing struggle for land rights, political representation, and self-determination more generally;
- using whatever power they do possess to limit competition from outside tourism companies (thereby giving local entrepreneurs favored access to the tourist dollar); and
- developing a local workforce capable of handling large flows of visitors.

Indigenous communities in the Arctic could potentially learn helpful lessons about managing large numbers of tourists from the experiences of Indigenous groups in other parts of the world, including those of Native Americans in the contiguous United States who have successfully directed profits from casino gaming into community development (Akee, Spilde & Taylor, 2015).

Minimum tourism

The premise of the minimum tourism scenario is that Arctic warming and associated biophysical changes occur so rapidly that the *idea of the north* is critically compromised in the minds of tourists. In this scenario, labeled here "The Chill is Gone", summer sea ice disappears completely, leading to the permanent disappearance of iconic ice-dependent animal species and ice-laden seascapes during the marine-tourism season. While access to Arctic destinations has improved, there is little demand. It turns out that the recent increase in Arctic tourism was driven largely by a last chance philosophy and, now that the opportunity to see the old Arctic is over, people have stopped coming. The state of related non-Arctic variables also play a role in decreasing visitor numbers. Due to increased travel costs, regional political instability, and greenhouse gas regulations, other destination have become relatively more attractive.

Even in the minimum tourism scenario, there are some visitors to the Arctic. In many ways, this is similar to how the Arctic tourism picture looked about 30 years ago—a small amount of independent adventurers spending large amounts of money to visit remote locations. The big difference is that the climatic and biotic characteristic of the destinations have been transformed. This situation provides both opportunities and challenges for the resilience of small communities to control and profit from tourism. With fewer tourists, the overall profit potential is less than in the other scenario. However, there is also less competition from outside tourism companies, which reduces the chance of Indigenous communities being overwhelmed and leaves a bigger slice of a smaller pie for local benefit.

In the minimum development scenario (easy to control, hard to profit), recommended adaptations revolve around new business strategies to maximize profit from limited visitors. This includes developing an *idea of the new-north*, which establishes the appeal of the Arctic once the ice is gone. Part of the new-north could be focused on environmental heritage (i.e., what used to be here and lessons learned) but could also focus on the present and future. Indigenous culture and power may play a larger role in the new-Arctic than it has in past (colonial?) Arctic imaginaries, and this could draw people to the region—in addition to being a positive step toward social equity. Indigenous communities may also benefit from developing vertically integrated tourism companies in order to minimize revenue leakage to outside operators.

Summary

This chapter's recommended adaptations aim to increase the political–economic sovereignty—or power—that Arctic Indigenous communities possess over regional tourism, and relative to other stakeholders with whom they compete for profit. The recommendations attempt to rectify historic power imbalances between social groups within the system—something resilience has been rightly criticized for ignoring. If it is accepted that Indigenous power over tourism is the appropriate subject of analysis for resilience-based Arctic tourism research, then researchers can begin conceptualizing Indigenous tourism governance as a key sub-system of the larger Arctic tourism system, and analyzing and advancing *its* resilience, specifically.

While this chapter takes a small step in that direction by considering future scenarios, the Stockholm Resilience Centre's seven principles for building resilience appear to be a promising framework to guide the next step (Stockholm Resilience Center, 2014). The seven principles are:

- maintain diversity and redundancy;
- manage connectivity;
- manage slow variables and feedbacks;
- foster complex adaptive systems thinking;
- encourage learning;
- broaden participation; and
- promote polycentric governance systems.

While some of the principles are likely less applicable than others (and less feasible given financial, social, and environmental constraints within the system), others would surely aid in building resilience in the Indigenous tourism governance system. The key message here, however, is the importance of choosing to study systems for which an increase in resilience would mean an increase in socio-economic equity.

Conclusion

As the environmental and social threats from climate change and globalizing capitalism continue to produce vulnerable communities, so too will resilience remain a popular paradigm for studying and steering change. Resilience, a framework predicated on adapting to disturbances, seems appropriate for a world full of disturbances, and both Arctic studies and tourism studies are embracing the resilience paradigm. This chapter contributes to resilience theory by making the case that, as resilience continues to expand in these fields, scholars and practitioners must not adopt its concepts and methods—which are founded in natural science—uncritically. It is essential that scholar/practitioners heed the critical work of social theorists that brings into question many of resilience's ontological and epistemological assumptions. Yes, change may be natural from an evolutionary standpoint, but this does not permit the social-ecological inequities seen around the globe today, which are much more a result of human agency than nature. Acceptable solutions must expand from *providing good science to policy makers* to eventually include *providing avenues for the politically marginalized to acquire more power*. Critical theorists have been saying this for a long time and it is still important.

Within tourism studies, there is a limited but vital history of critical political economy (e.g., Bianchi, 2011; Britton, 1991; Bunten, 2010). Ideally the field will further integrate insights from that work as it continues to adopt ideas from resilience. Integrating these approaches to create a critical resilience of tourism is a promising area for future research, and could advance resilience's critical turn

(Biermann, Hillmer-Pegram, Knapp & Hum, 2016). This chapter presents an early attempt at such integration by arguing that the power of Indigenous communities to control and profit from Arctic tourism ought to be the main concern of Arctic tourism research, and ought to be what is kept (or made) resilient to climate change and globalizing capitalism as the Arctic transforms, for the benefit of the Arctic's Indigenous inhabitants. More research on the political economy of tourism in particular communities is needed to advance such an agenda and would help further answer the questions, what should be made resilient to what, and for whom (Lebel et al., 2006)?

References

ACIA (2005). *ACIA Overview Report*. Arctic Climate Impact Assessment. New York: Cambridge University Press.

Akee, R. K. Q., Spilde, K. A. & Taylor, J. B. (2015). The Indian Gaming Regulatory Act and its effects on American Indian economic development. *The Journal of Economic Perspectives, 29*(3), 185–208.

AHDR (2004). *Arctic Human Development Report*. N. Einarsson, J. N. Larsen, A. Nilsson & O. R. Young (Eds.). Akureyri: Stefansson Arctic Institute.

AHDR II (2015). *Arctic Human Development Report: Regional Processes and Global Linkages*. J. N. Larsen & G. Fondahl (Eds.). Copenhagen: Nordisk Ministerråd.

Arctic Council (2013). *Arctic Resilience Interim Report 2013*. Stockholm: Stockholm Environment Institute and Stockholm Resilience Centre.

Bianchi, R. V. (2011). Tourism, capitalism and Marxist political economy. In J. T. Mosedale (Ed.), *Political Economy of Tourism: A Critical Perspective* (pp. 17–37). London: Routledge.

Biermann, M., Hillmer-Pegram, K., Knapp, C. & Hum, R. E. (2016). Approaching a critical turn? A content analysis of the politics of resilience in key bodies of resilience literature. *Resilience, 4*(2), 59–78

Britton, S. G. (1991). Tourism, capital, and place: towards a critical geography of tourism. *Environment and Planning D-Society & Space, 9*(4), 451–78.

Bunten, A. C. (2010). More like ourselves: Indigenous capitalism through tourism. *The American Indian Quarterly, 34*(3), 285–311.

Cameron, E. S. (2012). Securing Indigenous politics: a critique of the vulnerability and adaptation approach to the human dimensions of climate change in the Canadian Arctic. *Global Environmental Change, 22*(1), 103–14.

Carpenter, S. R., Bennett, E. M. & Peterson, G. D. (2006). Scenarios for ecosystem services: an overview. *Ecology and Society, 11*(1), 29.

Cerveny, L. K. (2007) *Sociocultural Effects of Tourism in Hoonah, Alaska*. Gen. Tech. Rep. PNW-GTR-734. Portland, OR: US Forest Service.

Chapin III, F., Kofinas, G. P. & Folke, C. (Eds.). (2009). *Principles of Ecosystem Stewardship: Resilience-based Natural Resource Management in a Changing World*. New York: Springer Science.

Chapin III, F. S., Sommerkorn, M., Robards, M.D. & Hillmer-Pegram, K. (2015). Ecosystem stewardship: a resilience framework for Arctic conservation. *Global Environmental Change, 34*, 207–17.

Conservation of Arctic Flora and Fauna (2010). *Arctic Biodiversity Trends 2010: Selected Indicators of Change*. Akureyri: CAFF International Secretariat.

Cote, M. & Nightingale, A. J. (2012). Resilience thinking meets social theory: Situating social change in socio-ecological systems (SES) research. *Progress in Human Geography,* 36(4), 475–89.

Cretney, R. (2014). Resilience for whom? Emerging critical geographies of socio-ecological resilience. *Geography Compass,* 8(9), 627–40.

Davidson, P. (2005). *The Idea of the North.* London: Reaktion Books.

Dawson, J., Stewart, E. J., Lemelin, H. R. & Scott, D. (2010). The carbon cost of polar bear viewing tourism in Churchill, Canada. *Journal of Sustainable Tourism,* 18(3), 319–36.

Enzenbacher, D. J. (2011). Polar tourism development: who benefits? In A. A. Grenier & D. K. Müller (Eds.), *Polar Tourism: A Tool for Regional Development* (pp. 23–60). Quebec: Presses de l'Université du Québéc.

Fay, G. & Karlsdottir, A. (2011). Social indicators for Arctic tourism: observing trends and assessing data. *Polar Geography,* 34(1–2), 63–86.

Folke, C., Carpenter, S. R., Walker, B., Scheffer, M., Chapin, T. & Rockström, J. (2010). Resilience thinking: integrating resilience, adaptability and transformability. *Ecology and Society,* 15(4), 20.

Foster, J. B. (2002). *Ecology Against Capitalism.* New York, NY: Monthly Review Press.

Fraser, N. (2009). *Scales of Justice: Reimagining Political Space in a Globalizing World.* New York: Columbia University Press.

Gunderson, L. H. & Holling, C. S. (2002). *Panarchy: Understanding Transformations in Systems of Humans and Nature.* Washington, DC: Island Press.

Hall, C. M. & Saarinen, J. (2010). Tourism and change in polar regions: introduction—definitions, locations, places and dimensions. In C. M. Hall & J. Saarinen (Eds.), *Tourism and Change in Polar Regions: Climate, Environment, and Experience* (pp. 1–42). London: Routledge.

Hall, R. (2013). Diamond mining in Canada's Northwest Territories: a colonial continuity. *Antipode,* 45(2), 376–93

Harvey, D. (1982). *The Limits to Capital.* Oxford: University of Chicago Press.

Harvey, D. (2005). *A Brief History of Neoliberalism.* Oxford: Oxford University Press.

Hillmer-Pegram, K. (2014). Understanding the resilience of dive tourism to complex change. *Tourism Geographies,* 16(4), 598–614.

Hillmer-Pegram, K. (2016). Integrating Indigenous values with capitalism through tourism: Alaskan experiences and outstanding issues. *Journal of Sustainable Tourism,* 1–17.

Hornborg, A. (2013). Revelations of resilience: from the ideological disarmament of disaster to the revolutionary implications of (p)anarchy. *Resilience: International Policies, Practices and Discourses,* 1(2), 116–29.

Johnston, A. M. (2006). *Is the Sacred for Sale? Tourism and Indigenous Peoples.* New York: Earthscan.

Johnston, M. (2011). Arctic tourism introduction. In P. T. Maher, E. J. Stewart & M. Luck (Eds.), *Polar Tourism: Human, Environmental and Governance Dimensions* (pp. 17–33). Putnam Valley, NY: Cognizant Communications.

Kaján, E. (2013). An integrated methodological framework: engaging local communities in Arctic tourism development and community-based adaptation. *Current Issues in Tourism,* 16(3), 286–301.

Kirchhoff, T., Brand, F. S., Hoheisel, D. & Grimm, V. (2010). The one-sidedness and cultural bias of the resilience approach. *Gaia-Ecological Perspectives for Science and Society,* 19(1), 25–32.

Larsen, J. N., Anisimov, O. A., Constable, A., Hollowed, A. B., Maynard, N., Prestrud, P. ... Stone, J. M. R. (2014). Polar Regions. In V. R. Barros, C. B. Field, D. J. Dokken, M. D.

Mastrandrea, K. J. Mach, T. E. Bilir … L.L. White (Eds.) *Climate Change 2014: Impacts, Adaptation, and Vulnerability. Part B: Regional Aspects. Contribution of Working Group II to the Fifth Assessment Report of the Intergovernmental Panel on Climate Change* (pp. 1567–612). Cambridge and New York, NY: Cambridge University Press.

Lebel, L., Anderies, J. M., Campbell, B., Folke, C., Hatfield-Dodds, S., Hughes, T. P. & Wilson, J. (2006). Governance and the capacity to manage resilience in regional social-ecological systems. *Ecology and Society,* 11(1), 19.

Lemelin, R. H., J. Dawson & E. Stewart (Eds.) (2012). *Last chance Tourism: Adapting Tourism Opportunities in a Changing World.* London: Routledge.

Lemelin, R. H., Koster, R. & Youroukos, N. (2015). Tangible and intangible indicators of successful aboriginal tourism initiatives: A case study of two successful aboriginal tourism lodges in Northern Canada. *Tourism Management,* 47, 318–28.

Lew, A. A. (2014). Scale, change and resilience in community tourism planning. *Tourism Geographies,* 16(1), 14–22.

Lew, A. A., Ng, P. T., Ni, C.-C. & Wu, T.-C. (2016). Community sustainability and resilience: similarities, differences, and indicators. *Tourism Geographies,* 18(1), 18–27.

Lück, M., P.T. Maher & E.J. Stewart (Eds.) (2010). *Cruise Tourism in Polar Regions: Promoting Environmental and Social Sustainability?* London: Earthscan.

McDowell Group Inc. (2016) *Economic Impacts of Alaska's Visitor Industry: 2014–2015 Update.* Juneau, AK: State of Alaska Department of Commerce, Community, and Economic Development.

MacKinnon, D. & Derickson, K. D. (2013). From resilience to resourcefulness: a critique of resilience policy and activism. *Progress in Human Geography,* 37(2), 253–70.

Maher, P.T. (2010). Cruise tourist experiences and management implication for Auyuittuq, Sirmillik, and Quttinirpaaq National Parks, Nunavut, Canada. In C. M. Hall and J. Saarinen (Eds.) *Tourism and Change in Polar Regions.* New York: Routledge.

Maher, P. T. (2013). Looking back, venturing forward: challenges for academic, community, and industry in polar tourism research In D. K. Müller, L. Lundmark & R. H. Lemelin (Eds.) *New Issues in Polar Tourism: Communities, Environments, Politics* (pp. 19–36). New York: Springer.

Maher, P.T., Gelter, H., Hillmer-Pegram, K., Hovgaard, G., Hull, J., Jóhannesson, G. … Pashkevich, A. (2014). Arctic tourism: realities and possibilities. In L. Heininen (Ed.), *Arctic Yearbook 2014: Scholarly Articles, Section II: Regional Economy and Property.*

Malcom, C. D. & Penner, H. C. (2011). Behavior of Belugas in the presence of whale-watching vessels in Churchill, Manitoba and recommendations for local Beulga-watching activities. In P. T. Maher, E. J. Stewart & M. Luck (Eds.), *Polar Tourism: Human, Environmental and Governance Dimensions* (pp. 54–79). Putnam Valley, NY: Cognizant Communications.Mikkelsen, A. & O. Langhelle (Eds.). (2008). *Arctic Oil and Gas: Sustainability at Risk?* New York: Routledge.

Nadasdy, P. (2007). Adaptive co-management and the gospel of resilience. In D. Armitage, F. Berkes & N. Doubleday (Eds.), *Adaptive Co-Management: Collaboration, Learning and Multi-level Governance* (pp. 208–27). Vancouver: UBC Press.

Nelson, S. H. (2014). Resilience and the neoliberal counter-revolution: from ecologies of control to production of the common. *Resilience,* 2(1), 1–17

Nuttall, M. (2000). Indigenous peoples, self-determination, and the Arctic environment. In M. Nuttall & T. V. Callaghan (Eds.), *The Arctic: Environment, People, Policy* (pp. 377–409). Amsterdam: Hardwood Academic Press.

PAME (2009). *Arctic Marine Shipping Assessment 2009 Report.* Protection of the Arctic Marine Environment. Akureyri: Arctic Council.

Pashkevich, A. & Stjernström, O. (2014). Making Russian Arctic accessible for tourists: analysis of the institutional barriers. *Polar Geography,* 37(2), 137–56.

Scheffer, M. (2009). *Critical Transitions in Nature and Society.* Princeton, NJ: Princeton University Press.

Stewart, E. J., Draper, D. & Dawson, J. (2011). Coping with change and vulnerability: a case study of resident attitudes toward tourism in Cambridge Bay and Pond Inlet, Nunavut Canada. In P. T. Maher, E. J. Stewart & M. Luck (Eds.), *Polar Tourism: Human, Environmental and Governance Dimensions* (pp. 33–54). Putnam Valley, NY: Cognizant Communications.

Stockholm Resilience Center (2014). *Applying Resilience Thinking: Seven Principles for Building Resilience in Social-Ecological Systems.* Stockholm: Stockholm Resilience Center.

Strickland-Munro, J. & Moore, S. (2013). Indigenous involvement and benefits from tourism in protected areas: a study of Purnululu National Park and Warmun Community, Australia. *Journal of Sustainable Tourism,* 21(1), 26–41.

Stroeve, J. C., Kattsov, V., Barrett, A., Serreze, M., Pavlova, T., Holland, M. & Meier, W. N. (2012). Trends in Arctic sea ice extent from CMIP5, CMIP3 and observations. *Geophysical Research Letters,* 39(16).

Tuulentie, S. & Rantala, O. (2013). Will "free entry into the forest" remain? In D. K. Muller, L. Lundmark & R. H. Lemelin (Eds.), *New Issues in Polar Tourism: Communities, Environments, Politics* (pp. 177–88). New York: Springer.

Tzekina, M. (2014). Estimation of tourism potential of Russian Far North. PhD, Economic, social, political and recreational geography. Moscow: Moscow State University.

Turner, M. D. (2013). Political ecology I: an alliance with resilience? *Progress in Human Geography,* 38(4), 616–23.

Walsh, J. E., Overland, J. E., Groisman, P. Y. & Rudolf, B. (2011). Ongoing climate change in the Arctic. *Ambio,* 40(1), 6–16.

Weaver, D. (2010). Indigenous tourism stages and their implications for sustainability. *Journal of Sustainable Tourism,* 18(1), 43–60.

Zeppel, H. (2006). *Indigenous Ecotourism: Sustainable Development and Management.* Oxford: Cabi.

17 Conceptualizing destinations as a *vanua*

The evolution and resilience of a Fijian social and ecological system

Apisalome Movono

Introduction

Tourism has become an invaluable tool that has helped many small island developing states (SIDS) to realize their national aspirations, and it is promoted in the Pacific region as an industry with great potential for community development (Prasad, 2014). In Fiji, tourism stands out as the most logical of economic alternatives because it fits well within the needs typical of island states in the region (Movono, Harrison, & Pratt, 2015; Prasad, 2014). However, there remains much ambiguity as the bulk of literature about Pacific communities and tourism is fragmented and rarely acknowledges the complex roles played by indigenous people as actors in their own development. Previous studies in Fiji have overlooked the complex and adaptive nature of its indigenous communities, ignoring their unique relationships with the biosphere. Issues related to resilience and to complex adaptive systems, as key features of indigenous Fijian society, have seldom been discussed, and this is an area to which this chapter will contribute.

This chapter will first provide a review of the literature before offering some background to Fiji, its tourism industry and the study area. The chapter will then outline the research methods employed and present the findings and conclusions of this study. The findings of this research will unpack the elements of the *vanua* as a complex and adaptive social and ecological system, highlight its historical developments and perceived changes, and propose the use of the *vanua*, SES and the resilience model as a tool to operationalize resilience in related indigenous Fijian communities.

Resilience and complex adaptive systems

Founded in the natural sciences and ecology, the notion of resilience emerged as 'a measure of the persistence of systems and their ability to absorb change, disturbance and still maintain the same relationships between populations and state variables' (Holling, 1973, p. 14). Recent academic literature has shown the development of *resilience* from a predominantly scientific term to one that shows increasing recognition for the complex relationships between society and the environment (Cretney, 2014; Folke, 2006; Gaillard, 2010; Holland, 2006). More recent definitions have focused on the varieties of elements and their capacities in a communal system to resist or change so that they may obtain an acceptable level of functioning and structure (UN, 2005). However, there is still much debate regarding its definition, measurement and application, brought about by the non-homogeneous nature of communities and the reality that each community and each system is different (Neely, 2015). Each system is exposed to different perturbations and is, therefore, subject to varying levels of socio-political, economic and ecological conditions that determine their levels of resilience (Manyena, 2006). Hence, it is important that the contextual components of a complex adaptive system be identified, and its non-homogenous social and ecological components examined to better understand resilience within indigenous communities involved in tourism.

There are many classifications of resilience in different disciplines that have emerged from the theoretical breakthrough of Holling (1973). Holling's work challenged the previous, dominant engineering-based definitions of resilience that imply a return to an original equilibrium. In doing so, Holling (1973) established that ecological systems do not have one static point of equilibrium, but rather a zone of stability that allows for the reorganization of a system to maintain survival. This process of reorganization is of particular interest to the current study because it examines an indigenous community as it adjusts to more than 40 years of tourism involvement. The literature reveals that most scholars have neglected to separate the notion of stability from resilience, causing confusion between the two terms as being part of one holistic definition (Coetzee, Niekerk & Raju, 2016; Rose, 2007; Paton, 2006). Gaillard (2010) points out that a resilient system is not always stable, and, in the case of communities, they cannot return to their former, equal state. Unlike engineered structures, societies are changing, continually responding to both internal and external pressures, rendering it with a safer equilibrium. Coetzee *et al.* (2016) assert that stability is not necessarily needed to attain resilience, but adaptability is. Despite this, many scholars still maintain that the underlying idea behind resilience is not necessarily 'bouncing back' to the same state, but that resilience must focus on adaptation and the processes of change that a system and its components can undertake while maintaining critical thresholds (Walker, Gunderson, Kinzig, Folke, Carpenter & Schultz, 2006).

It is, therefore, important to consider that there are crucial elements essential to understanding the different interpretations of resilience theory within communities, which include notions of adaptive capacity, transformation and social capital (Neely, 2015; Hammer, Edwards & Tapinos, 2012; Gallopin, 2006; Walker *et al.* 2006). In resilience studies, adaptive capacity refers to the processes and patterns of behaviour that change in order to maintain a system within critical thresholds (Holling, 1973). Such changes have been modelled in the form of the resilience cycle (Holling, 2001). Also referred to as the Holling Loop, the model illustrates the ability of systems to flip between different domains and regain stability. The cycle typically begins with the reorganization that leads to exploitation (new systems are created), conservation (building for a more stable state) and release (next disturbance event) (Holling, 2001). The resilience cycle makes it possible to focus on specific elements and zoom in on community capacities. This includes the capacity to learn, adapt and prepare for future perturbations, and it normally entails self-organization and taking action to ensure that the system is able to cope with any unforeseen circumstances (Folke *et al*. 2003; Gunderson & Holling, 2002). On the other hand, transformation consists of a more serious path involving a shift or collapse of a system and causing it to be transformed into an entirely new system (Walker, Holling, Carpenter & Kinzig, 2004; Timmerman, 1981). Cretney (2014) adds that this potential for change (dependant on willingness, capacity and ability to change) demarcates resilience from general capacities, suggesting that community capacities or social capital should be empirically examined in communities to measure and better understand resilience. He argues that social capital is part of a wider framework that includes other forms of capital but focuses on the social aspects of communities (Cretney, 2014). Furthermore, community capacities, as an alternative to social capital, recognize all aspects of social life and are well suited to this study. Cretney (2014, p, 61) emphasizes that, despite differences in the way resilience is understood or applied, 'the concepts of community capacities, alongside the integration of social and ecological systems, adaptive capacity and transformation are important to the theoretical base of resilience' and must therefore be incorporated in empirical research.

Gallopin (2006) reviews the concept of resilience in detail and elaborates on the differences and interrelationships between vulnerability and adaptive capacity as academic concepts. Both Gallopin (2006) and Folke (2006) hint that, if adaptability is important to attaining improved resilience, then emerging frameworks must endeavour to holistically encompass the prevalent ecological and socio-political contexts. Emphasis on specific contexts may address the shortfalls of resilience and increase focus on the intimate connections between components of social and ecological systems that affect it (Walker *et al.* 2004). Folke (2006) provides further clarity to the notion of social-ecological systems by emphasizing that the resilience approach should promote non-linear dynamics, thresholds and uncertainty, and examine how such dynamics interact across spatial and temporal scales. The subsequent increased awareness on the complex relationships between society and environment has helped catapult resilience to the mainstream via the significant theoretical advancements in social and ecological resilience. Adger (2000) acknowledges social and ecological systems as being interrelated, linked and dependant on one another through connections between livelihoods, well-being and environmental conditions. Folke *et al.*(2003) agree that these links between system elements can either work against one another or for mutual benefit and, in the process, affect the overall resilience of a system. As such, when studying resilience within communities, adopting a systems approach is essential.

In considering complexities within indigenous Fijian society and the dynamic nature of resilience, it is imperative that the case study, an indigenous Fijian community, be considered as a system. Essentially, a systems approach focuses on the human and environment interactions as part of an interrelated and interacting system. Systems thinking provides a pathway for adoption by this chapter through a specific variation of systems thinking, referred to as complex adaptive systems

(CAS) theory. CAS presents an interesting tool for studying dynamic systems and concepts such as resilience (Coetzee *et al.* 2016; Neely, 2015; Manyena, 2006). With roots in the natural sciences and ecology, CAS focuses on understanding non-linear dynamics and attempts to show how simple interactions at the micro level can lead to very complex implications at the macro level (Holland, 2006; Gunderson & Holling, 2002). Buckley's (1968) work opens opportunities for communities to be examined in their totality, allowing for the examination of specific system elements and focusing on how micro-level activities such as tourism employment can have a wider ranging impact on the broader elements and resilience of a human-driven system.

Hammer, Edwards and Tapinos (2012) consider any human system as a CAS because the systems are a set of diverse, interacting individuals that have the ability to self-reorganize and respond organically to any influence to the system. For this particular study, the indigenous Fijian community of Vatuolalai is presented as a CAS that has a specific set of social and ecological components which interact with internal and external systems including tourism. CAS theory is based on complex behaviour that emerges as a result of interactions among system components and between system components and the environment (Zhou, Wang, Wan & Jia, 2010). Through interacting with and learning from its environment, a complex adaptive system modifies its behaviour to adapt to changes in the environment (Coetzee *et al.*, 2016). CAS has also been referred to as complex systems, complex responsive processes, complex evolving systems and intelligent complex adaptive systems. It is also said to be characterized by panarchy, it can be dynamically influenced internally or externally and it is an ideal tool to examine communities that are always changing (Gunderson & Holling, 2002). By examining the resilience of a community as a CAS places emphasis on understanding individual components and capacities and how they interact to generate resilience. CAS concepts such as non-linearity, aggregation, emergent behaviour, feedback loops, adaptation and contextual-based responses become relevant for this study. The use of CAS paves the way for a more targeted and holistic approach to be taken when assessing the resilience of indigenous Fijian communities that are involved in tourism.

Methods

Research in indigenous Fijian communities is a multi-dimensional activity requiring immersion, careful engagement and the use of carefully selected methods and localized paradigms (Nabobo-Baba, 2008; Walsh, 1996). A case study approach and ethnography determine the nature of data collection (Veal, 2006). Fieldwork for this study was conducted in two phases, the first phase of 11 weeks and a second phase of 5 weeks, with the total data collection stretching over a period of 16 weeks (120 days). The researcher fully immersed himself during fieldwork and through access to all parts of the community engaged in close, long-term interaction with villagers to gain an understanding of their beliefs, motivations and behaviour (Tracy, 2013; Tedlock, 2000).

A total of 32 household survey questionnaires were administered, one to each household, with questions covering socio-economic variables focussing on resilience. These were used as a means to triangulate and quantify some of the parameters of this study. This research also employed conversations (or *talanoa*) as a principal tool to harness information and complement the ethnographic nature of this study (Nabobo-Baba, 2008). A *talanoa* (or *veitalanoa*) is an interview, but more. The techniques of semi-structured or unstructured interviewing are employed, encouraging consistent flow and deep discussions through conversation, adding more layers of detailed responses to the key research questions (Nabobo-Baba, 2008). *Talanoa*, rather than traditional interviews, were used to request the responses being sought by the researcher which essentially provided challenges that required the researcher to be adaptive and self-reflective. Respondents were observed to be more forthcoming when interviews were conducted in a relatively unstructured manner and respondents preferred that interviews were conducted in natural and comfortable settings such as a *talanoa*. As a result, over the course of the study, the researcher developed a routine that contextually employed the Pacific Island art of massaging. This 'massaging' of conversations, involved skilful and dynamic engagement with respondents through boundary-setting and probing that established the platform for rich dialogue which was gradually built on multiple encounters and progressive questioning on issues related to system components and community resilience.

This study interactively engaged with a total of 25 individuals who were its key informants. These individuals included the village headman, chairman of the village meeting, heads and elders of respective clans, current and retired hotel workers, heads of respective village committees, graduates with diplomas and degrees, and church representatives. The informants varied in age, experience and status. They were considered authorities on village issues and were relied on for advice and follow-up interviews. Their views and perceptions are presented in the findings section of this chapter using pseudonyms to maintain anonymity

Background: tourism in Fiji

The Republic of Fiji is an archipelago of some 333 islands with a total land mass of 18,274 sq. km located 175 degrees east longitude and 18 degrees south latitude (Fiji Bureau of Statistics, 2015). Fiji has a population of some 843,000 spread across its 14 provinces, over a quarter of whom are concentrated on its main island of Viti Levu (Fiji Bureau of Statistics, 2007). The island of Vanua Levu and Taveuni are smaller (5,538 sq. km), more sparsely populated and economically less developed.

Tourism began in Suva in 1920 with the establishment of the White Settlers League, a body comprised of early European settlers who marketed Fiji to passengers disembarking from ships that crossed the Pacific (Scott, 1970). The White Settlers League evolved into the Suva Tourism Board, later becoming the Fiji Visitors Bureau and then Tourism Fiji, the national tourism organization. Despite its early beginnings, tourism only really developed after the Second

World War, prompted by the rise of disposable incomes in Australia and New Zealand (Fiji's main source markets), and by developments in hotel infrastructure and transportation that made Fiji more accessible (Movono *et al.*, 2015). By 1982, tourism was coined the 'new sugar' when it replaced sugar production as the main source of foreign exchange, and it continued to grow into Fiji's most significant industry (Narayan, 2000, p. 15).

Despite numerous coups, floods and cyclones, tourism remains Fiji's most important industry, main foreign exchange earner and the largest employer (Fiji Bureau of Statistics, 2015; Ministry of Tourism, 2009; Rao, 2002). In 2014, a total of 680,000 tourists visited Fiji, of which about 53 per cent were from Australia, 22 per cent from New Zealand, and the rest from the United States, Europe and Asia (Fiji Bureau of Statistics, 2015). Data on tourism's economic contributions varies depending on the source. According to figures released by the FMITT (2014), the tourism sector contributed to 32 per cent of Fiji's GDP, employing directly and indirectly around 60,000 workers in 2013 and providing much-needed relief for Fiji's large appetite for foreign earnings.

The research area

The current study is set in Vatuolalai village, which is located along Fiji's Coral Coast, Fiji's oldest tourism area where resort-based tourism was pioneered in 1952. Located in the south-western part of Viti Levu, the Coral Coast (one hour from Nadi International Airport) is an area of unique cultural attributes and is one of Fiji's most environmentally diverse and sensitive coastal regions (Movono *et al.*, 2015). Comprised of five principal districts and made up of villages and resorts, the Coral Coast remains a leading tourism region in Fiji, receiving 18 per cent of visitors annually (Tourism Fiji, 2014).

The case study: Vatuolalai village

Vatuolalai village was specifically selected because the researcher has a long-term relationship with the community that began with visits in 2007 as an undergraduate student, and then later as part of a Masters of Arts research project (Movono, 2012). The researcher is an indigenous Fijian with cultural ties and long-term links to the village and has accumulated a good depth of knowledge about Vatuolalai, its people and their stories. There are 32 households in Vatuolalai, which equate to about 224 people, the majority of whom are below the age of 35. The 32 households are divided into two tribes (or *yavusa*): *Jubai* (48 per cent) and *Davutukia* (52 per cent). Each tribe has one *mataqali* (clan) and two *tokatoka* (sub-clans). The gender distribution is relatively balanced with a slightly higher percentage of females (54 per cent) compared to males (46 per cent), and a high number of people (over 94 per cent) are involved or have been involved in one form of tourism activity or another.

Vatuolalai village is relatively well endowed in terms of housing and infrastructure, with all homes built of concrete, with flush toilets, and with around

86 per cent having access to modern goods such as sofas, flat screen TVs, refrigerators and gas stoves (Movono *et al.*, 2015). About 92 per cent of households rely on paid work as a means of a living, of which around 88 per cent is directly related to the resort and tourism business. People rely on the supermarkets and urban centres for their daily food requirements and, because they have access to regular wages, they regularly participate in the formal economy. Also, 14 individual, small to medium tourism enterprises provide waterfall tours, jet ski rentals, handicrafts and massage services for tourists, 11 of which are owned by women. Collectively, the villagers run and share the benefits from village tours, hosting international student groups and providing entertainment in the form of *kava* ceremonies and dances at the resorts. Vatuolalai village has its own marine park, protected through the Fiji Locally Managed Marine Areas Network (FLMMA) and villagers are well versed in issues of conservation and sustainability. There are some internal institutions such as a youth group, a women's club, development and religious committees that meet regularly to discuss village-based issues.

Vanua: a complex and adaptive social and ecological system

Unlike other indigenous communities in the Pacific, Fijians have legally recognized systems that not only reflect their traditional patterns of social organization but also indicate the interdependence between humans and the biosphere. The *Vanua* SES Model (Figure 17.1) illustrates these realities and proposes the conceptualization of a destination as a CAS in order to better understand community resilience in Fiji. The term *vanua* means 'land' in the Fijian vernacular. For indigenous Fijians, however, the term means much more, and refers to a sacred overarching structure that has multiple dimensions (Ravuvu, 1983). Classical Fijian scholars have defined the *vanua* as an amalgamated entity having physical, social, economic and ecological components that are interrelated (Ravuvu, 1983; Nayacakalou, 1975). Ravuvu, for example, stressed that:

> It (*vanua*) does not mean only the land (*qele*) area one is identified with, and the vegetation, animal life, waters and coasts (*qoliqoli*) and other objects on it, but it also includes the social and cultural system. The people, the traditions and customs, beliefs and values, and the various other institutions established with the aim of achieving harmony, solidarity, and prosperity within a particular social order. Its social and cultural dimensions are a source of security and confidence as it provides a sense of identity and belonging. To most Fijians, the idea of parting with one's vanua or land is tantamount to parting with one's life.
>
> (Ravuvu, 1983, p. 70)

The *Vanua* SES and Resilience Model illustrates the case study as a CAS that has complex social, ecological and economic components. The model shows that the three major components of the system are dependent on one another and are linked through the wide combinations of livelihood activities which in turn affects levels of

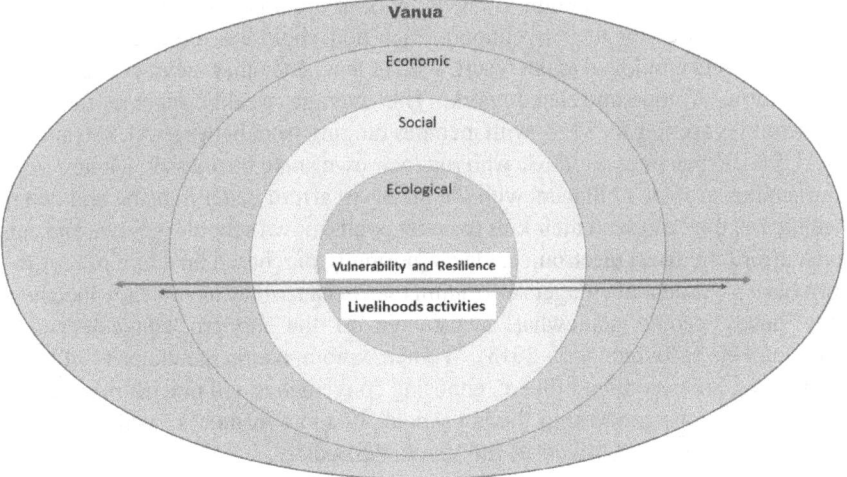

Figure 17.1 The *Vanua* social-ecological systems and resilience model
Source: Author

vulnerability and resilience. The following sections will unpack elements of each system component and highlight how an emphasis on tourism livelihoods creates disturbances and shifts within the system, ultimately affecting levels of resilience.

Economic system

The people of Vatuolalai were first introduced to the formal economy through copra and sugar, which later included fisheries and then tourism in 1952 when the first resort hotel was constructed some 5 km from the village. What seemed like an initial, or temporary intrusion into the *vanua*, the tourism system became cemented as part of the Vatuolalai SES in 1972 when the Naviti Resort was constructed on land right next to the village. The construction of the resort (and its subsequent influences) is highlighted in this chapter as an initial disturbance to the organic and traditional system that had prevailed prior to this time. The construction of the resort introduced new opportunities such as employment and new forms of livelihoods for the villagers. The positioning of the economic system at the outermost layer of the SES components reflects CAS thinking, which indicates how participation in the form of tourism employment (at the micro level) creates adjustments in society (macro level), ultimately influencing resilience. A period of over four decades of tourism employment, diversification and adaptation have elapsed, and the results indicate a significant level of participation, focus and reliance on tourism and its economic outputs.

Today, although land lease payments are received exclusively by the *Davutukia* tribe, all villagers have access to employment at the resort, assured through a lease arrangement that guarantees first preference is given to them. As a result, the majority (92 per cent) of adult villagers work at the hotel and in other tourism-related businesses for their livelihood. Each household has, on average, two to three members employed at the resort, with at least one other member engaged in some form of tourism-related work. The average weekly income for each household is around AU$125, with incomes ranging from between AU$80 and up to AU$1,100 per week for those who run their own small businesses. Money from tourism has provided villagers with the means to afford better housing and better healthcare; they can send their kids to better, well-known schools in Suva, and can even afford the latest electronic gadgets (normally purchased on a hire or rent-to-own basis). Vatuolalai villagers also admit that, because they have steady incomes, they have become somewhat accustomed to the modern convenience of supermarkets (Movono *et al.* 2015). Women, who make up the majority of hotel employees, are becoming affluent, choosing convenience and practicability when making decisions concerning the household. In some instances, women are the sole breadwinners and drivers of the Vatuolalai society.

The findings above indicates that the old internal complex adaptive system (*vanua*) is complemented by a new, spatially larger complex adaptive system (global tourism economy), which in this case is shown as part of the economic system within the *vanua*. This inclusion of tourism as a key livelihood activity within the *vanua* creates adjustments within system components, in the process affecting levels of resilience. Resilience, in this case, may be weaker in the long term for individuals participating in the new tourism-driven system compared to the old system. Indigenous Fijian communities are observed to be quick to discard the more 'traditional' activities because of the social status that comes with having a paid job. This, along with the promise of wealth and acquired elevated standards has encouraged people to adapt and rely more on tourism as a source of livelihood. Ayres (2000) notes that, without proper planning, communities with little economic diversification can easily fall into the trap of tourism dependency. Overdependence on a single source of livelihood is counterproductive to the ability of the system to respond to the volatilities exclusively attached with tourism (Cochrane, 2010; Strickland-Munro *et al.*, 2009). Furthermore, a heavy dependence on economic activities reduces participation and interest in traditional activities that not only adds to the diversity of livelihood portfolios of villagers but also enforce connections between humans and their biosphere. The data has shown that the majority of villagers are 'placing all their eggs in the tourism basket', reaching a point of possible over-dependence on tourism, causing shifts and adjustments in the natural and social setting, and creating increased vulnerabilities (Zhou *et al.*, 2010; Walker *et al.*, 2004).

Social system

In Vatuolalai, the social system can be defined as having two key components:

- observance of kinship relations; and
- the practice of traditional norms.

Over the years, the adherence and conduct of specific cultural protocols (such as the *kava* ceremonies) and the practice of kinship taboos (such as sibling taboos and marriage between kin) are quickly fading, due in part to modernization and the extensive social exchanges that have taken place between villagers and tourists. Traditional norms, such as the banning of hats, alcohol, and shorts worn by women on village grounds are loosely adhered to by youths. These changing practices are now, to an extent, acceptable to other villagers. Another example is the disappearance of authentic Vatuolalai *mekes* or traditional dances; villagers now perform the mekes from other parts of Fiji (and the Pacific) that are more popular with tourists but void of the authentic significance, chants and costumes unique to Vatuolalai. When asked, villagers state that the 'loss of appreciation' for cultural norms, 'adapting to tourists behaviour', 'having access to money' and 'modernization' are key reasons for the breakdown in these social norms. Respondents also raised diminishing respect for elders, disobedience and the reluctant participation of youths in traditional activities as other key changes that have occurred in society.

In Fijian culture, the observance of customs and traditions are conducted through the presentation of gifts of food, *kava*, mats, oil and other gifts for events such as births, deaths and weddings (Ravuvu, 1983). These cultural observances are what 'makes a Fijian a Fijian' and are considered essential to the preservation of Fijian culture (Nayacakalou, 1975, p. 32). Any cultural event will involve the presentation of traditional gifts, followed by a feast also involving consumption of large quantities of *kava*. According to Etonia, a clan elder:

> over the past four decades, the size and yield of farms have shrunk considerably; this is the same for our fisheries, which means we have to buy what is needed for a traditional event … sometimes we have two funerals one after the other, followed by another wedding, it gets quite difficult to bear.

Kado (2007) recognized that employment opportunities came at the cost of participation by locals in traditional activities, community work and cultural practice. Highly valued crops such as yams, *dalo* (taro) and *kava* are seldom planted because it is considered 'easier' or 'cheaper' to purchase depending on the alternatives (cassava, noodles, imported potatoes and rice) available at the Maui Bay Supermarket, some five minutes walk from the village. In effect, the data can be interpreted as showing that hotel work, by replacing traditional activities, has compounded the cost of carrying out traditions. Because villagers are unable to produce a greater variety of food and traditional items, they have to purchase what they need to not only survive, but also to meet their traditional obligations.

SES and CAS theory become relevant in identifying the interconnected and non-liner relationships between elements. The empirical evidence discussed above indicates how initial tourism employment and continued involvement in the tourism system has drawn attention away from conformance to cultural practices and traditions (Coetzee *et al.*, 2016; Cretney, 2014). To an extent, the findings prove how, 40 years since its introduction as a livelihood alternative, tourism has influenced how villagers interact with each other and with their natural environment. In essence, the connectivity between villagers, and their culture and environment, have become loose since embracing the tourism system. Despite the improvement of economic resilience, the resilience of culture and indigenous knowledge is in many ways diminished as society adapts to tourism and economically driven lifestyles.

The ecological system

The Vatuolalai ecological system can be broadly classified as having a land and marine component referred to as '*qele*' (all soil or dry land) and '*qoliqoli*' (waterways, coasts and springs) (Ravuvu, 1983). Land and marine resources are communally owned along patrilineal lines by *mataqali* or clans. Each villager is registered to a *mataqali* in a national register (the *Vola ni Kawa Bula*: VKB) which guarantees legal access to resources (Ravuvu, 1983). In Fijian society, ecological resources such as *qele* and *qoliqoli* are referred to as one's '*kanakana*', meaning from where sustenance is derived.

Indigenous Fijians interact with their biosphere through culturally established livelihoods practices as well as totemic connections which are a source of traditional knowledge, pride and identity. People of the same tribe are united through their totemic affiliations with each other (through the sharing of a totem tree, totem fish and totem bird), forming a cultural bond that links people to each other, links people to the *vanua*, and links the *vanua* to the people. Totems are also geographically unique for each tribe. For example, totems represent plants, trees and fish that are unique to Vatuolalai, adding a uniqueness of the place. These unique totemic connections enforce links between an individual and their natural surroundings, imparting custodianship, responsibility and confidence in relying on the *vanua* as a system in which they can coexist with nature.

Nowadays, villagers describe their relationship with their natural environment as 'distant' because of the minimal interaction and decreasing dependence on their '*kanakana*'. Villagers attribute this 'distancing' from their natural environment to their newfound wealth, tastes and commitment to tourism-related work, allowing villagers to lead more affluent lifestyles. This creates further isolation from their traditional knowledge and skills, which are often only transferred through these traditional activities. Surveys reveal that 32 per cent of the population have stopped planting altogether; others (38 per cent) have small gardens either for their consumption or are planting for a specific function. In Vatuolalai, only 11 per cent of the population own livestock and only 12 per cent still go fishing regularly, indicating an increasing reliance on non-traditional sources of protein (sausages, chicken, tinned meat and fish) and cassava (a staple). The absence of sweet potatoes, yams and plantains are also cause for concern as these crops are traditionally grown for their longevity and suitability during the cyclone seasons. Alivate, a lay preacher, said, 'people have adapted to the taste of "easy" food because we don't have time to work the land because our priority has become work and not living as nature intended.' Villagers were observed to be relying primarily on cassava, which historically was reserved as an animal (pig) feed and only consumed during adverse weather conditions, such as droughts and cyclones. The lack of variety in the crops being planted is in direct contradiction to traditional ideologies not only affecting traditional knowledge but affecting nutrition and increasing risks associated with a lack of food security.

The people of Vatuolalai often refer to themselves as '*kwai baravi*', meaning coastal people. Being so close to the sea, villagers have always used the ocean and its resources as a source of food, medicine and economic income for villagers. However, because of the resort, villagers now have to share their land and marine resources with tourists who use the same *qoliqoli* for water sports and snorkelling. Avid fishers, such as Epiuta, said that 'it's challenging to catch fish because the activities of tourists chase fish away.' Conversations with fishers suggest that construction of a causeway and a man-made island in the late 1990s had also caused the disappearance of fish. Of particular interest is the noticeable disappearance of seasonal bait fish, last seen in 1997 before the causeway was constructed. Another key ecological change noted by villagers is the scarcity of prawns in their streams, which they believe was caused by the construction of a water dam used to supply water to the resort. Villagers perceive these three physical constructs (dam, causeway and man-made islands) as the cause of changes experienced in their *qoliqoli* and as demonstrations of how coastal tourism affects the ecology of an area causing further adjustments within a system.

Respondents from this study value their links to the natural environment and confirm that changes, such as loss of fish species and a decline in ecological productivity, are having serious social implications on food security. Any reduction in connectivity and loss of traditional knowledge and learning systems diminishes diversity and redundancy and, in the longer term, may be highly detrimental to the resilience of the system (Stockholm Resilience Centre, 2014). Ultimately, as suggested by Veitayaki (2001), Fijians need to reinvigorate their links with the natural environment to better manage their natural resources effectively. Then they may be able to practice their customs, live with strengthened capacities and fewer vulnerabilities, and cope with external stressors. The findings above (p. 293) have shown the main components of the *Vanua SES* and has revealed how tourism, a CAS in its own respect, has become an integral part of the *vanua* (Neely, 2015). Its introduction as a means to diversify opportunities has stimulated adjustments and reorganizations within the various system components, and, in the process, affected the levels of economic, ecological and social aspects of community resilience (Gunderson and Holling, 2002). Table 17.1, adapted from the Stockholm Resilience Centre (2014) principles of resilience, provides a summary of this discussion as it relates to the principles of resilience, offering insight into the changes in levels of resilience brought about by involvement with and subsequent dependence on tourism.

Table 17.1 Resilience characteristics of the Vatuolalai SES before and after tourism involvement

Resilience characteristic	Pre-tourism system	Post-tourism system	Resilience implications
Diversity and Redundancy	• Broad range of livelihoods activities • Diverse range of traditional skills and knowledge • Wide range of known contextual connections between system elements enforced through cultural practice	• Heavy dependence and focus on tourism-based livelihoods • Loss of traditional skills and knowledge • Reduced connectivity between system elements due to mono focus on tourism activities	Diminished resilience
Learning and traditional knowledge	• Preservation of traditional skills and knowledge pertaining to fishing, farming, hunting and cultural etiquette • Continual practice of traditions that connect people with the biosphere • Transference of knowledge through practice, dances and crafts • Firm understanding and practice of seasonal agriculture and fisheries, ensuring preparedness during cyclone season	• Increased attention on tourism-based skills. • Increasing shifts away from traditional activities and skills • Loss of cultural knowledge about arts and dances. • Erosion of traditional skills and techniques have eroded due • Disruptions of systems of traditional knowledge transfer • Total loss of skills such as building traditional Fijian structures and canoe building	Diminished resilience
Managing connectivity	• Highly connected system • Interrelationships and interdependencies between system elements are strong. • Knowledge about system connections and relationships are clear • Sustainable resource management techniques are employed.	• Loss of connectivity between villagers and their natural environment • Increased affluence has increased dependency on purchased goods as opposed to locally sourced components. • Reduced connectivity between villagers	Diminished resilience
Broadening participation	• Participation in leadership and politics is generally gender based and participation in community and governance issues are male dominated • Participation in a broad range of activities, from fishing to multi-cropping by a broad spectrum of the population	• Empowerment of women and youths has led to increased participation in newly established committees and clubs. • Greater participation in community wide development issues • Greater participation of women in business • There is broader participation in economic work, particularly by women and youths • Increased participation of women in women led institutions such as the women's club	Improved resilience
Fostering complex systems thinking	• The pre-tourism system is one that emulates CAS theory. • Culture reinforces connections between communities and the biosphere • Slow variables and feedback through internal channels	• Greater emphasis and participation on the tourism system • Reduced complex systems thinking • Rapid changes and significant disruptions to pre-existing links between elements.	Diminished resilience

Source: Author

Conclusions

This chapter has shown that destination communities such as Vatuolalai are complex social and ecological systems that have diverse and interrelated non-homogeneous components. The *Vanua* SES and Resilience Model captures the fluidity and complexity of indigenous communities by incorporating social and ecological systems concepts within the bounds of the *vanua* to allow for a broader range of focused analysis to be conducted on this community (Folke, 2006; Gallopin, 2006; Holling, 1973). This chapter has defined a clear pathway where resilience is operationalized by adopting SES as a platform for empirical studies in tourism-based indigenous Fijian communities. It has also generated outcomes beyond purely academic discussions, providing empirical evidence to support resilience theory. This study incorporates resilience by assessing human interaction with the biosphere, identifying resource use patterns, issues of vulnerabilities, and providing focus on the elements of a system that need to be strengthened to increase adaptive capacities.

This chapter has also shown that there are two adaptive cycles: the *vanua* system and the tourism system. It can be argued that tourism grew from within the economic system through 'exploitation' to a stage of 'consolidation' where the traditional *vanua* system went through a stage of 'collapse' and is now in the midst of 'reorganization' (Holling, 2001). Based on the resilience Adaptive Cycle model, the tourism system is likely to experience a 'collapse' at some time in the future, while the *vanua* system is likely to experience 'reorganization'. A resilience approach would be aware of this and plan for both of these future stages to develop. However, this study has shown that society is highly reflexive, and its non-homogeneous components have the potential to self-organize as it adjusts to the initial disturbances brought about by tourism. More importantly, this chapter acknowledges that tourism itself does not lead to a loss of resilience. The community's resilience in this particular study was reduced by the diminishing attention to traditional values and knowledge over the long term, which could have been prevented with proper forethought, and might be recoverable with proper restoration. CAS theory hence becomes useful in indicating how micro-level influences such as tourism employment can stimulate changes and adjustments in the overall system, therefore affecting levels of resilience.

The findings of this empirical study fill a significant void in the literature and provide clear evidence of the complex, multilayered and interrelated nature of indigenous Fijian society. This chapter has shown, that, because villagers have adapted to tourism as a primary livelihood source, issues of resilience and vulnerability have arisen (Movono *et al.* 2015). Ultimately, this study has shown that the *Vanua* SES and Resilience Model may be useful in identifying vulnerabilities and has immense potential as a tool for initiating targeted action and achieving strengthened economic, social and ecological capacities. Actions such as reinvigoration of cultural practices, integrative planning, reducing economic dependence, diversifying livelihoods activities and efficient use of resources are some essential recommendations that must be considered in reducing vulnerabilities in a tourism-related destination community. Despite its findings, this case study (although a fair representation of Fijian communities) is nowhere near enough to adequately understand resilience in Fiji. Much yet remains to be done.

References

Adger, W. N. (2000). Social and ecological resilience: are they related? *Progress in Human Geography*, 24, 347–64.

Ayres, R. (2000). Tourism as a passport to development in small states: reflections on Cyprus. *International Journal of Social Economics*, 27(2), 114–33.

Becken, S. (2013). Developing a framework for assessing resilience of tourism sub-systems to climatic factors: *Annals of Tourism Research*, 43, 506–28

Buckley, W. F. (1968). Society as a complex adaptive system, in Buckley, W. F. (Ed). *Modern Systems Research for the Behavioural Scientist: A Source Book*, (pp. 77–125). Chicago: Aldine Publishing,

Cochrane, J. (2010). The sphere of tourism resilience. *Tourism Recreation Research*, 35(2), 173–85.

Coetzee, C., Niekerk, D. V., & Raju, E. (2016). Disaster resilience and complex adaptive systems theory. *Disaster Prevention and Management*, 25(2), 196–211.

Cretney, R. (2014). Resilience for whom? Emerging critical geographies of social-ecological resilience. *Geography Compass*, 8(9), 627–40.

Dahles, H., & Susilowati, T. B. (2015). Business resilience in times of growth and crisis. *Annals of Tourism Research*, 51, 34–50.

Fiji Bureau of Statistics (2007). *Fiji Census of Population and Housing*. Retrieved from www.statsfiji.gov.fj/statistics/2007-census-of-population-and-housing.

Fiji Bureau of Statistics (2015). *Key Statistics*. Suva: Fiji Bureau of Statistics.

FMITT (2014). *Fiji International Visitor Survey Report*. Suva: Fiji Ministry for Industry, Trade and Tourism.

Folk, C., Colding, J., & Berkes, F. (2003). Synthesis: building resilience and adaptive capacity in social-ecological systems. In: *Navigating Social-Ecological Systems; Building Resilience for Complexity and Change* (pp. 352–83). Cambridge: Cambridge University Press.

Folke, C. (2006). Resilience: the emergence of a perspective for social and ecological systems analysis. *Global Environmental Change*, 16, 253–67.

Folke, C., Carpenter, S., Elmqvist, T., Gunderson, L., Holling, C. S. & Walker, B. (2003). Resilience and sustainable development: building adaptive capacity in a world of transformations. *Ambio*, 31(5), 437–40.

Gaillard, J. (2010). Vulnerability, capacity and resilience: perspectives for climate development and policy. *Journal of International Development*, 22(2), 218–32.

Gallopin, C. G. (2006). Linkages between vulnerability, resilience, and adaptive capacity. *Global Environmental Change*, 16. 293–303.

Gunderson, L. & Holling, C. S. (Eds) (2002). *Panarchy: Understanding transformations in human and natural systems*. Washington, DC: Island Press.

Gunderson, L. H. (2000). Ecological resilience: in theory, and application. *Annual Review of Ecological Systems*, 31, 425–39.

Hammer, R. J., Edwards, J. S, & Tapinos, E. (2012). Examining the strategy development process through the lens of complex adaptive systems theory. *Journal of the Operational Research Society*, 63, 909–19

Holland, J. (2006). Studying complex adaptive systems. *Journal of System Science & Complexity*, 19(1), 1–8.

Holling, C, S. (1973). Resilience and stability of ecosystems. *Annual Review of Ecology and Systematic*, 4, 1–3.

Holling, C. S. (2001). Understanding the complexity of economic, ecological and social systems. *Annual Review of Ecology and Systematic*, 4, 390.

Kado, M. (2007) Tourism and poverty alleviation in Fiji: examining the impact of Coral Coast tourism on village livelihoods, the case of the villages of Namatakula and Vatuolalai. MA Thesis, School of Development Studies, University of the South Pacific, Suva.

Lew, A. A. (2014) Scale, change and resilience in community tourism planning, *Tourism Geographies*, 16(1), 14–22, DOI: 10.1080/14616688.2013.864325

Lew, A.A., Ng., P.T., Ni, C-C. & Wu, T-C. (2016). Community sustainability and resilience: similarities, differences, and indicators. *Tourism Geographies* 18(1), 18–27

Manyena, S. B. (2006). The concept of resilience revisited. *Disasters*, 30(4), 434–50.

Ministry of Tourism (2009). *Fiji International Visitor Survey Report*. Suva: Ministry of Tourism.

Movono, A. (2012) Tourism's impact on communal development in Fiji: a case study of the socio-economic impacts of The Warwick Resort and Spa and The Naviti Resort on the indigenous Fijian villages of Votua and Vatuolalai. Unpublished Master's Thesis, School of Tourism and Hospitality Management, University of the South Pacific, Suva.

Movono, A., Harrison, D. & Pratt, S. (2015). Adapting and reacting to tourism development: a tale of two villages on Fiji's Coral Coast. In S. Pratt & D. Harrison (Eds.), pp. 100–114, *Tourism in Pacific Islands: Current Issues and Future Challenges*. London and New York: Routledge

Nabobo-Baba, U. (2008). Decolonising framings in Pacific Research: Indigenous Fijian Vanua research framework as an organic response. *AlterNative: An International Journal of Indigenous Peoples*, 4(2), 140–54.

Narayan, P. (2000). Fiji tourism industry: a SWOT analysis. *Journal of Tourism Studies*, 11(2), 15–24.

Nayacakalou, R. R. (1975). *Leadership in Fiji*. Suva: University of the South Pacific. South Pacific Social Sciences Association in association with Institute of Pacific Studies.

Neely, K. (2015). Complex adaptive systems as a valid framework for understanding community level development. *Development in Practice*, 25(6), 785–97.

Paton, D. (2006). Disaster resilience: building capacity to co-exist with natural hazards and their consequences. In D. Paton & D. Moore Johnston (eds.) *Disaster Resilience: An Integrated Approach* (pp. 40–67). Springfield: Charles C. Thomas Publisher Ltd.

Prasad, B. C. (2014). Why is Fiji not the 'Mauritius' of the Pacific? Lessons for small island nations in the Pacific. *International Journal of Social Economics*, 41(6), 467–81.

Rao, M. (2002). Challenges and issues for tourism in the South Pacific island states: the case of the Fiji Islands, *Tourism Economics*, 8(4), 401–29.

Rose, A. (2007). Economic resilience to natural and man-made disasters: multi-disciplinary origins and contextual dimensions. *Environmental Hazards*, 7(4), 383–98.

Scott, R. J. (1970). *The Development of Tourism in Fiji since 1923*. Suva: Fiji Visitors Bureau.

Stockholm Resilience Centre (2014) *Applying Resilience Thinking: Seven Principles for Building Resilience in Social-Ecological Systems*. Stockholm: Stockholm University.

Strickland-Munro, J. K., Allison, H. E. & Moore, S. A. (2009). Using resilience concepts to investigate the impacts of protected area tourism on communities. *Annals of Tourism Research*, 37(2), 499–519.

Tedlock, B. (2000). Ethnography and ethnographic representation. In N.K. Denzin & Y.S. Lincoln (Eds.), *Handbook of Qualitative Research* (2nd ed.) (pp. 455–86). Thousand Oaks, CA: Sage Publications, Inc.

Timmerman, P. (1981). *Vulnerability, Resilience and the Collapse of Society: A Review of Models and Possible Climatic Application*, Toronto: Institute for Environmental Studies, University of Toronto.

Tourism Fiji (2014). *Annual Report*. Suva: Tourism Fiji.

Tracy, S. J. (2013). *Qualitative Research Methods: Collecting Evidence, Crafting Analysis, Communicating Impact*. Chichester: John Wiley and Sons.

UN (2005). *Hyogo Framework for 2005–15: Building Resilience of Nations and Communities to Disasters*. United Nations International Strategy for Disaster Risk Reduction.

Veal, A. J. (2006). *Research Methods for Leisure and Tourism: A Practical Guide* (3rd Ed.). Harlow: Prentice Hall.

Veitayaki, J. (2001). *Empowerment and the Challenges of Involving Local Fijian Communities*. Suva: University of the South Pacific.

Walker, B., Carpenter, S., Rockstrom, J., Crepin, A. & Peterson, G. (2012). Drivers, slow variables, fast variables shocks and resilience. *Ecology and Society*, 17(3), 30–34.

Walker, B., Gunderson, L. H., Kinzig, A. P., Folke, C., Carpenter, S. R. & Schultz, L. (2006). A handful of heuristics and some propositions for understanding resilience in social-ecological systems. *Ecology and Society*, 11(1), 13

Walker, B., Holling, C. S., Carpenter, S. R. & Kinzig, A. (2004). Resilience, adaptability, and transformability in social and ecological systems. *Ecology and Society*, 9(2)

Walsh, J. P. (1996). *Research Methods in Social Sciences* (5th Ed.). London: Edward Arnold.

Zhou, H., Wang, J., Wan, J. & Jia, H. (2010). Resilience to natural hazards: a geographic perspective, *Natural Hazards*, 53(1), 21–41.

Part V
Conclusions

18 Lessons learned

Tourism and the Anthropocene

Joseph M. Cheer and Alan A. Lew

Introduction

In the early days of contemporary research into tourism, the geographer Walter Christaller (1963) extolled the virtues of unspoiled nature, opining that utmost in the touristic experience was the landscape in all its natural glory. Imbued in his appeals was the value put on environmental integrity and how the merits of this stood in stark contrast to the industrialized urban settings of the era. In particular, and apropos to this book, is Christaller's (1963, p. 103) conception that 'such a landscape is a whole complex: it is built up of various elements, more or less dependent on each other'. It is this interdependency between the human and the environmental that underlines resilience thinking insofar as environmental change is concerned and is principally what this volume addresses.

In contrast, and around the same time, Nancy Mitford's (1959) examination of tourism alluded to issues of carrying capacity and questioned what the limits of acceptable change might be. In referring to the Venice of the late 1950s, Mitford was uncertain in her embrace of tourism. When referring to the tourism-driven changes on Torcello, a neighbouring island, she observed: 'Torcello which used to be lonely as a cloud has recently become an outing from Venice. Many more visitors than it can comfortably hold pour into it' (Mitford, 1959, p. 5). Her observations were prescient in the sense that she tried to reconcile the economic and environmental costs of tourism before it had become a global concern: 'there is nothing to be done about it, impossible to stem the tide, but what will happen when it is swollen by the li, the rouble, the yen, and the rupee?' (Mitford, 1959, p. 7).

In making the leap from the 1950s to the present day, very different circumstances are encountered. At the end of 2016, the world's population was close to 7.4 billion, and over 1.2 billion international travellers were accounted for (UNWTO, 2016). These unprecedented contextual changes have major implications for natural areas whose values are shaped through demands for extractive resources, arable land and human settlement. The expansion of tourism is one of the many human-caused geophysical drivers of the Anthropocene epoch that have engendered debates over sustainability and, in the case of this book, resilience to environmental change and disasters. Climate change, global warming and the hole in the ozone layer, for example, did not feature in national and global conversations in the 1950s and 1960s, when Christaller and Mitford made their observations, and in all cases the circumstances today are vastly more intense.

Understanding how resilient people and places are to environmental change today holds greater importance than ever. For tourism-based communities, environmental change can have tremendous impacts on their ability to respond and survive. The low-lying islands in the Pacific Ocean are one illustration of how environmental change is leading to drastically transformed local contexts where 'there will be fundamental irreversible changes in island geography, settlement patterns, subsistence systems, societies and economic development' (Nunn, 2013, p. 143). In much the same way, climate change in alpine and ski regions is impinging on the length and quality of the ski season, placing business and community resilience under increasing strains and generating innovative adaptations (Hopkins & Maclean, 2014; Kaján, Tervo-Kankare & Saarinen, 2015).

Any type of development or effort that utilizes the natural environment necessitates the delivery of trade-offs. Ideally, such negotiations should avoid net loss positions to communities and maintain a steady state in the quality of ecosystem services through sustainability and resilience responses. Environmental change is often a result of a whole host of intertwined and complex variables, with climate change being one of the most widely acknowledged examples of this. As such, governance responses that underline adaptive capacities and adaptations to environmental change are best formulated through the lens of system resilience (Nelson, Adger & Brown, 2007).

Underscoring environmental change is the occurrence of disaster hazards and the links between the greater frequency of such events and environmental changes are detailed in Part II of this volume. Often (though not always) the more experienced communities are at dealing with disaster, the greater their resilience. This comes about because of robust social capital, cohesive community–industry–institution relationships and human social networks that pride reciprocity and benevolence. All of these are critical in natural hazard disaster events. Spiritual resilience, tapping into core beliefs systems, is especially pertinent given its links to sense of place, community identity and a personal sense of belonging.

In Part III of this book, environmental change and tourism contextualizes indigenous peoples by highlighting how resilience to environmental change is linked strongly to social justice, sovereignty and empowerment concerns, and the undermining of any of these can result in diminished adaptive capacities. Moreover, the ability to have agency over system shifts (responses) from one state to another is central to resilience in indigenous contexts, especially the ability to self-organize and reboot cultural ecosystems in times of crisis. In today's world, however, the scales and rates of change are amplified, which means that, for many indigenous cohorts, the old ways of responding are in need of change themselves. In general, this replies to all societal contexts where new approaches to old problems and innovative tactics to new problems are pressing.

Emergent themes

The question as to how tourism communities can respond to environmental change remains local, given that whatever the repercussions are, it is they who encounter them firsthand. In attempting to understand this question, the chapters in this volume have presented instructive lessons that characterize the experiences of local communities in developing adaptive capacities to environmental changes and disasters. The major themes generalized from these perspectives feature in the following sections.

Maintenance of robust human social networks

In almost all of the cases in this volume, the capacity of communities to maintain robust human social networks is paramount, especially in response to disasters or monumental changes. When it comes to responding to natural hazard disasters, while reliance on wider governance mechanisms (such as emergency services) can go some way to aiding a recovery, communities with pre-existing supportive and tight-knit networks are more assured of a successful and thriving rebound. For example, communities that have experienced past disasters tend to have formalized adaptive responses that can be rapidly mobilised in the event of a new crisis, based on their prior experience. Disregard for human social networks and the value they can bring to resilience building is one of the key undermining factors that prevents the development of efficacious adaptive capacities. (Lew, Ni, Wu and Ng, Ch. 3; Burns, Ch. 4; Nogueira de Moraes, Ch. 5; Holmes, Ch. 6; Lapointe and Sarrasin, Ch. 9; Xu, Chen and Dai, Ch. 11)

Community collaboration: bottom-up self-organization

The extent to which adaptive responses are effective also rests on the ability of stakeholders to mobilize and self-organize from the bottom-up in crisis situations. This is especially relevant where a diminished supply of essential ecosystems services (such as in a drought) has a widespread impact beyond smaller segments of the community. In such cases, individual responses become nullified given that the enormity of the crisis requires a whole community approach. Thus, enabling both individual and collective agency is critical to resilience building from below. (Holmes, Ch. 6; Bakti and Lew, Ch. 10; Xu, Chen and Dai, Ch. 11; Miller, Ch. 12; Movono, Ch. 17)

Government and industry leadership

Beyond community-driven initiatives to construct adaptive capacity, the extent to which government and industry respond in a consistent and collaborative manner to dealing with environmental change and crises frames the success or failure of resilience building efforts. In times of crisis, the tourism industry can offer the vital services of secure employment and investment in enhancing social capital.

Additionally, where economic and ecological resources are in short supply, it is critical that tourism and other industries band together with governmental entities to fund and support innovative initiatives to respond to wider community needs. (Holmes, Ch. 6; Lew and Wu, Ch. 7; Ooi, Ch. 8; Bakti and Lew, Ch. 10)

Social capital enhancement

The deliberate enhancement of social capital through formal and informal investments underlines almost all of the chapters in this volume. This is social capital that enables a closer alignment between various stakeholders, including community, civil society, industry and government. Social capital at a local level is considered vital to building communities of compassion and spiritual resilience. In much the same way, investing in social capital reinforces resilience building when it strengthens traditional values, while also allowing the capacity to maintain individual agency and ensuring that the community benefits from tourism. (Kato, Ch. 14; Hillmer-Pegram, Ch. 16; Movono, Ch. 17)

Measuring and articulating resilience

Unfortunately, like sustainability, resilience has increasingly become a catchword for which theory has not been matched by the application of effective efforts to enhance adaptive capacities in real world settings. This situation is a particularly damning indictment of industry and government stakeholders who should be taking the lead in community planning and resilience. This situation, however, is also compounded by the challenge of quantifying and measuring community resilience, and being clear about what and who (systems) are being made resilient to what drivers of change, and what the trade-offs are in such decisions. (Nogueira de Moraes, Ch. 5; Lapointe and Sarrasin, Ch. 9; Miller, Ch. 12; Orchiston and Espiner, Ch. 15; Hillmer-Pegram, Ch. 16)

Coordinated stakeholder response

From a resilience theory perspective, the modelling of system disturbance and responses is essential in understanding the multiplicity of potential stable states (scenarios) that can occur, and planning for them. The adaptive roles that government and tourism industry organizations play, as well as civil society groups, should also be clearly articulated in much the same way to ensure that communication flows concerning policy directions reaches all stakeholders in a timely and inclusive manner is advocated. Well-timed interventions that protect the public interest are also critical to ensure that private interventions are not conducted with short-termism in mind, but with a view to supporting initiatives that build social-ecological resilience is vital. (Duke, Cotterell and Cotterell, Ch. 2; Holmes, Ch. 6; Ooi, Ch. 8; Lapointe and Sarrasin, Ch. 9; Orchiston and Espiner, Ch. 15)

Well-being is physical, mental, and spiritual

Within resilience discourses, the emphasis on physical, mental and spiritual dimensions of resilience are often overlooked when they are fundamental to the adaptive capacities of all stakeholders. Especially in the face of monumental disaster hazards, well-being amidst traumatic circumstances requires more than just 'bricks and mortar' engineering initiatives. The sense of connection of people to place promotes feelings of mutual obligation and reciprocity. The connectedness between well-being and social justice concerns in times of crisis is also essential as it underlines dignity, identity and the adaptive capacities beyond the individual. (Herrschner and Honey, Ch. 13; Kato, Ch. 14; Hillmer-Pegram, Ch. 16)

Cultural ecosystems services enhancement

The management and enhancement of cultural ecosystems services (socio-cultural benefits derived from natural ecosystems) is considered a type of insurance in the event of a crisis and can reinforce adaptive capacities and sustainability. This is especially critical where different systems and subsystems exist, operating through different time scales, and where traditional governance systems and mechanisms are ineffectual. Maintaining optimum cultural ecosystems services requires a regard for systemic collective actions that, if mobilized, can shape socio-political discourses that enhance adaptation. (Burns, Ch. 4; Lew and Wu, Ch. 7; Lapointe and Sarrasin, Ch. 9; Bakti and Lew, Ch. 10)

Maintenance of core community identities, values and common goals

Where stakeholders are united in maintaining particular identities and fostering common goals, adaptation to environmental change can be more effective. These core values can provide a bedrock in times of crisis, when individual and group liminality opens new opportunities for personal and community adaptive capacities and transformability. Building resilience responses on top of core value systems is the most efficient path for individual and collective agency in changing times. (Burns, Ch. 4; Nogueira de Moraes, Ch. 5; Miller, Ch. 12; Herrschner and Honey, Ch. 13)

Tourism and the Anthropocene

All the chapters in this volume speak of adaptive capacities as being contingent on an awareness of an evolutionary resilience approach (adaptation to constant change) that is focused on building resilience from the bottom up. Indeed 'adaptation to environmental change is best formulated as an issue of system resilience' (Nelson et al., 2007, p. 396). The urgency for a system resilience response in tourism contexts have been emphatically demonstrated in numerous monumental disasters in recent years, including:

- the Indian Ocean earthquake and tsunami (2004);
- Hurricane Katrina in the US (2005);
- massive earthquakes in Sichuan, China (2008), Christchurch, New Zealand (2011), and Nepal (2015); and
- Cyclone Pam in Vanuatu (2015) and Hurricane Mathew in Haiti (2016).

These cases demonstrate that natural hazard disaster events, on both small and grand scales will continue to occur in the future, resulting in the question of how can communities in disaster-prone locations develop resilience and the capacity to adapt and bounce back assuredly? This will almost certainly be a hallmark issue for much of tourism in the Anthropocene epoch. All the stakeholders who draw from tourism are interconnected in this shared common future, and the future viability of destinations demands new ways of thinking about tourism beyond transactional economic perspectives (Hollenhorst, Houge-Mackenzie & Ostergren, 2014).

The Anthropocene epoch, in which human activity is a major driver of the planet's geophysical change, poses ethical and moral questions, especially where the conduct of tourism compromises ecosystem services or where the need to develop adaptive capacities to disaster and environmental change is ignored (Cole, 2014; Weeden & Boluk, 2014). Because tourism is, as Gren and Huijbens (2014, p. 18) put it, 'part of the Anthropocene environmental problem of global sustainability', it must therefore move beyond clichéd sustainable development thinking and question its complicity in contributing to the strain on critical ecosystems services. The World Tourism Organization celebrated the one billionth international travel in 2012 (UNWTO 2016), and its efforts to drive this number higher, while also preaching sustainability, is one of the starkest illustrations of the many contradictions that will prevail in the Anthropocene.

Many of the chapters in Part III of this volume describe resilience building as being conditional upon the interconnectedness of stakeholders and the establishment of robust frameworks that enable rapid self-organization in the face of crises, and lesson-building in their aftermath. Folke's (Folke, 2016, p. 14) assertion that '[i]n resilience thinking and social-ecological systems research, people are viewed as part of the planet, as part of the biosphere' is an essential paradigm upon which resilience building is predicated. For many indigenous peoples, this realization is not new and their traditional responses to change has been shaped by adaptations borne out of practice, rather than grounded on the cessation of disasters.

The Anthropocene is emblematic of the urgency for a paradigm shift in bolstering general community resilience to disasters and environmental change that is based on humans being in and of nature. Concern should be more about living with, rather than attempting to change, nature's course. This feeds into the conundrum faced by many tourism communities in clearly demarcating the extent to which their expanding tourism growth impacts their abilities to adapt to environmental change and disasters. For example, while seaside and alpine resorts may be desirous, how they are positioned in the context of environmental change requires forethought and reflection. In much the same way, shifting towards greater reliance on the tourism economy deserves the same considerations.

In building resilience, political–economy trade-offs and compromises to the various competing and at times asynchronous positions will be required. As Galafassi et al. (2017, p. 1) argue:

> Management of social-ecological systems necessarily involves trade-offs, and the consideration and resolution of trade-offs is likely to be influenced by the politics of decision making and the relative power of winners and losers to articulate and pursue their interests.

This is the reality and the politics that underpins tourism development where economic expansion exploits ecosystems services, and concomitantly reduces adaptive capacities. The desire for triple-bottom line effects is clear, and acknowledging the interconnectedness between the human, environmental and the economic should be an essential mantra in the Anthropocene.

If tourism communities are to build adaptive capacities, both bottom-up and top-down impetus are required that embraces 'whole of system resilience', avoids short-termism and instead builds frameworks that embody the essence of a 'common future', as invoked by the *Brundtland Report*. The implications for research into the nexus between tourism resilience and adaptation to environmental change and disasters are clear: for communities to be better able to confront change and deal with shocks, investing in social capital is critical. This in itself is not revelatory because in the aftermath of disaster or shocks, as demonstrated in this volume, communities that tend to rebound more effectively, also happen to exhibit robust social capital.

In general, research into resilience should place equal emphasis on the social dimensions that underpin adaptive capacities. This is what Davoudi et al. (2013, p. 319) refer to as evolutionary resilience: the 'pursuit of building capacity for envisaging and embracing transformation through creativity and imagination at institutional, community and individual levels and through cultivating flexibility, resourcefulness and cooperative networks at various scales'. Almost all of the chapters in this volume speak of evolutionary resilience as the key to adapting to environmental changes and recovering from disasters, and this signals a potentially rich seam of research on building resilience to environmental change and natural hazard disasters in tourism contexts.

References

Christaller, W. (1963). *Some Considerations of Tourism Location in Europe: The Peripheral Regions, Underdeveloped Countries, Recreation Areas*. Paper presented at the Regional Science Association; Papers XII, European Congress, Lund.

Cole, S. (2014). Tourism and water: from stakeholders to rights holders, and what tourism businesses need to do. *Journal of Sustainable Tourism, 22*(1), 89–106.

Davoudi, S., Brooks, E. & Mehmood, A. (2013). Evolutionary resilience and strategies for climate adaptation. *Planning Practice & Research, 28*(3), 307–22.

Folke, C. (2016). Resilience. In *Oxford Research Encyclopedia of Environmental Science*. Oxford: Oxford University Press.

Galafassi, D., Daw, T., Munyi, L., Brown, K., Barnaud, C. & Fazey, I. (2017). Learning about social-ecological trade-offs. *Ecology and Society,* 22(1), 2.

Gren, M. & Huijbens, E. H. (2014). Tourism and the Anthropocene. *Scandinavian Journal of Hospitality and Tourism,* 14(1), 6–22.

Hollenhorst, S. J., Houge-Mackenzie, S. & Ostergren, D. M. (2014). The trouble with tourism. *Tourism Recreation Research,* 39(3), 305–19.

Hopkins, D. & Maclean, K. (2014). Climate change perceptions and responses in Scotland's ski industry. *Tourism Geographies,* 16(3), 400–414.

Kaján, E., Tervo-Kankare, K. & Saarinen, J. (2015). Cost of adaptation to climate change in tourism: methodological challenges and trends for future studies in adaptation. *Scandinavian Journal of Hospitality and Tourism,* 15(3), 311–17

Mitford, N. (1959). The Tourist. *Encounter,* 13(3), 3–7.

Nelson, D. R., Adger, W. N. & Brown, K. (2007). Adaptation to environmental change: contributions of a resilience framework. *Annual Review of Environmental Resources,* 32, 395–419.

Nunn, P. D. (2013). The end of the Pacific? Effects of sea level rise on Pacific Island livelihoods. *Singapore Journal of Tropical Geography,* 34(2), 143–71.

UNWTO (2016). *UNWTO Tourism Highlights, 2016 Edition.* Madrid: United Nations World Tourism Organization. Retrieved from http://mkt.unwto.org/publication/ unwto-tourism-highlights-2016-edition.

Weeden, C. & Boluk, K. (2014). *Managing ethical consumption in tourism.* London: Routledge

Index

Page numbers in *italics* refer to illustrations. Page numbers in **bold** refer to tables.